T0146048

TUNAS and BILLFISHES of the WORLD

Bruce Collette and John Graves
Illustrated by Val Kells

TUNAS and BILLFISHES of the WORLD

Bruce Collette and John Graves

Illustrated by Val Kells

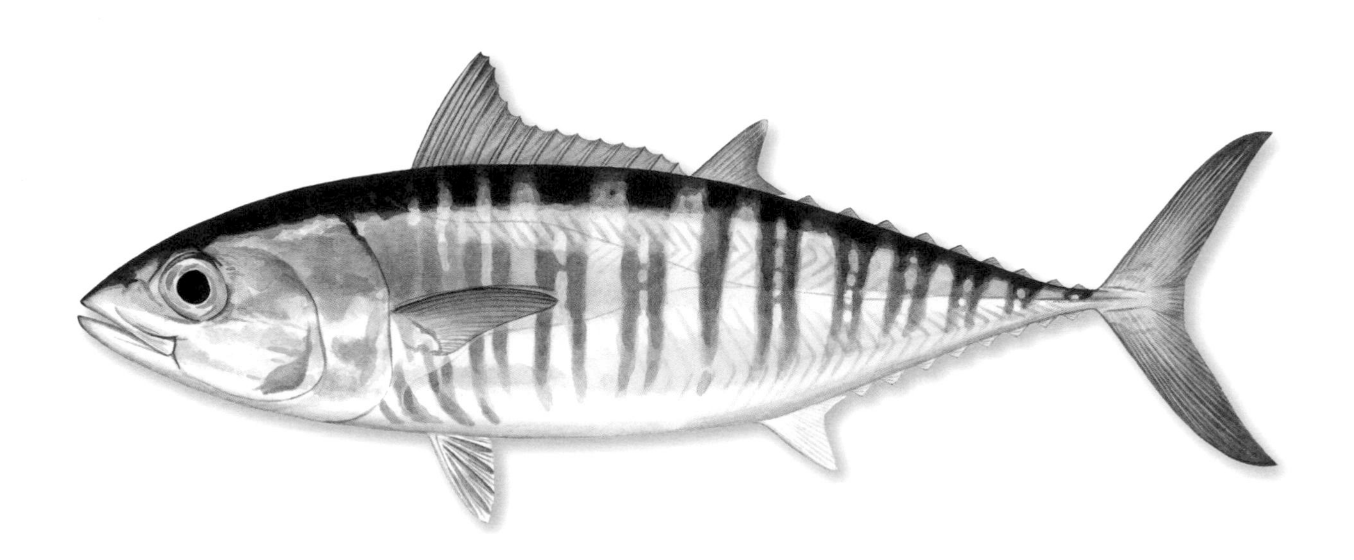

Johns Hopkins University Press
Baltimore

Printed in Canada on acid-free paper
2 4 6 8 9 7 5 3 1

Johns Hopkins University Press
2715 North Charles Street
Baltimore, Maryland 21218-4363
www.press.jhu.edu

Library of Congress Cataloging-in-Publication Data

Names: Collette, Bruce B., author. | Graves, John, 1953–, author.
Title: Tunas and billfishes of the world / Bruce Collette and John Graves ;
illustrated by Val Kells.
Description: Baltimore : Johns Hopkins University Press, 2019. | Includes
bibliographical references and index.
Identifiers: LCCN 2018055374 | ISBN 9781421431574 (hardcover : alk. paper) |
ISBN 1421431572 (hardcover : alk. paper)
Subjects: LCSH: Tuna--Identification. | Billfishes--Identification.
Classification: LCC QL638.S35 C624 2019 | DDC 597/.783--dc23
LC record available at https://lccn.loc.gov/2018055374

A catalog record for this book is available from the British Library.

Johns Hopkins University Press uses environmentally friendly
book material, including recycled text paper that is composed
of at least 30 percent post-consumer waste, whenever possible.

For Sara
—BC

For Sharon, Kyrie, and Courtney
—JG

For my father, David G. Kells Jr.,
and my grandfather, David G. Kells
—VK

Contents

ACKNOWLEDGMENTS

We would like to thank our dedicated editor Vince Burke for championing this book and seeing it through over six years of development. His unwavering support and thoughtful guidance helped make this book a reality. We also thank our production editor Debby Bors for her eagle eye. Her expertise and attention to detail gave this book a fine polish.

We send our deep gratitude to the following persons for providing assistance in the form of information, referrals, reference, and/or translations: Larry Allen, Martini Arostegui, Michel Bariche, Massimo Brogna, Daniel Golani, Gene Kira, Milton Love, Elvio Pennetti, Rosemary Robinson, Ross Robertson, Andrea Saldana, John Snow, and Steve Wozniak. Shane Griffiths, Anthony Lewis, Jason Schratwieser, Kurt Schaefer, and Mitchell Zischke provided valuable comments on several species.

Bruce thanks the collection managers and curators who provided access to collections of scombrids and the interns, assistants, and illustrators who helped with specimens and x-raying and particularly the collaborators who helped produce the taxonomic revisions of groups of scombrids, especially Chris Aadland, Labbish Chao, Gary Gillis, Bob Gibbs, and Joe Russo.

John would like to thank the many captains, mates, anglers, and fisheries scientists who not only shared their passion and knowledge of these incredible fishes, but also kindly provided many opportunities for him to catch and study them. He is particularly indebted to Guy Harvey, Julian Pepperell, Ken Neill, Jan McDowell, and his graduate students, from whom he has learned so much.

Val would like to thank her family, friends, and many fellow fish geeks for the support, encouragement, and appreciation that helped drive her desire to commit these beautiful and intriguing fishes to paint. Thank you, Dave and Drew! Each painting was sheer joy.

INTRODUCTION

Tunas and billfishes! Highly desired and avidly sought by big game fishermen. Important sources of food and income for many maritime nations whose fleets of vessels target them with longlines, which trail across the oceans for many miles, or huge purse seines. Canned tuna—a staple food for millions. Tuna was the fourth-most valuable globally traded fishery product in 2012. Fastest swimmers in the ocean. Peak predators of the oceans. All these statements pertain to some of the 61 species in 3 families of fishes: Scombridae, the mackerels and tunas; Istiophoridae, sailfish and marlins; and Xiphiidae, the Swordfish.

In this book we summarize all information available about these fishes: what they look like, accompanied by full-color illustrations by renowned illustrator Val Kells; and where they live, accompanied by distribution maps for each species. We also include summaries of their biology: size, food, habitat, reproduction, and early life history; fisheries interests; and conservation status. Much of the basic information comes from two Food and Agricultural Organization (FAO) catalogues, one by Bruce Collette and Cornelia Nauen on the mackerels and tunas (1983), the other by Izumi Nakamura on the billfishes (1985). Several changes in tuna and billfish taxonomy of scombrids during the past 30 years warranted an update of these references, and we included evaluations of the threat status for each of the 61 species developed in a series of workshops by the Tuna and Billfish Specialist Group of the International Union for the Conservation of Nature and published in their online Red List (2011).

The Scombridae comprises 15 genera and 51 species of epipelagic oceanic fishes. They possess many and varied morphological and physiological adaptations that are of great interest to physiologists and evolutionary biologists. They are divided into two subfamilies, the Gasterochismatinae, which contains only the Butterfly Kingfish, *Gasterochisma melampus*, and the Scombrinae, which contains the rest of the family. The Scombrinae consists of four tribes: Scombrini, the mackerels (*Scomber* and *Rastrelliger*); Scomberomorini, the Spanish mackerels (*Scomberomorus*), double-lined mackerels (*Grammatorcynus*) and the Wahoo (*Acanthocybium solandri*); Sardini, the four genera of bonitos (*Sarda*, *Cybiosarda*, *Orcynopsis*, and *Gymnosarda*); and the Thunnini, *Allothunnus*, frigate tunas, little tunas, Skipjack, and the higher tunas (*Thunnus*). The morphological evidence strongly indicates that there is a clear evolutionary trend within the Scombridae from the more primitive mackerels (Scombrini) and Spanish mackerels (Scomberomorini) to the more advanced bonitos (Sardini) and tunas (Thunnini). There are two families of billfishes, encompassing nine species of epipelagic oceanic fishes in the family Istiophoridae: Sailfish, *Istiophorus platypterus*; Black Marlin, *Istiompax indica*; Blue Marlin, *Makaira nigricans*; two species of smaller marlins, *Kajikia*; and four species of spearfishes, *Tetrapturus*; plus the Swordfish, the only member of the family Xiphiidae.

Scombrids are swift, epipelagic, or epi-mesopelagic predators; some species occur in coastal waters, others far from shore. Spanish mackerels, bonitos, and tunas feed on larger prey, including small fishes, crustaceans, and squids. The main predators of smaller scombrids are other predacious fishes, particularly large tunas and billfishes. Scombrids are dioecious (separate sexes), and most display little or no sexual dimorphism in structure or color pattern. Reports of courting behavior in scombrids are rare. Males of many species attain larger sizes than females. Batch spawning of most species takes place in tropical and subtropical waters, usually at least 24°C. The eggs are buoyant and pelagic and hatch into planktonic larvae. Larvae are characterized by large heads, triangular gut, and large jaws but are very difficult to identify to species. Mackerels and tunas support very important commercial and recreational fisheries as well as substantial artisanal fisheries throughout the tropical and temperate waters of the world. Catches in cold and warm temperate waters predominate over tropical catches, with more than half of the world catch being taken in the western North Pacific, the eastern North Atlantic, and the southeastern Pacific. Many species of tunas and mackerels are the target of distant-water and high-seas fisheries. The principal fishing methods used for fish schooling near the surface include purse seining, drift-netting, hook-and-line/bait boat fishing, and trolling; standard and deep longlining are used to target (usually bigger) fish, with gear set at depths ranging from near the surface to several hundred meters. Recreational fishing methods involve mostly surface trolling and pole-and-line fishing, while numerous artisanal fisheries deploy a great variety of gear including bag nets, cast nets, lift nets, gill (drift) nets, beach seines, hook and line, handlines, harpoons, specialized traps, and fish corrals.

Billfishes are at or near the apex of pelagic food webs, have broad diets, grow very rapidly, reach ages of 9–12 years or more, and undertake long-distance migrations. Females are usually larger than males. They are oviparous and highly fecund. Egg production increases with size, with annual fecundities of 0.75–19 million eggs. All but the smallest young billfishes are quite easily identified to family because the snout starts to elongate by 3 mm notochord length, although it does not take on the adult spear shape until a lower jaw fork length (LJFL) of 50 mm or longer. All are important sport fishes, and many are taken as incidental catch by longlining operations targeting tunas and swordfish and used for food. Several species are under intense fishing pressure and since the early 1980s, stock assessments have indicated that many stocks of billfish are overfished. Size limitations, encouragement of catch-and-release sport fishing, and recommendations for using circle hooks instead of J-hooks are measures designed to increase survival in catch-and-release sport fishing and may be instrumental in their successful management.

ECOLOGICAL SIGNIFICANCE: Large tunas and billfishes are at the top of the pelagic foodweb and contribute to pelagic ecosystem structure, functioning, and stability. Market tunas are some of the largest and fastest marine fishes. Atlantic and Pacific bluefins can exceed 2 m in length and have a maximum recorded weight of 685 kg. Tunas are capable of extreme speeds of up to

72 km per hour. Market tunas have extremely efficient metabolic systems, including a circulatory system that permits them to retain or disperse heat to achieve optimal performance, enabling them to tolerate a broad range of water temperatures, allowing these highly migratory species to make use of a range of oceanic habitats. For example, Atlantic Bluefin Tuna experience sea surface temperatures ranging from 3–20°C in northern waters during the summer. As they get larger, tunas produce more heat, enabling some species to inhabit higher latitudes and deeper waters. Tunas have strong schooling behavior and often associate with floating objects and large live animals (e.g. whale sharks, dolphins). They conduct large basin-scale migrations. For example, Bluefin and Albacore tunas can cross the Atlantic and Pacific in under 60 days. They are highly fecund and mature at a relatively young age. Bluefin tunas reach sexual maturity at about 4–7 years, while tropical tunas generally mature at 2–3 years. Temperate species can live over 20 years. Some nursery grounds are close to the coast, overlapping with the grounds of artisanal coastal fisheries.

Skipjack and Yellowfin tunas are largely tropical in distribution. Albacore and the three species of bluefin tunas occur in temperate waters. Bigeye Tuna are tropical in distribution but feed in cooler waters by feeding deeper than other tropical tunas. All of the principal market tunas spawn in warm tropical waters. The three bluefin tuna species spawn in limited areas: Pacific Bluefin spawn in an area off Taiwan, Atlantic Bluefin spawn in the Gulf of Mexico and the western Mediterranean, and southern Bluefin spawn between Australia and Indonesia. Bluefins also have relatively short spawning periods of of 1–2 months, whereas Bigeye, Yellowfin, Skipjack, Albacore, and many of the billfishes have long spawning seasons and extensive spawning grounds in tropical waters. Bluefin and Albacore make seasonal migrations between foraging grounds at higher latitudes and spawning grounds in tropical waters. Adult tunas prey primarily on small schooling fishes, such as anchovies and small squids, but also, when near shore, may eat crustaceans, plankton, and small shallow-water fishes. Tuna predators include marine mammals, sharks, billfishes, and other large predatory fishes, including larger tunas. With the increased fishing effort for tunas, bycatch (or incidental catch) of billfishes has also greatly increased.

SOCIOECONOMIC SIGNIFICANCE: Since ancient times, tuna and tuna-like species have been an important food source and are target species of fisheries worldwide. In the nineteenth century, most tuna fisheries were coastal and conducted by locally based fleets. Industrial tuna fisheries were initiated in the 1940s, and over the next few decades fishing grounds quickly expanded, as did the number of countries with large-scale coastal and distant-water tuna fleets. Demand for both canned and fresh tuna has been rapidly increasing, tripling from about 500,000 tons of tuna consumed in 1976 to about 1,500,000 tons in 2004.

INTRODUCTION

There are seven principal market species of tunas: Albacore (*Thunnus alalunga*), Atlantic Bluefin (*T. thynnus*), Pacific Bluefin (*T. orientalis*), Southern Bluefin (*T. maccoyii*), Bigeye (*T. obesus*), Yellowfin (*T. albacares*), and Skipjack (*Katsuwonus pelamis*). These seven species of market tunas and billfishes are all pelagic. Other tunas and some tuna-like species are primarily neritic, occupying waters primarily over continental shelves. Tuna-like species include, for example, Kawakawa (*Euthynnus affinis*), Frigate Tuna (*Auxis thazard*), Bullet Tuna (*A. rochei*), and Longtail Tuna (*T. tonggol*), Wahoo (*Acanthocybium solandri*), and Narrow-barred Spanish (*Scomberomorus commerson*) and King Mackerels (*S. cavalla*).

The seven principal market tuna species and several tuna-like species enter international trade as fresh, frozen, and canned products. Tuna was the fourth-most valuable globally traded fishery product in 2012, accounting for about 8% of the total $129 billion value of internationally traded fishery products, behind shrimp (15%), salmon (14%), and groundfish (10%). Reported landings of Skipjack and Yellowfin tunas represented the third and eighth top marine capture fishery landings by weight in 2012, making up 3.5% and 1.7% of total global landings by weight, respectively. Of the principal market tuna species, the three bluefin tuna species each make up less than 1% of total global reported landings by weight. The top six fishing nations supplying principal tunas are Japan, Taiwan, Indonesia, the Philippines, Spain, and the Republic of Korea, which in 2007 accounted for over half of the total weight of tuna landings. Since 2006, over half of principal market tunas have come from the western and central Pacific Ocean. In 2014, 60% of global total reported landings of combined Albacore, Bigeye, Skipjack, and Yellowfin tunas were from the western and central Pacific Ocean, 14% from the eastern Pacific Ocean, 9% from the Atlantic Ocean, and 17% from the Indian Ocean.

ORGANIZATION

ORGANIZATION OF THIS BOOK: The species accounts are arranged alphabetically by genera and species and summarize existing literature. Citations were not included to provide a more reader-friendly text. Short reference lists are included at the end of each species account, but the full bibliography (all 100+ pages) is available at www.press.jhu.edu. Each account is supported by a full-color illustration in life-like color and a range map. Information pertaining to each species is arranged as follows:

COMMON NAMES: English, Spanish, and French names used primarily by FAO.
ETYMOLOGY: The origin of the scientifc name and historical devolopment of its meaning.
SYNONYMS: Previous and different name combinations.
TAXONOMIC NOTE: A brief explanation about the taxonomy of the species.
FIELD MARKS: Primary identifying marks or features of the species.
UNIQUE FEATURE: A physical, geographic, or taxonomic feature unique to the species.
DIAGNOSTIC FEATURES: Detailed and distinctive characters of the species as an additional aid for identification.
GEOGRAPHIC RANGE: The entire known geographic range of the species, including seasonal occurrence and accompanied by a range map.
SIZE: The maximum known length and/or weight and the all-tackle angling record.
HABITAT AND ECOLOGY: Information about depth range, salinity, water temperature, migrations, school size, and predators.
MOVEMENTS: Seasonal migrations and displacements of individuals.
FOOD: Prey items, quantity and size of prey, and time of feeding when known.
REPRODUCTION: Age or size at sexual maturity, type of spawning, depth of spawning, and fecundity.
AGE AND GROWTH: Rate of growth, age at maturity, and longevity.
EARLY LIFE HISTORY: Size of newly hatched larva, feeding, and growth of larva.
STOCK STRUCTURE: Intraspecific biological and/or management units.
FISHERIES INTEREST: Where specie are targeted, nature of the gears and fisheries, annual landings, and utilization.
THREATS: Fisheries threats to stock stability, sustainability, and renewal and IUCN Red List status.
CONSERVATION: Current conservation and management measures.
REFERENCES: Full or select species account references.

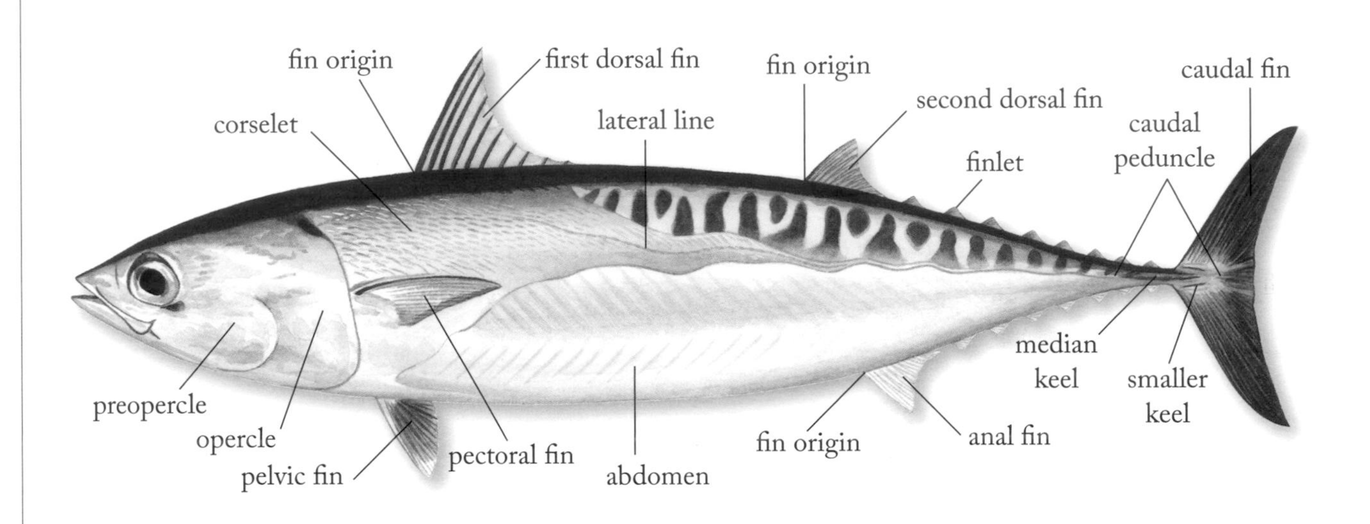
fin origin
first dorsal fin
fin origin
caudal fin
corselet
lateral line
second dorsal fin
caudal
peduncle
finlet
median
keel
smaller
keel
preopercle
opercle
pelvic fin
pectoral fin
abdomen
fin origin
anal fin

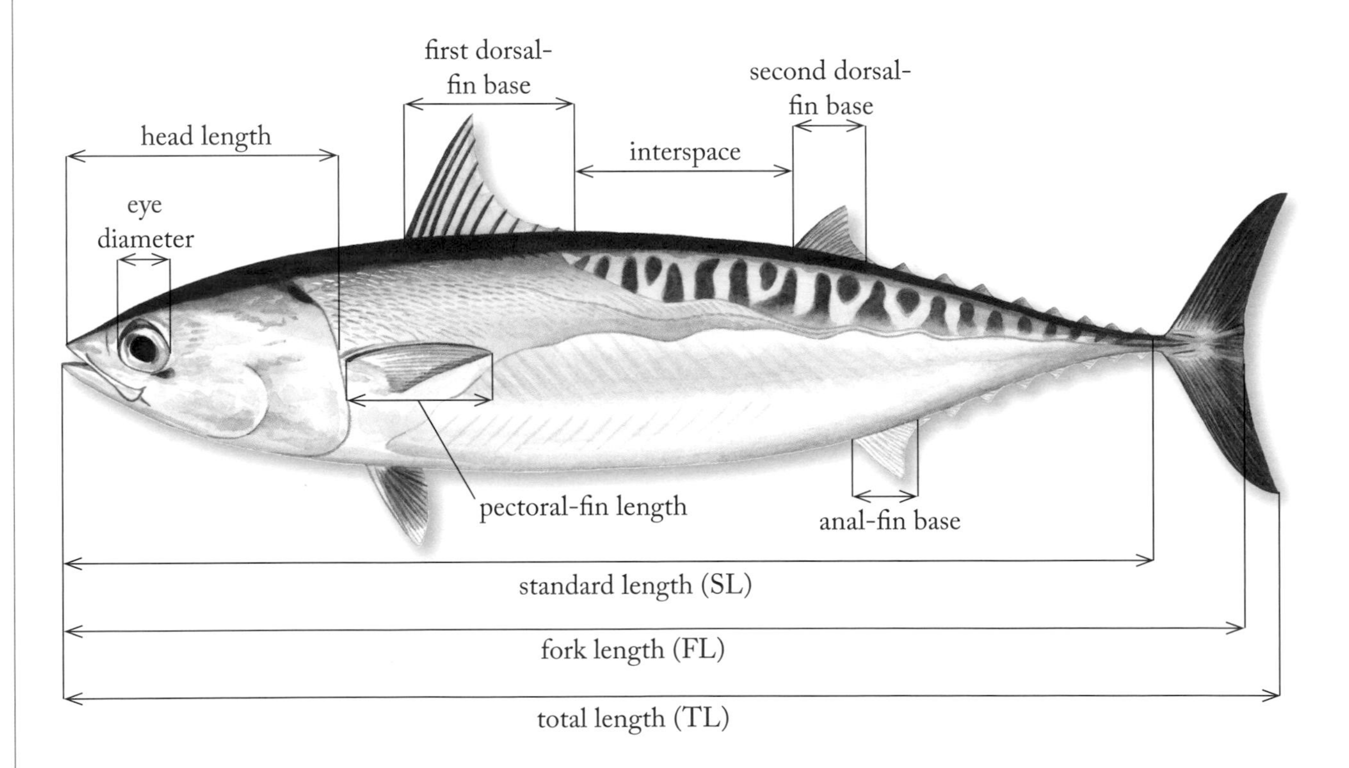
first dorsal-
fin base
second dorsal-
fin base
head length
interspace
eye
diameter
pectoral-fin length
anal-fin base
standard length (SL)
fork length (FL)
total length (TL)

ANATOMY and MEASUREMENTS

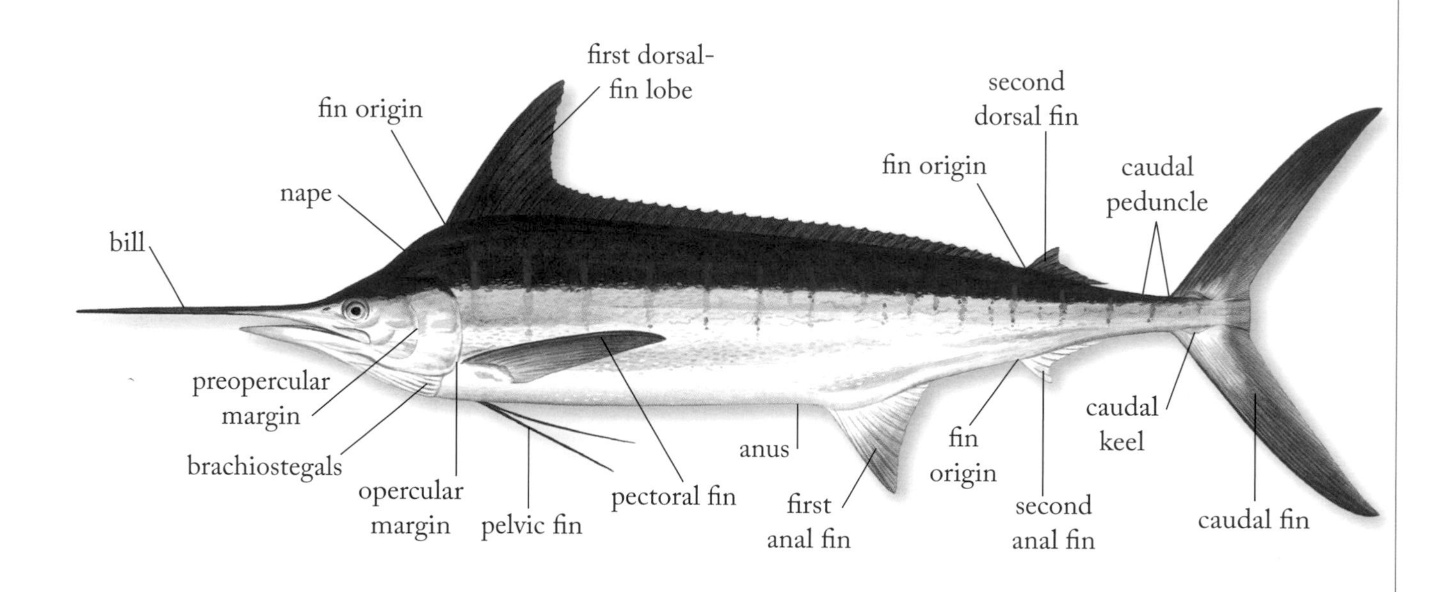

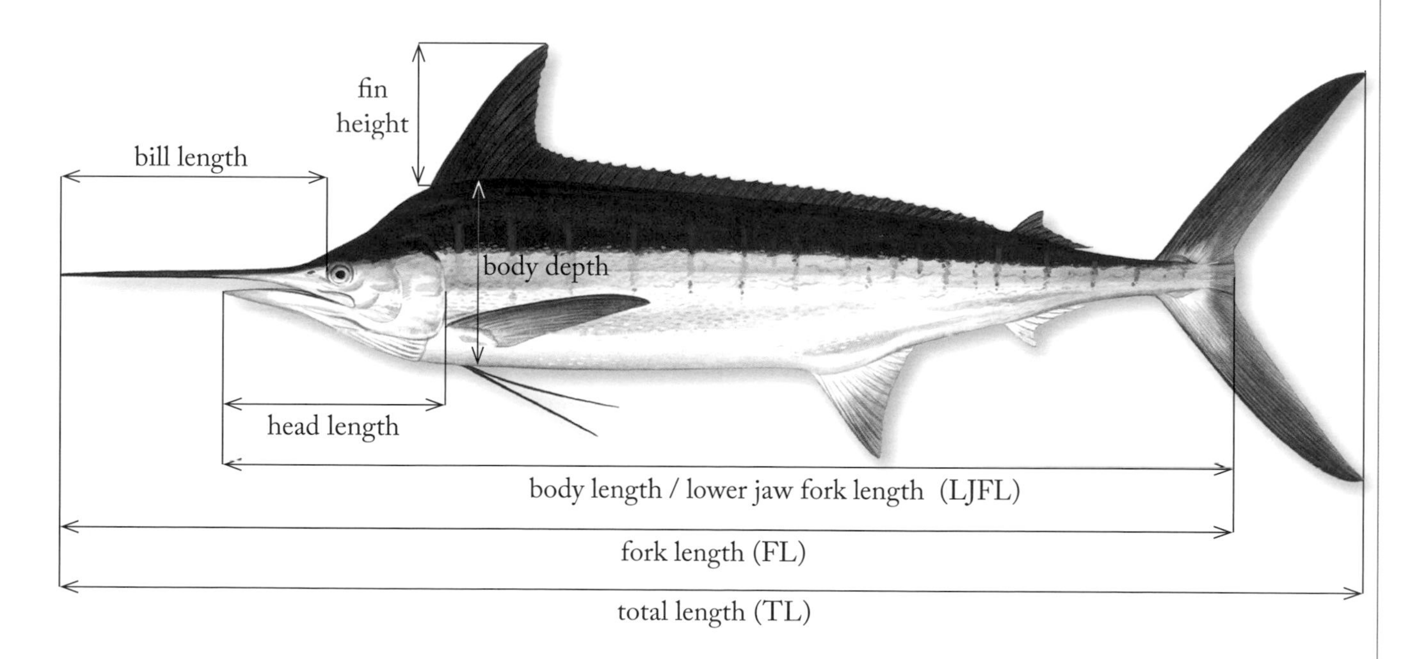

HOW TO USE THIS KEY

This key is broken into two sections. Each section consists of couplets and triplets of characters that, by process of elimination, guide the user to an identification. To use this key, start with the first alternative in the first couplet. If the description fits the specimen, follow the direction given at the right-hand margin. If a number is given, go to the couplet with that number for the next character. If a name is given, you have identified the species in question. If the first alternative does not fit the specimen, go to the second character in the couplet or triplet, which will lead you to another name or another couplet or triplet. Proceed down the key and through as many couplets or triplets as necessary to identify the species in question.

1a Upper jaw bluntly rounded to bluntly pointed in adults; pectoral fins high on the body; first dorsal fin with 9–10 spines; second dorsal fin followed by 5 or more finlets .... **SCOMBRIDAE**

1b Upper jaw elongate and flattened or round in cross section in adults; pectoral fins low on the body; first dorsal fin with 34 or more spines; finlets absent .... **XIPHIIDAE, ISTIOPHORIDAE**

SCOMBRIDAE

1a Body covered with large cycloid scales; pelvic fins huge in juveniles, proportionally smaller in adults and depressible into a groove at all sizes .... *Gasterochisma melampus*—p. 63

1b Body naked or covered with small to moderate-sized scales; pelvic fins small, groove into which they fit absent 2

2a Two small keels on either side of the caudal peduncle; 5 dorsal and 5 anal finlets present; adipose eyelids cover front and rear of eye 3

2b Two small keels and a large median keel between them present on either side of the caudal peduncle; 7–10 dorsal and 6–10 anal finlets present; adipose eyelids absent 8

3a Entire body covered with moderately large scales; gill rakers very long, longer than gill filaments and plainly visible through open mouth; teeth absent on vomer or palatine bones 4

3b Entire body covered with small scales; gill rakers shorter than gill filaments and barely visible through open mouth; teeth present on vomer and palatine 6

4a Gill rakers on lower half of first arch 21–26; body relatively slender with depth at posterior margin of opercle contained 4.9 to 6 times in fork length; length of intestine equal to or less than fork length .... *Rastrelliger faughni*—p. 93

4b Gill rakers on lower half of first arch 30–48; body relatively deep with depth at posterior margin of opercle contained 3.7 to 5.2 times in fork length; length of intestine 1.4 to 3.6 times the fork length 5

5a Body depth at posterior margin of opercle contained 4.3 to 5.2 times in fork length; length of intestine 1.4 to 1.8 times the fork length .... *Rastrelliger kanagurta*—p. 97

5b Body depth at posterior margin of opercle contained 3.7 to 4.3 times in fork length; length of intestine 3.2 to 3.6 times the fork length .... *Rastrelliger brachysoma*—p. 89

6a Space between end of first dorsal-fin groove and second dorsal-fin origin greater than total length of first dorsal-fin groove—about 1.5 times as long; swim bladder absent; belly unmarked; 13 precaudal plus 18 caudal vertebrae; 21–28 interneural bones under first dorsal fin .... *Scomber scombrus*—p. 139

6b Space between end of first dorsal-fin groove and second dorsal-fin origin about equal to or less than length of groove; swim bladder present; belly unmarked or marked by spots or broken wavy lines; 14 precaudal plus 17 caudal vertebrae; 12–20 interneural bones under first dorsal fin 7

7a First dorsal-fin spines 9–10; distance from last dorsal-fin spine to second dorsal-fin origin less than distance between first and last spine; belly unmarked or vaguely marked; 12–15 interneural bones under first dorsal fin .... *Scomber japonicus*—p. 132

7b First dorsal-fin spines 9–10; distance between end of first dorsal-fin groove and second dorsal-fin origin equal to or less than length of groove; belly spotted; 12–15 interneural bones under first dorsal fin .... *Scomber colias*—p. 126

7c First dorsal-fin spines 10–13; distance from tenth dorsal-fin spine to second dorsal-fin origin greater than distance between first and tenth spines; belly spotted; 15–20 interneural bones under first dorsal fin.............................. *Scomber australasicus*—p. 121

8a Two lateral lines, the lower joining upper behind pectoral fins and at caudal peduncle; interpelvic process single and small; vertebrae 31.............................. 9

8b One lateral line; interpelvic process single or double; vertebrae 41–64.............................. 10

9a Eye small (3–4% of fork length); 14 or 15 gill rakers on first arch .... *Grammatorcynus bicarinatus*—p. 67

9b Eyes large (7–9% of fork length); 19–24 gill rakers on first arch .... *Grammatorcynus bilineatus* —p. 70

10a Teeth in jaws strong, compressed, almost triangular or knife-like; corselet of scales obscure............ 11

10b Teeth in jaws slender, conical, hardly compressed; corselet of scales well developed.............................. 29

11a Snout as long as rest of head; gill rakers absent; first dorsal fin with 23–27 spines; posterior end of maxilla concealed under preorbital bone; vertebrae 62–64 .... *Acanthocybium solandri*—p. 24

11b Snout much shorter than rest of head; 1–27 gill rakers on first arch; first dorsal fin with 12–22 spines; posterior end of maxilla exposed; vertebrae 41–56.............................. 12

12a Lateral line with a deep dip below first or second dorsal fin; vertebrae 40–46.............................. 13

12b Lateral line straight or descending gradually to caudal peduncle; vertebrae 44–56 15

13a Dip in lateral line under first dorsal fin; total gill rakers on first arch 12–15; caudal vertebrae 21 or 22 .... *Scomberomorus sinensis*—p. 219

13b Dip in lateral line under second dorsal fin; total gill rakers on first arch 2–13; caudal vertebrae 23–27 14

14a Total gill rakers on first arch 7–13; precaudal vertebrae 16 or 17 ....*Scomberomorus cavalla*—p. 151

14b Total gill rakers on first arch 3–8; precaudal vertebrae 19 or 20 ...... *Scomberomorus commerson*—p. 157

15a Total gill rakers on first arch 21–27; no spots or bars on sides of body of males................................. ....*Scomberomorus concolor*—p. 164
15b Total gill rakers on first arch 1–18; spots, bars, or other markings on sides usually present, except in *S. multiradiatus* 16

16a Anal-fin rays 25–29; second dorsal-fin rays 21–25, usually 23 or more; gill rakers on first arch 1–4; total vertebrae 54–56; body unmarked.......................................*Scomberomorus multiradiatus*—p. 185
16b Anal-fin rays 15–24; second dorsal-fin rays 15–24; total gill rakers on first arch 3–18; total vertebrae 44–53; sides usually with spots or other markings.. 17

17a Dorsal-fin spines 19–22 18
17b Dorsal-fin spines 13–19, usually 18 or fewer.. 19

18a First dorsal fin black between first and fifth to seventh spine, white posteriorly; intestine straight with no folds; total vertebrae 48–50 ....*Scomberomorus niphonius*—p. 192
18b First dorsal fin black or almost black to posterior end; intestine with 2 loops and 3 limbs; total vertebrae 50–52 .... *Scomberomorus munroi*—p. 188

19a Lateral line with many small auxillary branches anteriorly.. 20
19b Lateral line without or with only a few auxillary branches anteriorly... 21

20a Dorsal-fin spines 15–18, usually 16 or more; intestine with 2 loops and 3 limbs; total vertebrae 47–52, usually 50 or 51; head long, 20.2–21.5% of fork length; body depth shallow, 22.8–25.2% of fork length... *Scomberomorus guttatus*—p. 168
20b Dorsal-fin spines 14–17, usually 14 or 15; intestine with 4 loops and 5 limbs; total vertebrae 46 or 47, usually 46; head short, 19.7–20.4% of fork length; body depth deep, 24.4–26.7% of fork length... .... *Scomberomorus koreanus*—p. 173

21a Sides with spots and at least one stripe that may be short, wavy, or interrupted.............................. 22
21b Sides without stripes; spots usually present 24

22a One long stripe on sides with spots or interrupted lines above and below; total gill rakers on first arch 12–18, usually 15 or 16; total vertebrae 47 or 48, usually 48 ....*Scomberomorus regalis*—p. 206
22b Sides with several rows of short to long stripes; total gill rakers on first arch 9–15, usually 14 or fewer; total vertebrae 44–47, usually 46... 23

23a Sides with a series of straight stripes and few if any spots; total gill rakers on first arch usually 11 or fewer; second dorsal-fin rays 15–19, rarely 21 or 22, usually 18 or fewer.. .... *Scomberomorus lineolatus*—p. 176
23b Sides with a series of short wavy lines plus many small spots; total gill rakers on first arch usually 12 or more; second dorsal-fin rays 19–21, usually 20 or more .... *Scomberomorus plurilineatus*—p. 197

24a Sides with bars or large spots, mostly larger than eye diameter 25
24b Sides with small round spots, at most about equal to eye diameter and orange in life...................... 27

25a Sides with relatively large, round, and regular spots or blotches; total gill rakers on first arch 3–9, usually 7 or fewer .... *Scomberomorus queenslandicus*—p. 201
25b Sides either with bars or with irregular, vertically elongate spots; total gill rakers on first arch 6–15, usually 9 or more 26

26a First dorsal-fin spines 13–15; second dorsal-fin rays 19–22, usually 20 or more; total gill rakers on first arch 6–13, usually 11 or fewer; total vertebrae 44–46, usually 45; sides with broad crossbars that tend to fade in adults .... *Scomberomorus semifasciatus*—p. 210
26b First dorsal-fin spines 15–18, usually 16 or more; second dorsal-fin rays 16–19, usually 17; total gill rakers on first arch 12–15; total vertebrae 46 or 47, usually 46; sides with irregular, elongate spots that tend to form crossbars in large adults .... *Scomberomorus tritor*—p. 222

27a Total vertebrae 51–53; second dorsal-fin rays 17–20, usually 18 or more .... *Scomberomorus maculatus*—p. 180
27b Total vertebrae 46–49; second dorsal-fin rays 15–19, usually 18 or fewer 28

28a Pectoral-fin rays 21–24, usually 22 or 23; pelvic fins short, 3.6–5.9% of fork length .... *Scomberomorus brasiliensis*—p. 146
28b Pectoral-fin rays 20–24, usually 21; pelvic fins comparatively longer, 4.7–6.4% of fork length .... *Scomberomorus sierra*—p. 215

29a Upper surface of tongue without cartilaginous longitudinal ridges 30
29b Upper surface of tongue with 2 longitudinal ridges 37

30a Jaw teeth tiny, 44–55 on each side of upper and lower jaws; gill rakers fine and numerous, total gill rakers on first arch 70–80; body elongate; distance from snout to second dorsal fin 61–65.4% of fork length; maxilla comparatively short, 35.4–37.9% of head length .... *Allothunnus fallai*—p. 30
30b Jaw teeth larger and more prominent, 10–30 on each side of upper and lower jaws; total gill rakers on first arch 8–27; body less elongate; distance from snout to second dorsal fin 48.1–61% of fork length; maxilla comparatively long, 43.1–55.7% of head length 31

31a Five to ten narrow, dark longitudinal stripes on upper part of body; teeth absent on tongue; spleen prominent in posterior third of body cavity in ventral view 32
31b Body either without stripes or with dark spots above lateral line and longitudinal dark stripes below; two patches of teeth present on tongue; spleen either concealed or located in anterior third of body cavity in ventral view 35

32a First dorsal fin with 20–23 spines; total vertebrae 50–55 .... *Sarda sarda*—p. 116
32b First dorsal fin with 17–19 spines; total vertebrae 43–46 33

33a Total gill rakers on first arch 8–13; supramaxilla narrow .... *Sarda orientalis*—p. 111
33b Total gill rakers on first arch 19–27; supramaxilla comparatively wider 34

34a Total gill rakers on first arch 19–21; pectoral rays 25–27, usually 26; teeth sometimes present on vomer; length of first dorsal-fin base 31.5–34.3% of fork length; maxilla 50.3–53.9% of head length. .... *Sarda australis*—p. 102

34b Total gill rakers on first arch 23–27; pectoral rays 22–26, usually 24 or 25; teeth absent on vomer; length of first dorsal-fin base 26.7–31.4% of fork length; maxilla 46–50.3% of head length........ *Sarda chiliensis*—p. 105

35a Body with dark spots above lateral line and dark longitudinal stripes below; first dorsal-fin spines 16–18........ *Cybiosarda elegans*—p. 44

35b Body without prominent pattern of stripes or spots; first dorsal-fin spines 12–15 36

36a Pectoral-fin rays 21–23; small conical teeth in jaws; total gill rakers on first arch usually 14 or more; interpelvic process bifid; spleen not visible in ventral view; laminae in olfactory rosette 25–28; interorbital width 23.9–31% of head length .... *Orcynopsis unicolor*—p. 85

36b Pectoral-fin rays 25–28; jaw teeth very large and conspicuous; total gill rakers on first arch usually 13 or fewer; interpelvic process single; spleen visible on right side of body cavity in ventral view; laminae in olfactory rosette 48–56; interorbital width 32.1–40% of head length .... *Gymnosarda unicolor*—p. 74

37a First and second dorsal fins widely separated, the space between them at least equal to length of first dorsal-fin base; first dorsal-fin spines 10–12; interpelvic process single and long, at least as long as longest pelvic-fin ray 38

37b First and second dorsal fins barely separated, at most by a space equal to eye diameter; first dorsal-fin spines 12–16; interpelvic process bifid and short, much shorter than pelvic-fin rays........ 39

38a Posterior extension of corselet narrow, only 1 to 5 scales wide under second dorsal-fin origin; pectoral fin extends to a point beyond anterior margin of dorsal scaleless area .... *Auxis thazard*—p. 39

38b Posterior extension of corselet wide, 10–15 scales wide under second dorsal-fin origin; pectoral fin does not reach past anterior margin of dorsal scaleless area........ *Auxis rochei*—p. 34

39a Three to five prominent dark longitudinal stripes on belly; total gill rakers on first arch 53–63; total vertebrae 41 .... *Katsuwonus pelamis* —p. 78

39b No dark longitudinal stripes on belly; total gill rakers on first arch 19–45; total vertebrae 37–39.... 40

40a Body naked behind corselet; several black spots usually present between pectoral- and pelvic-fin bases; back with a complex striped pattern under dorsal-fin bases; pectoral-fin rays 25–29 41

40b Body covered with very small scales behind corselet; no black spots on body; back lacks a striped pattern; pectoral-fin rays 30–36........ 43

41a Vomerine teeth absent; total gill rakers on first arch 37–45........ *Euthynnus alletteratus*—p. 53

41b Vomerine teeth present; total gill rakers on first arch 29–39 42

42a Total gill rakers on first arch 29–33; total vertebrae 39; bony caudal keels on vertebrae 33 and 34; no trace of vertebral protuberances.......... *Euthynnus affinis*—p. 48
42b Total gill rakers on first arch 33–39; total vertebrae 37; bony caudal keels on vertebrae 31 and 32; conspicuous protuberances on 31st and 32nd vertebrae .... *Euthynnus lineatus*—p. 58

43a Ventral surface of liver with prominent striations; center lobe of liver equal to or longer than left or right lobes.......... 44
43b Ventral surface of liver lacks striations; right lobe of liver much longer than left or central lobes 47

44a Total gill rakers on first arch 31–43; pectoral fin short, less than 80% of head length.......... 45
44b Total gill rakers on first arch 23–31; pectoral fin moderate or long, greater than 80% of head length.......... 46

45a Pectoral fin 17.0–21.7% of fork length; median caudal keel dark; first ventrally directed parapophysis on eighth vertebra.......... *Thunnus thynnus*—p. 265
45b Pectoral fin16.8–20.8% of fork length; median caudal keel dark; first ventrally directed parapophysis on eighth vertebra.......... *Thunnus orientalis*—p. 259
45c Pectoral fin 20.2–23% of fork length; median caudal keel yellow; first ventrally directed parapophysis on ninth vertebra .... *Thunnus maccoyii*—p. 246

46a Caudal fin with narrow white posterior margin; pectoral fin very long, reaching well past end of second dorsal-fin base; greatest body depth at or slightly before second dorsal-fin level.......... *Thunnus alalunga*—p. 225
46b Caudal fin lacks white margin; pectoral fin short or moderate in length, not reaching end of second dorsal-fin base (except in small specimens); greatest body depth about middle of body, near middle of first dorsal fin .... *Thunnus obesus*—p. 252

47a Total gill rakers on first arch 26–34, usually 27 or more; second dorsal and anal fins of larger specimens (120 cm fork length or larger) elongate, more than 20% of fork length; maximum size is over 200 cm fork length .... *Thunnus albacares*—p. 232
47b Total gill rakers on first arch 19–28, usually 26 or fewer; second dorsal and anal fins never greatly elongate, less than 20% of fork length at all sizes; maximum size less than 110 cm fork length....... 48

48a Lower sides of body with a pattern of pale streaks and spots either horizontally oriented or without obvious orientation; swim bladder absent or rudimentary; vertebrae 18+21=39.......... *Thunnus tonggol*—p. 274
48b Lower sides lack pale streaks and spots, or with such markings at least partially in vertical rows; swim bladder present; vertebrae 19+20=39.......... *Thunnus atlanticus*—p. 240

KEY to SPECIES

XIPHIIDAE, ISTIOPHORIDAE

1a Pelvic fins absent; a large median keel on each side of the caudal peduncle region; bill depressed in cross section of adults; first and second dorsal fins well separated (much greater than body depth); no scales on body or teeth in jaws of adults .... *Xiphias gladius*—p. 280

1b Pelvic fins present; a pair of caudal keels on each side of the caudal peduncle region; bill rounded in cross section; distance between first and dorsal fins short (less than 1/2 body depth); body covered with small, elongate bony scales on body and small, rasp-like teeth present in jaws of adults 2

2a First dorsal fin sail-like and much higher than body depth at mid-body; pelvic-fin rays very long, tips nearly reaching to origin of anal fin .... *Istiophorus platypterus*—p. 294

2b First dorsal fin not sail-like and lower than body depth at mid-body; pelvic-fin rays short, tips well separated from origin of anal membrane 3

3a Height of anterior lobe of first dorsal fin less than body depth in adults; nape highly elevated; body not compressed laterally 4

3b Height of anterior lobe of first dorsal fin slightly greater than body depth in adults; nape slightly or not elevated; body compressed laterally 5

4a Pectoral fins flexible, capable of being folded back against sides of body; origin of second dorsal fin slightly posterior to insertion of second anal fin .... *Makaira nigricans*—p. 317

4b Pectoral fins rigid, not capable of being folded against sides of body; origin of second dorsal fin slightly anterior to insertion of second anal fin .... *Istiompax indica*—p. 289

5a Anus situated slightly anterior to origin of first anal fin, the distance being less than half the height of first anal fin 6

5b Anus situated far anterior to origin of first anal fin, the distance between them greater than half the height of first anal fin 7

6a Tip of first dorsal fin, pectoral fins, and first anal fin rounded .... *Kajikia albida*—p. 303

6b Tip of first dorsal fin, pectoral fins, and first anal fin pointed .... *Kajikia audax*—p. 310

7a Bill short, its length usually equal to or shorter than head length; pectoral fins narrow and short, less than 18% of body length 8

7b Bill long, its length usually equal to or longer than head length; pectoral fins wide and long, more than 18% of body length 9

8a Bill very short, less than 15% of body length .... *Tetrapturus angustirostris*—p. 324

8b Bill moderately short, less than 18% of body length .... *Tetrapturus belone*—p. 328

9a The distance between the anus and the anal-fin origin is nearly equal to anal-fin height; first anal fin triangular in shape with pointed tip .... *Tetrapturus pfluegeri*—p. 336

9b Anus situated closer to origin of first anal fin, the distance between the anus and the anal-fin origin equal to more than half the anal-fin height; first anal fin triangular in shape with a truncated tip .... *Tetrapturus georgii*—p. 332

Family SCOMBRIDAE

Tunas and Mackerels

Wahoo

Acanthocybium solandri (Cuvier, 1832)

COMMON NAMES: English – Wahoo
French – Thazard-bâtard
Spanish – Peto
Hawaiian – Ono

ETYMOLOGY: Named by the great French anatomist and ichthyologist Georges Cuvier (1832) after Daniel Solander, who provided much of the description for Cuvier in his unpublished papers.

SYNONYMS: *Cybium solandri* Cuvier, 1832; *Cybium sara* Lay and Bennett, 1839; *Cybium petus* Poey, 1860; *Cybium verany* Döderlein, 1872; *Acanthocybium forbesi* Seale, 1912; *Scomber amarui* Curtiss, 1938; *Acanthocybium solanderi* (Cuvier, 1832)

TAXONOMIC NOTE: Recent genetic information suggests that this species consists of one highly connected population worldwide.

FIELD MARKS:

1 Wahoo resemble an elongate Spanish mackerel but the snout is much longer and about as long as the rest of the head.
2 Gill rakers are absent.
3 The posterior end of the upper jaw (maxilla) is concealed under the preorbital bone instead of being exposed as in the Spanish mackerels.
4 There are more dorsal spines (23–27) than in any of the Spanish mackerels (12–22).

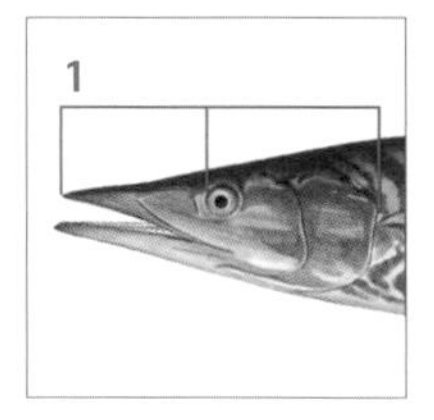

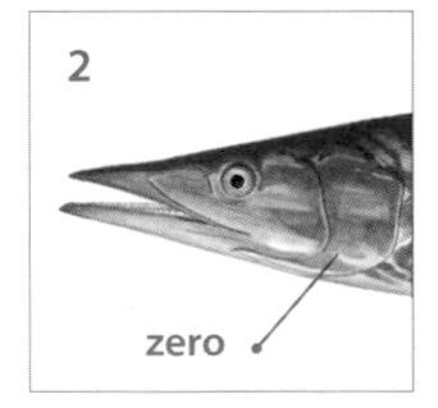

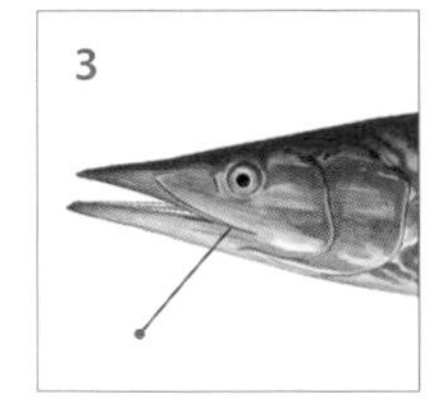

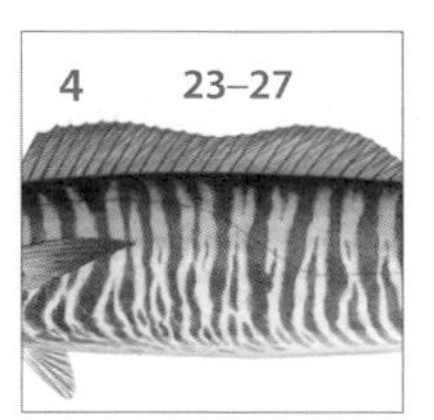

UNIQUE FEATURE: The scombrid with the most vertebrae (62–64) and the only scombrid with no gill rakers.

DIAGNOSTIC FEATURES: The body is very elongate, fusiform, and only slightly laterally compressed. The mouth is large with strong triangular, compressed, and slightly serrate teeth closely set in a single series in both jaws. The snout is about as long as the rest of the head. **Gill rakers:** Unlike all other scombrids, no gill rakers are present. The gills are modified for increased oxygen uptake by having thin-walled lamellae and reduced water-blood barrier distances, similar to higher tunas. Gill filaments and filament fusions on both the leading and trailing edges of the gill are covered by bony epithelial toothplates as in the billfishes and unlike the condition in other scombrids. The posterior part of the maxilla is completely concealed under the preorbital bone. **Fins:** The first dorsal fin has 23–27 spines, usually 26, separated by a short distance from the second dorsal fin, which has 11–16 rays, usually 13, followed by 7–10 finlets. The anal fin has 11–14 rays followed by 7–10 finlets. The pectoral fins have 22–26 rays. The interpelvic process is small and bifid. **Caudal peduncle:** The caudal peduncle is slender with a well-developed lateral keel between the two smaller keels on each side. **Swim bladder:** A swim bladder is present. **Vertebrae:** Vertebrae number 30–32 precaudal plus 31–33 caudal, totaling 62–64. Descriptions and illustrations of the osteology and soft anatomy are presented by Collette and Russo (1984). **Color:** The body is iridescent blue-green; the sides are silvery with 24–30 cobalt-blue vertical bars that extend to below the lateral line.

GEOGRAPHIC RANGE: Wahoo are present in the Atlantic, Indian, and Pacific oceans in tropical and subtropical waters, including the Caribbean and Mediterranean seas. There are records from around many Atlantic islands, Bermuda, Saint Peter and Saint Paul Archipelago, Ascension, Madeira, Cape Verdes, and São Tome and Principe. There are only three verified records from the Mediterranean. In the eastern Pacific, it occurs from southern California south to northern Peru, including all of the oceanic islands.

SIZE: The maximum recorded length is at least 210 cm from St. Lucia, West Indies. The maximum recorded weight is 213 lb (96.4 kg) while the IGFA all-tackle game fish record is of a 184 lb (83.46 kg) fish taken off Cabo San Lucas on the Pacific coast of Mexico in July 2005.

HABITAT AND ECOLOGY: This is an oceanic, epipelagic species frequently found solitarily or forming small, loose aggregations rather than compact schools. It spends the vast majority of its time both day and night above the thermocline, at least around topographic features, but occasionally dives to at least 340 m. In the western North Atlantic, pop-up satellite tags showed Wahoo spent more than 90% of their time in water less than 200 m at temperatures between 17.5°C and 27.5°C. The probability of occurrence of Wahoo in the eastern Pacific Ocean has been predicted to be the highest when sea surface temperatures are 20–25°C and chlorophyll-a concentrations are <2 mg m^{-3}.

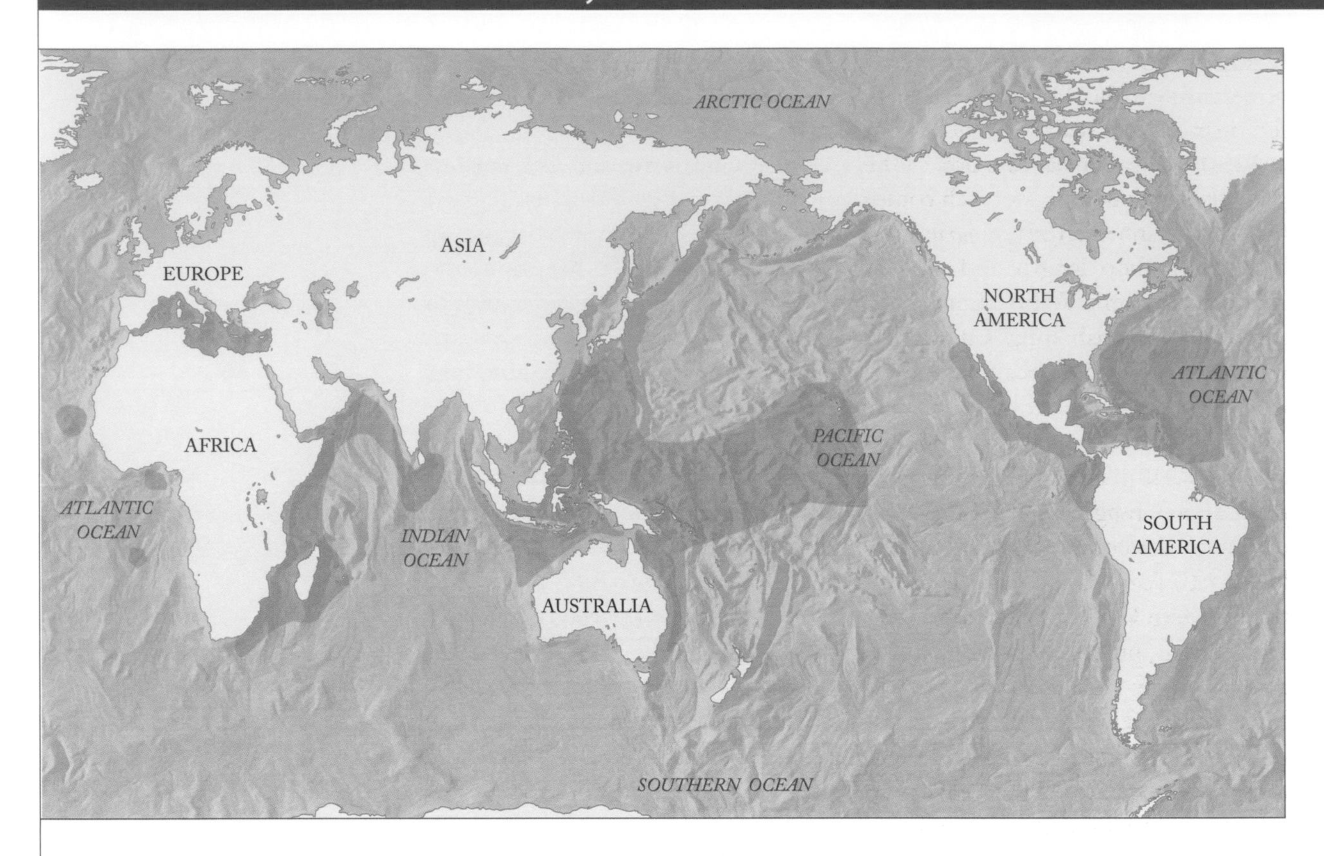

Above: Wahoo range map.

MOVEMENTS: Relative to other large scombrids, there is little information on horizontal movements of Wahoo. Individuals have undertaken extensive movements, traveling straight-line distances of almost 2,000 km in 31 days; however, studies in the eastern North Pacific demonstrated net displacements of less than 20 km after times at liberty of up to 68 days. In the western central Atlantic, there are shifts in seasonal abundance, with Wahoo moving into northern waters during summer months.

FOOD: Wahoo feed primarily on pelagic and epipelagic fishes but occasionally consume squids. Along the Atlantic and Gulf coasts of the United States, Frigate Tuna (*Auxis*), porcupinefish (*Diodon*), and flyingfishes occurred most frequently among the approximately 30 different species of fishes found in the stomach contents of 466 Wahoo. In the Gulf of Mexico, 27 species were found in 127 Wahoo stomachs with the most important identifiable prey being a jack, the Blue Runner (*Caranx crysos*), and dolphinfishes (*Coryphaena* sp.), followed by small tunas and flyingfishes. Small Wahoo appear to feed on smaller flyingfishes and jacks while larger Wahoo fed on Blue Runner, Dolphinfish, and small tunas. Around the São Pedro and São Paulo Archipelago, Brazil, squids and the flyingfish *Cypselurus cyanopterus* are important food items, together with other fishes, such as *Oxyporhamphus micropterus* and the Flying Gurnard, *Dactylopterus volitans*. In the Pacific Ocean, 11 different prey species were found in Wahoo in the Galapagos Marine Reserve, with the flyingfish *Prognichthys tringa* and the Humboldt squid *Dosidicus gigas* representing 40% of the diet. In the Line Islands of the central Pacific, Wahoo fed on mackerel scads (*Decapturus* spp.), squids, Skipjack Tuna (*Katsuwonus pelamis*) and other fishes. Also in the central Pacific, the most frequent fish prey items recorded in stomach contents of 43 Wahoo were Chiasmodontidae (23%) and lancetfish, *Alepisaurus* (14%). This study did not find a relationship between Wahoo length and prey length and noted a fast digestion rate. Nearshore-caught Wahoo from the central North Pacific fed largely on pre-settlement reef fish species during the summer followed by jacks in autumn months, while snake mackerels, remoras, and scombrids were dominant prey in Wahoo caught in the offshore fishery.

REPRODUCTION: Previous studies have reported a skewed sex ratio in Wahoo populations, ranging from 1.3 to 3.4 females to every male. Wahoo exhibit early sexual maturity and live for at least 5–6 years, and some may live as long as 10 years. Generation length is estimated to be between 3–5 years. In the northern Gulf of Mexico, 50% sexual maturity in males is reached before 93.5 cm FL, probably at an age of 1, and for females, size at 50% maturity is approximately 102 cm FL, at an estimated age of 2 years. For the Bahamas and Florida, 50% female maturity is reached at 92.5 cm FL, 0.6 years of age. Much greater age at first maturity was reported for Saint Peter and Saint Paul's off the coast of Brazil, males at 102 cm and females at 110 cm, corresponding to a length of approximately 3.8 years. Off eastern Australia, 50% of females reach maturity by 104.6 cm and approximately 7 months of age. Wahoo have an extended spawning season during warmer months in both the northern and southern hemispheres. The spawning season of Wahoo in the Gulf of Mexico, Bahamas, and Bermuda extends from at least May to October. Similarly, a spawning season from April–October was observed for female Wahoo off the Atlantic coast of Florida and the Bahamas. At Saint Peter and Saint Paul's rocks off Brazil, spawning occurs in April and May. Off the Pacific coast of Australia, female Wahoo spawn in October–February. Females are multiple batch spawners and are highly fecund. Mean spawning frequency has been reported as 2–9 days during a spawning season. Predicted mean weight at length in Hawaii was highest at the beginning of the spawning season in June and lowest after the spawning season in September. Mean batch fecundity in the northern Gulf of Mexico is 1.1 million eggs, for Florida and the Bahamas 0.44–1.57 million eggs, and relative fecundity was estimated as 57.7 (+/- 5.4) oocytes per gram of ovary-free body weight. Average batch fecundity in Saint Peter and Saint Paul's Rocks is about 1.3 million eggs. Fish in different maturity stages are frequently caught at the same time. Batch fecundity is positively correlated with fish size in eastern Australia, ranging between 0.65 and 5.12 million oocytes. Relative fecundity was estimated at 122 (+/- 9.7) oocytes per gram of ovary-free body weight.

AGE AND GROWTH: Wahoo are short lived, grow rapidly in their first year, achieve a large size, and have high mortality rates. Accounting for differences in body size, Wahoo appear to grow faster than other scombrids and most istiophorid billfishes, with a growth rate and maximum size most similar to dolphinfishes.

EARLY LIFE HISTORY: The eggs of Wahoo were identified using genetic techniques, then they were described and illustrated. The eggs ranged in diameter from 1.02–1.04 mm. Hatching is at 2.5 mm and the length at flexion is about 6 mm. Larval Wahoo were identified from off Hawaii using DNA barcoding, a molecular technique. Matsumoto (1967) described and illustrated 8 of 38 larvae 2.8–23.7 mm SL from the central Pacific Ocean.

Above: Juvenile Wahoo, approx. 30 mm.

FISHERIES INTEREST: Wahoo are targeted in artisanal and recreational fisheries and are captured as bycatch in pelagic longline and purse seine fisheries throughout their range. In many areas, the abundance of Wahoo is seasonal, and during times of peak abundance Wahoo can constitute a significant fraction of local fish landings. Targeted artisanal fisheries typically employ handlines, troll lines, drift nets, and small surface longlines. Landings are most certainly underreported for this species. Reported worldwide catches show a gradual increase from 100 metric tons (mt) in 1960 to more than 12,000 mt in 2012. It is possible that this increase may represent an improvement in documentation and/or increase in fishing effort. As with other large pelagic species, Wahoo accumulate mercury in their flesh and there is a positive linear relationship between mercury and length and age with a mean of 0.50 mg/kg.

Atlantic Ocean: Over the past 30 years, reported catches in the Atlantic and Caribbean have gradually increased from less than 1,000 mt to almost 5,000 mt, with considerable variability year to year. It is important to note that variability in catches might be more related to reporting rate than to actual fluctuations in the population. Between 1990 and 2015, the US Atlantic and Gulf of Mexico recreational landings fluctuated between 22,000 and 138,000 fish/year without discernible trend. Landings from the US Atlantic pelagic longline fishery averaged 3,608 fish per year from the mid 1990s to mid 2000s, but dropped to 749 in 2010 and increased to 3,325 in 2015. Wahoo became the single most important species in commercial fisheries in Bermuda in 1956; in 1986 the reported catch was 65,406 kg, which increased to 104,591 kg in 2000. In the western South Atlantic this species is caught by artisanal handline and trolling in the northeast and central Brazil, as bycatch in industrial longliners and game fisheries, and around the Saint Peter and Saint Paul Archipelago.

Pacific Ocean: Reported annual landings have recently increased to more than 1,000 mt in the Pacific Ocean. In the Hawaiian commercial longline, troll, and handline fisheries for highly migratory pelagic fishes, Wahoo and Dolphinfish are incidentally caught and are of secondary importance to tunas and billfishes. Wahoo are an important pelagic species to artisanal and subsistence fisheries in many South Pacific island nations. In the Eastern Pacific, Wahoo are often caught as bycatch in purse seines, especially in sets associated with Fish Aggregating Devices (FADs).

Indian Ocean: Only minimal catches of Wahoo (well below 1,000 mt) have been reported for the Indian Ocean.

THREATS: Wahoo constitute an economically important fishery for many coastal nations but reporting, especially from artisanal fisheries, is incomplete. Assessment of this resource is further hampered by a lack of basic life history information. Increased purse seine fishing effort on FADs is resulting in higher bycatch rates for Wahoo. FAO fisheries statistics suggest that global landings have increased over the past 20 years, as they have in the western Atlantic and the western and central Pacific. Wahoo is fast growing and early maturing, and there is no current evidence of it being significantly impacted by current fishing effort, although local depletions may occur. This species was evaluated as Least Concern on the IUCN Red List in the eastern tropical Pacific, the Atlantic, the Gulf of Mexico, and on a global basis.

CONSERVATION: Wahoo are widespread, with genetic evidence indicating that it has high genetic connectivity globally. The status of Wahoo resources in the western central Atlantic remains unclear. Two per-recruit stock assessments have been undertaken on Wahoo populations in the Pacific, in waters off Taiwan and in the western South Pacific Ocean. Results indicated that fishing mortality is likely below benchmark levels. Catch rates have not declined in the eastern Caribbean, but data are insufficient for a benchmark assessment. The recommendation was that there should be a precautionary approach to the development of the fishery. In contrast to the Caribbean, the catch per unit effort (CPUE) for Wahoo off south and southeast Brazil drastically declined from 1971–2009. There are limited species-specific conservation measures in place for this species. In the US Atlantic, commercial harvest is restricted to hook-and-line gear and there is commercial trip limit of 500 lb. The recreational fishery is regulated with a bag limit of two fish per person per day. Federally managed commercial fishers in Queensland, Australia, may retain a maximum of 20 Wahoo per trip. The Queensland fisheries agency imposes a recreational size restriction of 75 cm TL (ca. 71.4 cm FL) but fish of this size are infrequently caught in either the commercial or recreational fishery.

SELECTED REFERENCES: Collette and Russo 1985; Franks et al. 2007; Garber et al. 2005; Maki Jenkins and McBride 2009; Manooch and Hogarth 1983; Matsumoto 1967; McBride et al. 2008; Oxenford et al. 2003; Perelman et al. 2017; Theisen et al. 2008; Vitek 2014; Zischke 2012; Zischke et al. 2013a.

Slender Tuna

Allothunnus fallai Serventy, 1948

COMMON NAMES: English – Slender Tuna
French – Thon élégant
Spanish – Atún Lanzón

ETYMOLOGY: The genus was named *Allothunnus* by the Australian biologist David Serventy (1948) because of its clear anatomical affinity with the higher tunas such as *Thunnus*. The species name *fallai* was named after Dr. R.A. Falla, then Director of the Canterbury Museum, Christchurch, New Zealand, who had communicated with Dr. Serventy about a bonito-like fish and then in 1942 sent him a specimen that became the type of the new species.

SYNONYMS: No synonyms. However, early records of other scombrids, *Cybium guttatus* and *Auxis thazard* from New Zealand, have been shown to be misidentifications of the Slender Tuna.

TAXONOMIC NOTE: The Slender Tuna is the most primitive of the higher tunas (tribe Thunnini).

FIELD MARKS:

1 Slender Tuna have more gill rakers than any other scombrid, 70–80 on the first gill arch.
2 Their teeth are very small and conical, with 40–55 on each side of the upper and lower jaws.

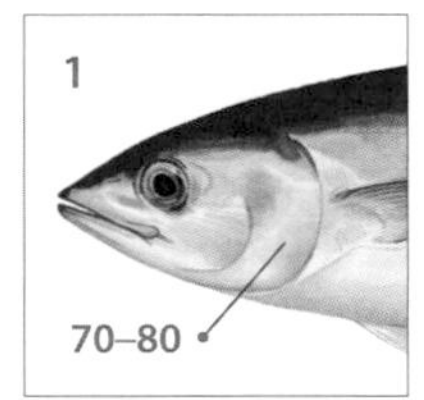

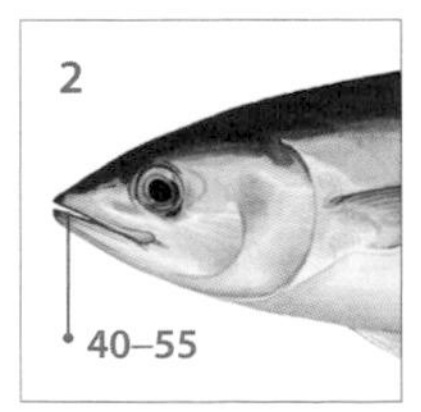

UNIQUE FEATURE: The scombrid with the most gill rakers, 70–80.

DIAGNOSTIC FEATURES: The body is robust, elongate, and rounded. The teeth are very small and conical, with 40–55 on each side of the upper and lower jaws. There are more gill rakers than in any other species of scombrid, with 70–80 on the first arch. **Fins:** The dorsal fins are close together, the first with 15–18 spines, the second with 12 or 13 rays, followed by 6 or 7 finlets. The anal fin has 13 or 14 rays and is also followed by 6 or 7 finlets. The pectoral fins have 24–26 rays. The interpelvic process is small and bifid. **Caudal peduncle:** The caudal peduncle is slender with a well-developed lateral keel between the two smaller keels on each side. **Scales:** The body is naked ventrally behind the long anterior corselet. The dorsal half of the body down to the lateral line is covered with small scales. **Swim bladder:** A swim bladder is present. **Vertebrae:** Vertebrae number 20 precaudal plus 19 caudal, totalling 39. Descriptions and illustrations of the osteology and soft anatomy of the Slender Tuna are included in the revision of the bonitos by Collette and Chao (1975). **Otoliths:** The sagittal otolith is distinguishable from that of all other eastern Pacific tunas and mackerels. **Color:** The body is dark blue dorsally, silver laterally and ventrally, with no distinct dorsal, lateral, or ventral markings.

GEOGRAPHIC RANGE: Slender Tunas are distributed circumglobally in the Southern Ocean between 20° and 50° S. There are records from around New Zealand and Tasmania. Slender Tuna have been reported from off Uruguay and Tierra del Fuego in the South Atlantic and from Santos, São Paulo, Brazil. One individual was taken in Los Angeles Harbor and another from the North Pacific subarctic gyre at 44°01′ N, 151°13′ W; both were probably vagrants.

Below: Slender Tuna range map.

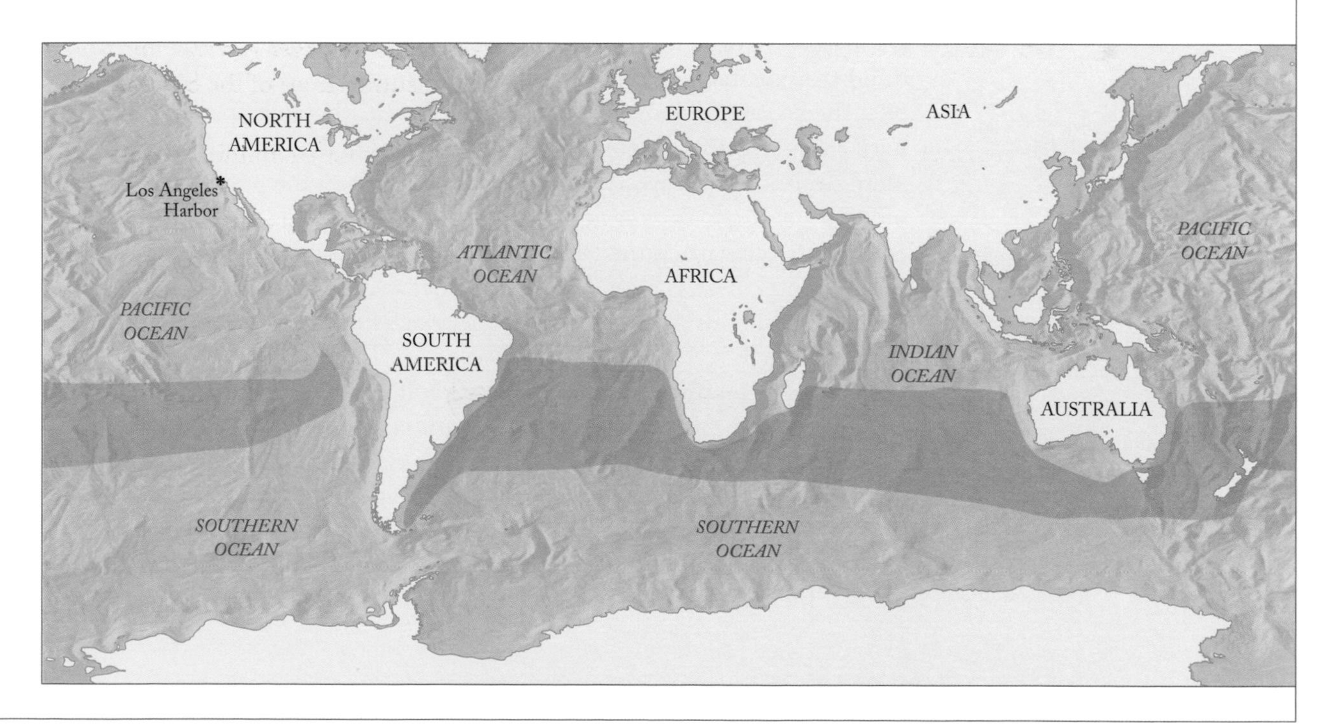

SIZE: 105 cm FL. The IGFA all-tackle game fish record is a 26 lb, 1 oz (11.85 kg) fish taken off Taiaroa Heads, Otago, New Zealand, in May 2001. Japanese longline catch sizes range between 65 and 96 cm FL.

HABITAT AND ECOLOGY: Slender Tuna are pelagic and oceanodromous and occasionally schooling. Juveniles are found between 20° and 35° S at surface temperatures ranging from 19–24°C. With increasing size, they gradually move into higher latitudes where water temperatures are lower. Slender Tuna have all four extraocular eye muscles fused into a distinct tissue complex that may function to warm the brain and eye region, and they also have vascular modifications of the central circulatory system that allow them, like other tunas, to elevate the temperature of the lateral musculature. Measurements on 30 *A. fallai* (75–85 cm FL) captured by hook and line off the coast of southern New Zealand revealed that their lateral red muscle temperatures are elevated by 6.7–10.0°C above the ambient sea surface temperature of 14.3–16.4°C.

FOOD: Slender Tuna are plankton feeders concentrating on crustaceans, particularly krill (euphausiids), but they also feed on squids and small fishes. The individual from Los Angeles harbor had eaten 38 small anchovies. The lipid composition of their flesh differs from other tunas and reflects the concentration of krill in their diet.

REPRODUCTION: Spawning is presumed to take place during the southern summer months (October–December) over a wide range of the temperate Indian and South Pacific oceans north of 31° S. Both sexes have a reported length at first maturity of 71.5 cm FL in Tasmania.

FISHERIES INTEREST: There is no directed fishery for Slender Tuna, but they are caught incidentally, largely by fisheries for Southern Bluefin Tuna south of 38°. There was a lack of reported catches for this bycatch species worldwide until the mid-2000s. Reported catches increased to 272 mt in 2013, mainly by New Zealand. In the western South Atlantic, 57% of the purse seine landings off Mar del Plata, Argentina, in January 2010 were Slender Tuna, and small catches have been reported from the Falkland Islands. The protein and oil content of the flesh is high. The flesh is paler than that of most true tunas and is very oily, but Tasmanians consider the cooked meat to be good.

THREATS: Slender Tuna are taken as bycatch in longline fisheries for Southern Bluefin Tuna. The impact of incidental take is not known but it is not considered a major threat at this time. It is listed as Least Concern on the IUCN Red list. More information is needed on their biology and population trends.

CONSERVATION: This species is widespread and locally abundant in the Southern Ocean. There are no species-specific conservation measures in place.

REFERENCES: Amorin et al. 2011; Bishop et al. 1976; Collette 2010; Collette and Chao 1975; Collette and Diaz de Astarloa 2008; Collette and Nauen 1983; Collette et al. 2011; Cousseau and Figueroa 1989; Fitch and Craig 1964; Garciarena 2011; Graham and Dixon 2000; Horn et al. 2013; IGFA 2017; Mori 1967, 1972; Nakamura and Mori 1966; Olsen 1962; Roberts, C.D. 1989; Roberts, P.E. 1975; Roberts et al. 1977; Schaefer and Childers 1999; Sepulveda et al. 2007, 2008; Serventy 1948; Tominaga 1966; Warashina and Hisada 1972; Watanabe et al. 1966; Wolfe and Webb 1975; Yatsu 1995; Yatsu and Watanabe 1987.

Bullet Tuna

Auxis rochei (Risso, 1810)

COMMON NAMES: English – Bullet Tuna
French – Bonitou, Auxidem
Spanish – Melva

ETYMOLOGY: Described by the Italian naturalist Joseph Antoine Risso (1810:165–167) as Scombre de Laroche, *Scomber Rochei*, apparently a patronymic.

SYNONYMS: *Scomber rochei* Risso, 1810; *Scomber bisus* Rafinesque, 1810; *Thynnus rocheanus* Risso, 1827; *Auxis vulgaris* Cuvier, 1832; *Auxis thynnoides* Bleeker, 1855; *Auxis ramsayi* Castelnau, 1879; *Auxis maru* Kishinouye, 1915; *Auxis rochei eudorax* Collette and Aadland, 1996.

TAXONOMIC NOTE: Most authors, such as Fraser-Brunner (1950), thought that there was only one worldwide species in the genus *Auxis*, which was recorded as *Auxis thazard* even though other authors, as early as Kishinouye (1915), clearly differentiated two species. After the review by Fitch and Roedel (1963), most authors agreed that there were two species, the Bullet Tuna *Auxis rochei* and the Frigate Tuna *Auxis thazard*, and this has been corroborated with subsequent analyses of morphological and molecular characters (Collette and Aadland 1996, Kumar et al. 2013, Habib and Sulaiman 2016).

FIELD MARKS:

1 Both species of *Auxis* have a single large fleshy process between the pelvic-fin bases.
2 The corselet is wider in the Bullet Tuna than in the Frigate Tuna. It is at least six scales wide in the Bullet Tuna and is under the origin of the second dorsal fin.
3 The bars in the naked area above the lateral line are nearly vertical in the Bullet Tuna, whereas they are distinctly oblique in the Frigate Tuna.

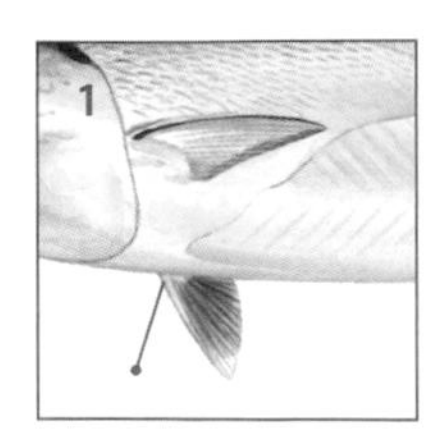

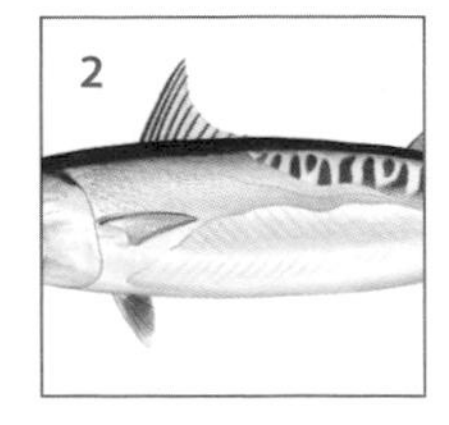

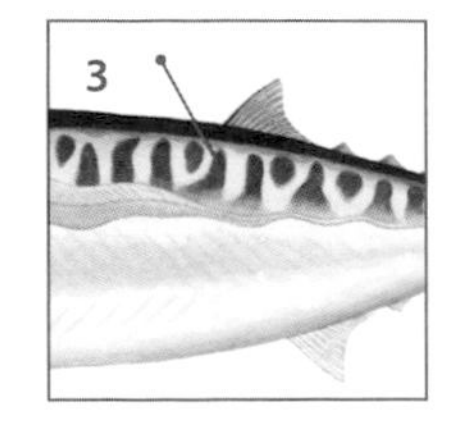

UNIQUE FEATURE: The Bullet Tuna has a very wide corselet under the second dorsal fin, much wider than in the Frigate Tuna.

DIAGNOSTIC FEATURES: Body robust, elongate, and rounded. The teeth are small and conical and in a single series on each jaw. The palatine and vomer lack teeth. **Gill rakers:** Gill rakers on the first gill arch number 39–49, usually 40–44 in the western Atlantic, 41–47 in the eastern Atlantic and Mediterranean, and 44–47 in the eastern Pacific subspecies. Juveniles of both species of *Auxis* are particularly difficult to identify. Gill raker counts are informative; Bullet Tuna have 37–43 while Frigate Tuna have 44–47 in Philippine waters; however, the number of gill rakers increases with size. Molecular markers can unambiguously distinguish the two species, and the complete mitochondrial DNA sequences of both the Bullet Tuna and the Frigate Tuna have been published. **Fins:** There are two dorsal fins, the first with 10–12 spines, separated from the second by a long interspace (at least equal in length to the length of the first dorsal-fin base). The second dorsal fin has 10 or 11 fin rays and is followed by 8 finlets. The anal fin has 12 or 13 fin rays and is followed by 7 finlets. The pectoral fins are short. There is a single, long pelvic process between the pelvic fins. **Caudal peduncle:** A strong central keel is present on each side of the caudal fin base between two smaller keels. **Scales:** The body is naked except for the corselet, which extends posteriorly along the lateral line past the origin of the second dorsal fin and is 6–30 scales wide under the origin of the second dorsal fin. **Vertebrae:** Vertebrae number 20 precaudal plus 19 caudal for a total of 39. The osteology and soft anatomy was described and illustrated in the synopsis of the genus by Uchida (1981) and in the generic revision by Collette and Aadland (1996). **Color:** Upper sides of body bluish, turning to deep purple or almost black on the head; belly white, usually without stripes or spots, but occasionally a few black spots may be present on the belly between the pectoral and pelvic fins. There are 14–16 or more fairly broad, nearly vertical bars in the naked area above the lateral line. The pectoral and pelvic fins are purple, with their inner sides black.

GEOGRAPHIC RANGE: Bullet Tuna are cosmopolitan in warm waters of the Atlantic, Indian, and Pacific oceans, including the Mediterranean Sea. Presence of both Bullet Tuna and Frigate Tuna in Hawaii was confirmed in 1965, and the Frigate Tuna was added to the Atlantic fauna by Richards and Randall in 1967. The eastern Pacific population is recognized as a distinct subspecies, *Auxis rochei eudorax*, and it ranges from California and the mouth of the Gulf of California to Peru, including the Galapagos, Cocos, and Malpelo islands.

SIZE: The maximum fork length is 50 cm in Japanese catches, commonly to 35 cm FL. The IGFA all-tackle game fish record is of a 4 lb, 1 oz (1.84 kg) fish taken at Santa Margherita, Ligue, Italy, in October 2014. Length-weight information has been presented for Bullet Tuna in the Mediterranean Sea.

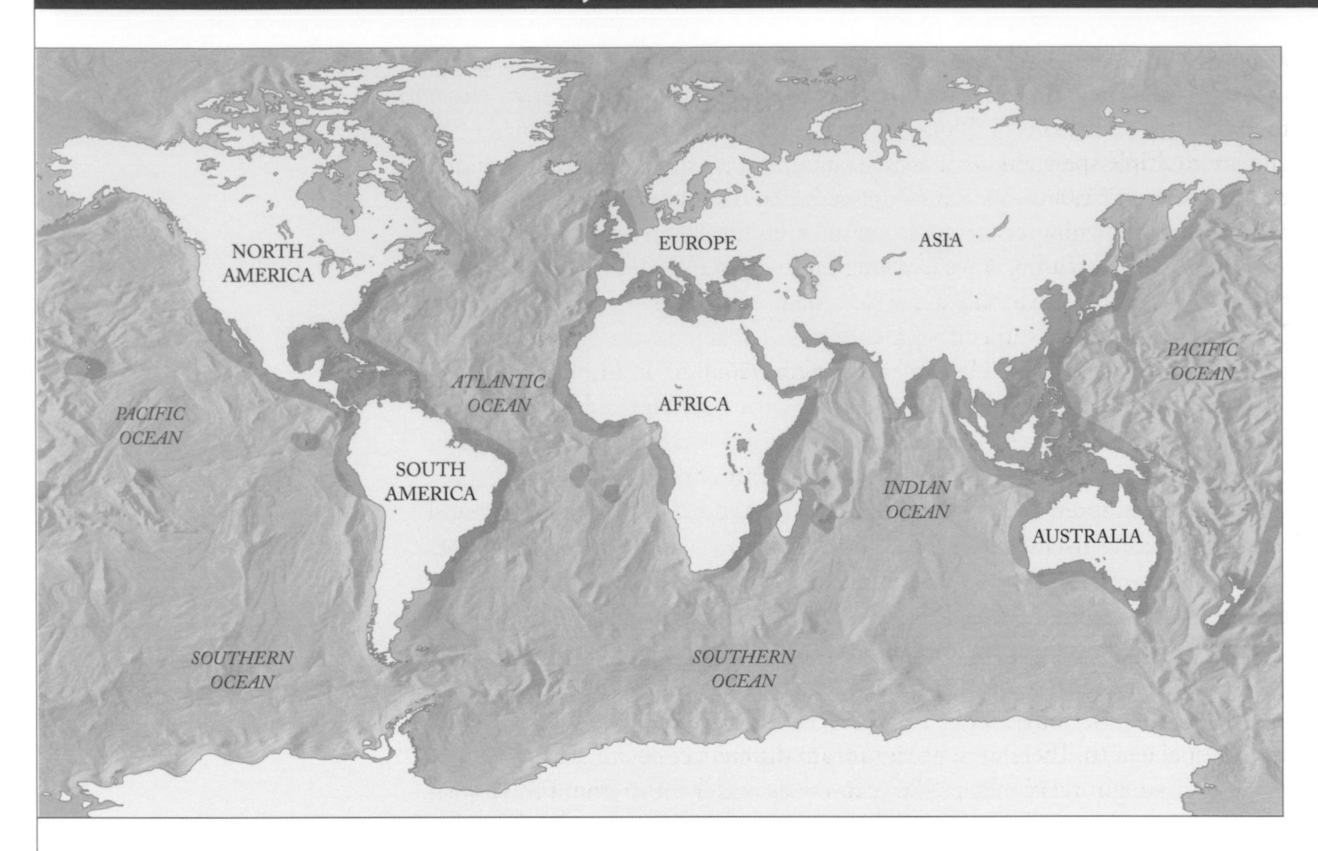

Above: Bullet Tuna range map.

HABITAT AND ECOLOGY: Bullet Tuna are pelagic and oceanodromous species and form schools. The lower limit of their depth range is about 200 m. Bullet Tuna frequently school with Frigate Tuna in some parts of the Indo-Pacific such as Ceylon, but they are smaller than the Frigate Tuna in a school. Because of their abundance, Bullet Tuna are an important element of the pelagic food web, particularly as forage for other species of commercial interest. In eastern Taiwan, Sailfish are a major predator of Bullet Tuna, and in Japan and the western Atlantic, Bullet Tuna are a major component of the food of Blue Marlin.

FOOD: Bullet Tuna feed on a wide variety of fishes, crustaceans, and mollusks. In the Philippines, they show a preference for fishes, planktonic crustaceans (shrimps, copepods, crab larvae, and amphipods), and small cephalopods. Among the crustaceans consumed in the Tyrrhenian Sea, hyperiidean amphipods are often the most important, followed by euphausiids. The fish component of the diet in the Philippines and along the Indian coast was represented by anchovies, sardines, mackerel, and jacks.

REPRODUCTION: Length at first maturity for Bullet Tuna in the Philippines is 17 cm FL, generally before 1 year of age. Length at 50% maturity of both sexes is 24 cm off India and 18.8 cm in the Philippines. Age of first maturity in Japan is 1.25 years. Length at first maturity in Indian waters is 23.6 cm FL at an age of about 2 years, or for females 23.8 cm FL and 24 cm FL for males. Bullet Tuna spawn year-round in tropical waters. They spawn closer to shore than Albacore or Bluefin Tuna in the western Mediterranean Sea and are multiple spawners with asynchronous oocyte development. Fecundity ranges between 31,000 and 103,000 eggs per spawning (according to the size of the fish). Spawning occurs at sea surface temperatures of 24°C or higher, and the spawning season varies from region to region. In the western Atlantic, spawning begins in the Straits of Florida in February and peaks in the Gulf of Mexico from March to April and from June to August while in coastal waters from Cape Hatteras to Cuba. In the eastern Atlantic, high percentages of ripe males and females were found from May through September, and Bullet Tuna larvae were collected in the eastern Mediterranean in early June and in the northwestern Mediterranean between July and September when the sea surface temperatures were 23.7–25.4°C. Spawning occurs between July and September in Tunisian waters. Off Cape Ashizuri, Japan, hydration of eggs occurred between 1100 and 1300 hours and ovulation at about 1500 hours, followed by spawning. Batch fecundity estimates in Spanish Mediterranean waters ranged between 150,000 and 350,800 oocytes.

AGE AND GROWTH: Longevity of Bullet Tuna is 5 years, and males and females attain equal length. There were no significant differences between the slopes of the length-weight relationships between males and females from the Turkish Mediterranean coasts. Bullet Tuna from southern Spain reached 25 cm FL at 1 year, 33.1 cm at 2 years, 37.4 cm at 3 years, and 41.2 cm at 4 years. Mediterranean Bullet Tuna aged by reading cross sections of the first dorsal spine averaged 35.1 cm FL at age 2, 39.2 cm at age 3, 41.2 cm at age 4, and 42.1 cm at age 5. Length-weight curves for Mediterranean Bullet Tuna have been presented by de la Serna et al. (2005). Longevity in Japan has been reported to be only 2–5 years.

EARLY LIFE HISTORY: *Auxis* are the most abundant juvenile tunas in the world and are widely distributed in tropical and temperate waters. Bullet Tuna larvae are found in shallower waters than Atlantic Bluefin and Albacore larvae, close to the shelf break in the Balearic Sea, western Mediterranean. Larvae were collected off the Catalan coast of Spain in the northwestern Mediterranean between July and September when the sea surface temperatures were 23.7–25.4°C. Larvae feed mostly during daylight hours, and larvae 3–5 mm selectively feed on cladocerans and appendicularians, while those longer than 5 mm prefer appendicularians and fish larvae. The positive selection of fish larvae as food at sizes of 5 mm is concomitant with greater tooth development, appearance of specialized teeth, and flexion of the urostyle. In the Balearic Sea, Bullet Tuna larvae grew faster in waters of Mediterranean origin than those in water of Atlantic origin, indicating a greater trophic specialization in their food. There are illustrations of six western Atlantic Bullet Tuna, 3.5 mm notochord length to 39.0 mm SL, 3 Mediterranean larvae 2.9–7 mm SL, and 5 larvae 4.5–7.7 notochord length from Korean seas.

FISHERIES INTEREST: Bullet Tuna are widespread and abundant in many parts of their range. They are caught in directed fisheries using several gear types, including troll lines, hand line, gill nets, and coastal purse seines. They are also taken incidentally in pelagic purse seine and a variety of trawl fisheries. In southern Spain, and particularly in Andalucía, both species of frigate tunas have been exploited due to the excellent properties of the meat, with its mild taste and low cholesterol content. Differentiating canned frigate tuna from other tunas to protect the label "Melva de Andalucía" led to development of a novel fluorescent-based multiplex-PCR analysis. Bullet Tuna have high regional commercial importance in southern Spain and many other areas, but landings are often mixed with the Frigate Tuna. FAO reports statistics for Bullet Tuna, Frigate Tuna, and for *Auxis* spp., not identified to species. Worldwide reported landings for *Auxis* spp. show a gradual increase from 22,278 mt in 1950 to 480,971 in 2014. In the Atlantic, ICCAT has recorded landings for both species for the past 30 years, which have ranged from 3,000 to 9,000 mt for Bullet Tuna and 4,500 to 15,800 mt for Frigate Tuna.

THREATS: Since 1991, the use of FADs by tropical purse seiners may have led to an increase in fishing mortality of small tropical tuna species. However, there is a general lack of information, including incomplete reporting from directed artisanal fisheries and incomplete reporting of bycatch, both of which are exacerbated by the confusion regarding species identification. The present rate of exploitation in Indian waters indicates that the Indian stock is under high fishing pressure and measures need to be taken to reduce the fishing pressure. Bullet Tuna are common and abundant, with landings showing fluctuation but no serious declines in most regions. It is considered of low value in many regions and is not especially targeted by large commercial fisheries. It is listed as Least Concern on the IUCN Red List.

CONSERVATION: There are few conservation measures in place for Bullet Tuna. It is a highly migratory species, listed on Annex I of the 1982 Convention on the Law of the Sea. No species-specific fishery management measures are in place, although the species benefits from an EU prohibition on drift nets and general time/area closures in Turkish waters. It has been suggested that the closed fishing season off Taiwan and the East China Sea be extended two months earlier, from April to June, to protect spawning of the 72 most important commercial fishes in the region, including Bullet Tuna.

SELECTED REFERENCES: Catanese et al. 2008; Collette and Aadland 1996; Fitch and Roedel 1963; Habib and Sulaiman 2016; Jasmine et al. 2013; Ji et al. 2011; Kumar et al. 2013; Niiya 2001a, 2001b; Rodríguez-Roda 1983; Rohit et al. 2014; Uchida 1981; Yesaki and Arce 1994; Yoshida and Nakamura 1965.

Frigate Tuna

Auxis thazard (Lacepède, 1800)

COMMON NAMES: English – Frigate Tuna
French – Auxidem
Spanish – Melva

ETYMOLOGY: The Frigate Tuna was described by the French ichthyologist Bernard Lacepède (1800:599), based on an earlier manuscript description by Philibert Commerson.

SYNONYMS: *Scomber thazard* Lacepède, 1800; *Scomber taso* Cuvier, 1832; *Auxis tapeinosoma* Bleeker, 1854; *Auxis hira* Kishinouye, 1915; *Auxis thazard brachydorax* Collette and Aadland, 1996.

TAXONOMIC NOTE: Most authors thought that there was a single worldwide species in the genus *Auxis*, which was recorded as *Auxis thazard* even though other authors clearly differentiated two species. After the 1963 review by Fitch and Roedel, most authors agreed that there were two species, the Bullet Tuna *Auxis rochei* and the Frigate Tuna *Auxis thazard*, and this has been corroborated using both morphological and molecular characters (Collette and Aadland 1996, Paine et al. 2006, Kumar et al. 2013, Habib and Sulaiman 2016).

FIELD MARKS:
1 Both species of *Auxis* have a single large fleshy process between the bases of the pelvic fins.
2 The corselet is narrower in the Frigate Tuna than in the Bullet Tuna. It is five scales or less in the Frigate Tuna and is under the origin of the second dorsal fin.
3 The bars in the naked area above the lateral line are distinctly oblique in the Frigate Tuna, whereas they lie nearly vertical in the Bullet Tuna.

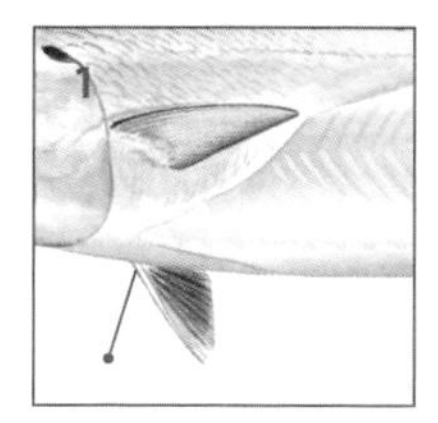

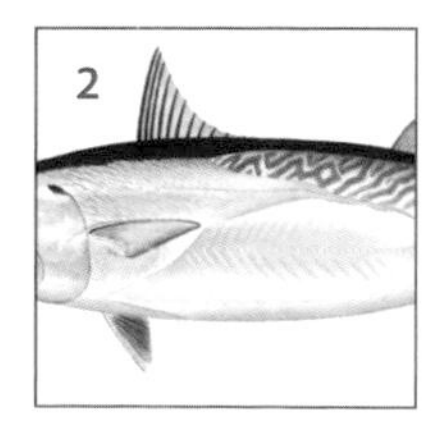

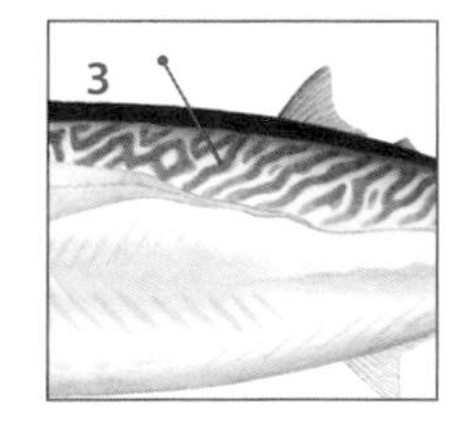

UNIQUE FEATURE: The Frigate Tuna has a very narrow corselet extension under the second dorsal fin, much narrower than in the Bullet Tuna.

DIAGNOSTIC FEATURES: Body robust, elongate, and rounded. The teeth are small and conical and in a single series on each jaw. The palatine and vomer lack teeth. **Gill rakers:** Gill rakers on the first gill arch number 36–44, usually 38–41 in the Atlantic and Indo-West Pacific and 40–48 in the eastern Pacific subspecies. Juveniles of both species of *Auxis* are particularly difficult to identify. Gill raker count helps; Bullet Tuna have 37–43, while Frigate Tuna have 44–47 in Philippine waters, but the number of gill rakers increases with size, so genetic analyses are more definitive. Complete mitochondrial DNA sequences of both the Frigate Tuna and the Bullet Tuna have been published. **Fins:** There are two dorsal fins, the first with 10–12 spines, separated from the second by a long interspace (at least equal in length to the length of the first dorsal-fin base). The second dorsal fin has 10 or 11 fin rays and is followed by 8 finlets. The anal fin has 12 or 13 fin rays and is followed by 7 finlets. The pectoral fins are short. There is a single, long pelvic process between the pelvic fins. **Caudal peduncle:** A strong central keel is present on each side of the caudal-fin base between two smaller keels. **Scales:** The body is naked except for the corselet, which extends posteriorly along the lateral line past the origin of the second dorsal fin and is no more than 5 scales wide under the origin of the second dorsal fin. **Vertebrae:** Vertebrae number 20 precaudal plus 19 caudal for a total of 39. The osteology and soft anatomy were described and illustrated in the generic revision by Collette and Aadland (1996). **Color:** Upper sides of body bluish, turning to deep purple or almost black on the head; belly white, usually without stripes or spots, but occasionally a few black spots may be present on the belly between the pectoral and pelvic fins. There are 15 or more narrow, oblique to nearly horizontal dark wavy lines in the naked area above the lateral line. The pectoral and pelvic fins are purple, with their inner sides black.

GEOGRAPHIC RANGE: Frigate Tuna are cosmopolitan in warm waters of the Atlantic, Indian, and Pacific oceans but are considered a vagrant in the Mediterranean Sea. There are relatively few records of this species in the Atlantic, as most of the *Auxis* in the Atlantic are Bullet Tuna. Presence of both Frigate and Bullet Tuna in Hawaii was confirmed by Matsumoto (1960), and the Frigate Tuna was confirmed as part of the Atlantic fauna by Richards and Randall (1967). Frigate Tuna recently extended their range northward in the western Pacific to Peter the Great Bay, Russia. The eastern Pacific population is recognized as a distinct subspecies, *Auxis thazard brachydorax* Collette and Aadland 1996, and it ranges from California and the mouth of the Gulf of California south to Peru, including the Galapagos and all the other oceanic islands except Clipperton.

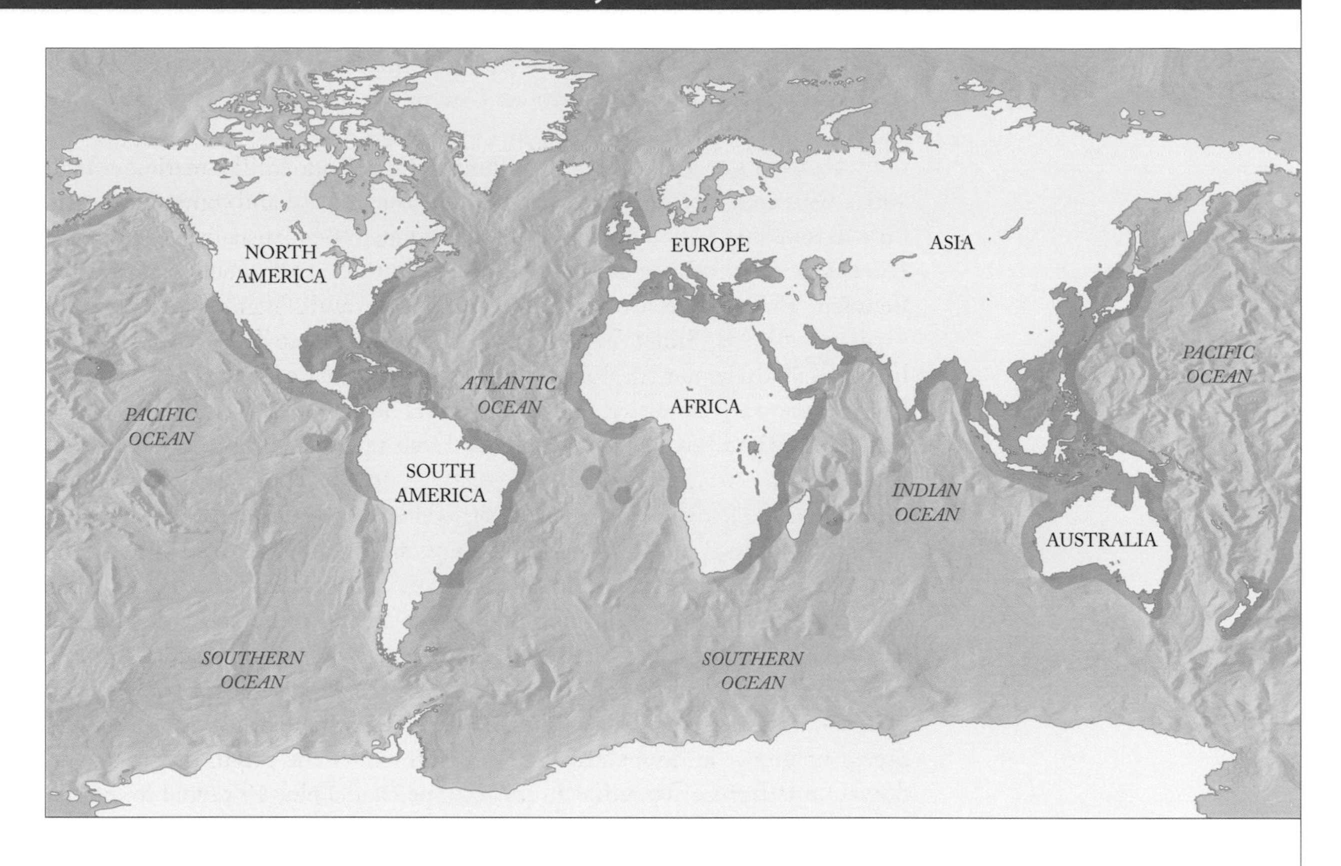

Above: Frigate Tuna range map.

SIZE: Frigate Tuna reach at least 62 cm FL. The IGFA all-tackle game fish record is of a 3 lb, 8 oz (1.6 kg) fish caught off the Shima Peninsula, Mie, Japan, in November 2015.

HABITAT AND ECOLOGY: Frigate Tuna are pelagic, oceanodromous, and epipelagic in neritic and oceanic waters. Adults are coastal or near coastal, while juveniles are more widely distributed throughout the world's ocean. As with other members of the tribe Thunnini, Frigate Tuna have countercurrent heat exchangers, which conserve metabolic heat and enable them to maintain body temperatures several degrees (3.5–9.5°C) warmer than the waters they inhabit. Bullet Tuna frequently school with Frigate Tuna in some parts of the Indo-Pacific, such as Hawaii and Ceylon. Because of their abundance, they are considered an important element of the food web, particularly as forage for other species of commercial interest. They are preyed upon by larger fishes, including other tunas, like Skipjack Tuna, and billfishes such as Sailfish and Blue Marlin.

FOOD: Frigate Tuna feed on small fishes, squids, planktonic crustaceans (megalops), and stomatopod larvae. Sardines, anchovies, mackerels, scads, and juvenile tunas were dominant among finfishes in the diet in Indian waters.

REPRODUCTION: Frigate Tuna spawn near shore and in oceanic waters. Young have been collected as far north as Cedros Island, Baja California, in the eastern tropical Pacific south to off Point Santa Elena, Ecuador. Correlated with temperature and other environmental changes, the spawning season varies with area, but in some places it may even extend throughout the year. Gravid and ripe females were found in Indian waters in all months except December, with peak occurrences during February and again during July–October. The smallest maturing female Frigate Tuna off the west coast of Thailand were 31–33 cm FL. Length at 50% maturity was 34–37 cm FL in the Gulf of Thailand and 38.5 cm for females and 36.7 cm for males in India. However, a much smaller size at maturity has been reported in India, 30.5 cm FL. Total fecundity ranged from 6.97 to 11.63 million eggs in Indian females.

AGE AND GROWTH: The longevity of Frigate Tuna is approximately 4 years.

FISHERIES INTEREST: Frigate Tuna are of high commercial interest regionally. They are caught with beach seines, drift nets, purse seines, hook and line, gill nets, and by trolling. Landings are often mixed with the other species of *Auxis*, the Bullet Tuna. In southern Spain, and particularly in Andalucía, both species of frigate tunas have been exploited due to the excellent properties of the meat, with its mild taste and low cholesterol content. Differentiating canned frigate tuna from other tunas to protect the label "Melva de Andalucía" led to development of a novel fluorescent-based multiplex-PCR analysis. FAO reports statistics for Bullet Tuna, Frigate Tuna, and for *Auxis* spp., not identified to species. Worldwide reported landings for *Auxis* spp. show a gradual increase from 22,278 mt in 1950 to 480,971 in 2014. In the Atlantic, ICCAT has recorded landings for both species for the past 30 years, which have ranged from 3,000 to 9,000 mt for Bullet Tuna and 4,500 to 15,800 mt for Frigate Tuna. Almost all the catch from Venezuela in the Atlantic and from countries in the Mediterranean is thought to be Bullet Tuna.

THREATS: Since 1991, the use of FADs by tropical purse seiners may have led to an increase in fishing mortality of small tropical tuna species, including Frigate Tuna. In the Taiwan Strait, the mean fork length, body weight, and age of Frigate Tuna from 1982–84 decreased from 349.0 mm, 662.9 g, and 1.88 years to 321.5 mm, 622.3 g, and 1.79 years due to overfishing. There is a general lack of information in most regions, including incomplete reporting from directed artisanal fisheries and incomplete reporting of bycatch, both of which are exacerbated by the confusion regarding species identification. Frigate Tuna are widespread and abundant, with no current major threats from fishing in most regions, and are listed as Least Concern on the IUCN Red List.

CONSERVATION: Frigate Tuna are highly migratory, listed on Annex I of the 1982 Convention on the Law of the Sea. There are no known conservation measures except for a prohibition on drift nets in EU countries. It has been suggested that the closed fishing season off Taiwan and the East China Sea be extended two months earlier, from April to June, to protect spawning of the 72 most important commercial fishes in the region, including Frigate Tuna. Gill nets with a mesh size of 84 mm are recommend for use in the Indian gill net fishery, not those with 60 mm mesh size, because they catch juveniles. Present levels of fishing in India should be maintained to keep the stock at near optimum fishing pressure.

REFERENCES: Ankenbrandt 1985; Catanese 2006; Catanese et al. 2008; Chow et al. 2003; Collette and Aadland 1996; Collette and Nauen 1983; Collette et al. 2011; Fitch and Roedel 1963; Fraser-Brunner 1950; Ghosh et al. 2012; Habib and Sulaiman 2016; ICCAT 2009; IGFA 2018; Infante et al. 2004; Iswarya Deepti and Sujatha 2012; Jude et al. 2002; Klawe 1963; Kumar et al. 2013; Lacepède 1800; Matsumoto 1960; Muthiah 1985; Pimenta et al. 2005; Richards and Randall 1967; Robertson and Allen 2006; Schaefer 1985; Shih et al. 2009; Shimose et al. 2013; Sivasubramaniam 1973; STECF 2009; Uchida 1981; Wade 1949; Yesaki and Arce 1994; Yoshida and Nakamura 1965; Zemnukhov and Epur 2011.

Leaping Bonito

Cybiosarda elegans (Whitley, 1935)

COMMON NAMES: English – Leaping Bonito
French – Bonite à Dos Tacheté
Spanish – Bonito Saltador

ETYMOLOGY: As one of the most attractive species of scombrids, the Leaping Bonito was appropriately named *elegans* by the Australian ichthyologist Gilbert Whitley and placed in a new subgenus, *Cybiosarda,* of Spanish mackerels, *Scomberomorus*, in the family Sardidae, the bonitos.

SYNONYMS: *Scomberomorus (Cybiosarda) elegans* (Whitley, 1935), *Gymnosarda elegans* (Whitley, 1935)

TAXONOMIC NOTE: Except for the original description where it was considered a species of Spanish mackerel and inclusion in *Gymnosarda* by Fraser-Brunner (1950), the Leaping Bonito has been considered a valid species of bonito with no synonyms.

FIELD MARKS:

1 As its specific name, *elegans*, implies, the Leaping Bonito is one of the most colorful and strikingly patterned scombrids, with stripes on its belly and spots on its upper surface.
2 There are two tooth patches on the upper surface of the tongue, as in the Dogtooth Tuna and the Plain Bonito.
3 The first dorsal fin is large and high.

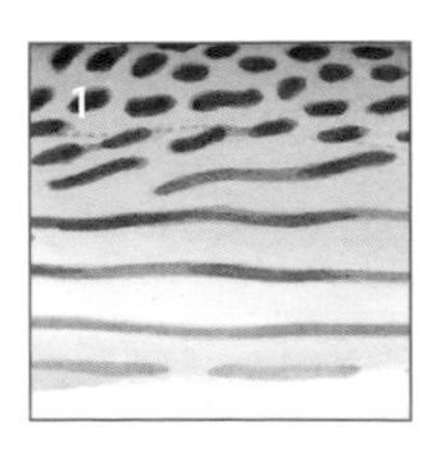

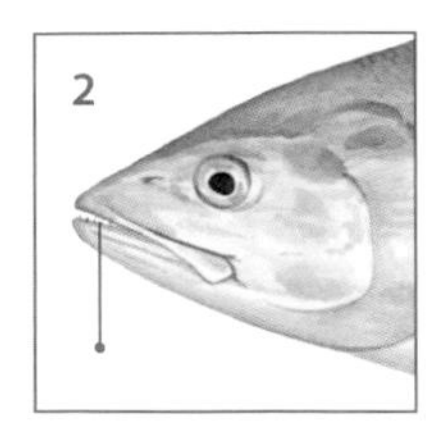

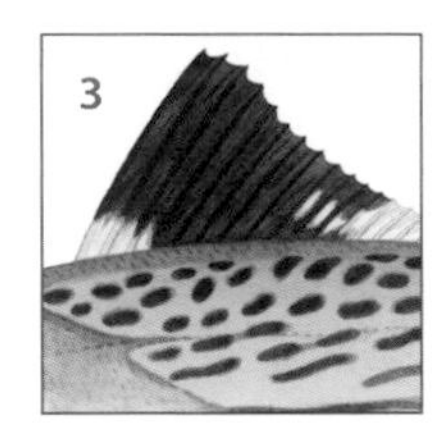

UNIQUE FEATURE: The most colorful and boldly patterned scombrid.

DIAGNOSTIC FEATURES: The body is relatively short and deep and strongly compressed. The mouth is rather large, with the upper jaw extending back to the posterior end of the eye. There are two tooth patches on the upper surface of the tongue. There are 13–22 large conical teeth on each side of the upper jaw and 10–17 teeth on the lower jaw. The olfactory rosette contains 28–33 laminae. The interorbital distance is 23.9–31% of the head length. **Gill rakers:** There are 12–15 gill rakers on the first gill arch. **Fins:** The dorsal fins are close together; the first is high anteriorly, with 16–18 spines. The second dorsal fin has 17–19 rays followed by 8–10 finlets. The anal fin has 13–17 rays followed by 6 or 7 finlets. The pectoral fin is short and has 22–24 rays. The interpelvic process is small and bifid. **Caudal peduncle:** The caudal peduncle is slender, with a well-developed lateral keel between two small keels on each side. **Scales:** The body is mostly naked behind the well-developed anterior corselet except for a band of scales along the bases of the dorsal and anal fins and patches of scales around the bases of the pectoral and pelvic fins. **Organs:** No swim bladder is present. The spleen is concealed under the liver and is not visible in a ventral dissection. The liver has an elongate right lobe and a short left lobe, which tends to fuse with the middle lobe. **Vertebrae:** The vertebrae number 22–24 precaudal plus 23–26 caudal for a total of 47 or 48 vertebrae. Descriptions and illustrations of the osteology and soft anatomy of the Leaping Bonito are included in the revision of the bonitos (Collette and Chao 1975). **Color:** The belly is light, with several stripes reminiscent of those of the Skipjack Tuna. In life, the back is bright turquoise to deep blue and covered with elongate black spots. The first dorsal fin is jet black anteriorly and white on the last few posterior membranes. The anal and second dorsal fins are yellow.

GEOGRAPHIC RANGE: Leaping Bonito are found in the western Pacific restricted to the northern three-quarters of Australia (absent from the south coast) and the southern coast of Papua New Guinea. They have been recorded as far south on the east coast as Green Island 35.27° S, 150.51° W, but anecdotal accounts exist of rock fishers catching them as far south as Greencape near the NSW/Victoria border, which usually defines the southernmost extremity of the East Australia Current during summer/autumn.

SIZE: Maximum size is 45 cm FL, 6.6 lb (3 kg). The IGFA all-tackle game fish record is of a 2 lb, 2 oz (0.96 kg) fish taken in the Macleay River, Australia, in May 1995. The largest fish caught by hook and line was a 1.15 kg specimen (length not available) off Shellharbour, New South Wales (records of the NSW Fishing Clubs Association, affiliated with the Australian Anglers Association).

HABITAT AND ECOLOGY: Leaping Bonito are epipelagic and neritic, forming schools of several hundred individuals. During midwinter months, sizable schools move inshore to feed on surface aggregations of clupeids (herrings and sardines) and anchovies. A 12-month dietary study of the species in eastern Australia revealed that the species has a low diversity of prey (six taxa), primarily consuming clupeids and anchovies in terms of both biomass and frequency of occurrence. Limited tagging information and seasonal catch trends in commercial fisheries suggests the species moves south during summer and autumn with the southward expansion of the East Australia Current.

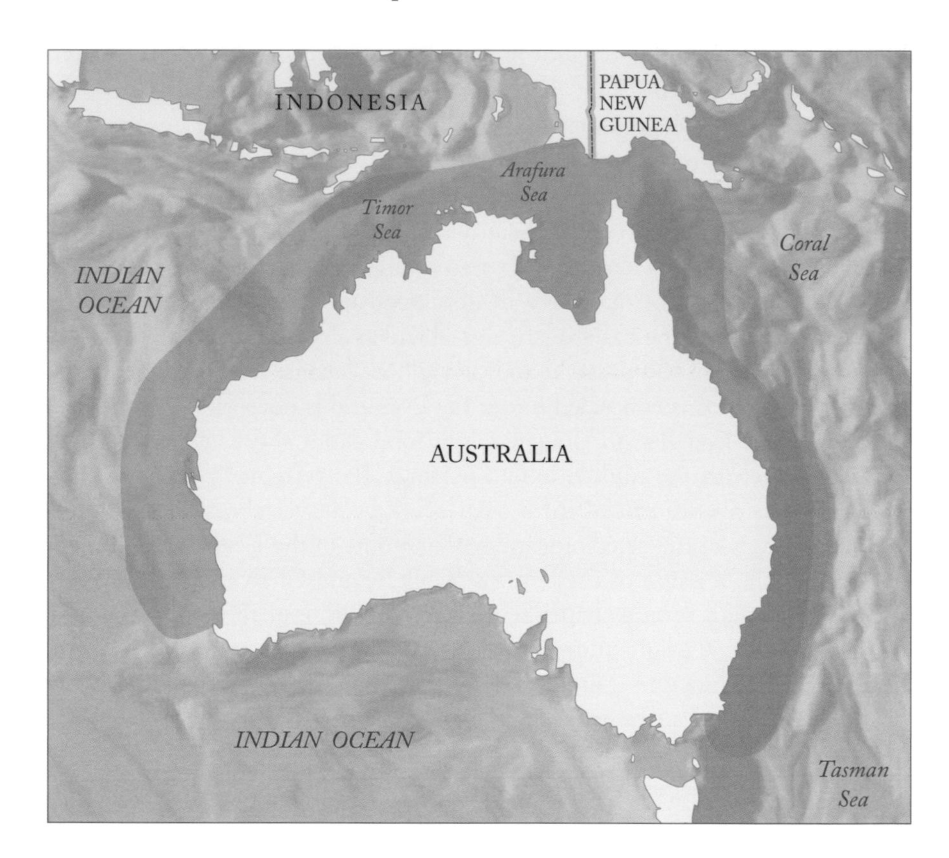

Right: Leaping Bonito range map.

AGE AND GROWTH: Leaping Bonito show allometric growth, indicating that they grow proportionally faster in terms of weight relative to length compared to the nearly isometric growth of closely related bonitos of the genus *Sarda*. Maximum age is unknown.

EARLY LIFE HISTORY: Three small juvenile Leaping Bonito were taken with several other species of scombrids with light traps in coastal waters of the central Great Barrier Reef. A sample of 15 juvenile fish (101–55 mm FL) were collected using a small seine net in shallow sand flats of Hervey Bay, Queensland, mixed with large schools of hardyheads. It is likely that shallow sand flats provide a nursery area for the species.

FISHERIES INTEREST: Leaping Bonito are occasionally taken in coastal commercial fisheries, primarily in large marine embayments and sheltered beaches using seine nets. They are commonly taken by anglers trolling lures at the surface but are usually sought as bait for larger game fish. They are rather dark-fleshed fish that are better eating than popular opinion will admit.

THREATS: The Leaping Bonito has little commercial importance in eastern Australia but is taken as bait for a range of reef species by commercial fishers and for marlins and sharks by sport fishers. It is occasionally targeted by commercial net fisheries when abundant or caught as bycatch when targeting Spanish mackerels and is frequently caught by recreational fishers. It is listed as Least Concern on the IUCN Red List. However, more research is needed on the biology and population trends of this species.

CONSERVATION: The Leaping Bonito is widespread and relatively common, and there is no directed fishery for it. There are no species-specific conservation measures for it.

REFERENCES: Collette and Chao 1975; Collette and Nauen 1983; Collette et al. 2011; Grant 1987; Griffiths et al. 2017; Robertson et al. 2007; Thorrold 1993; Whitley 1935.

Kawakawa

Euthynnus affinis (Cantor, 1849)

COMMON NAMES: English – Kawakawa, Black Skipjack
French – Thonine Orientale
Spanish – Bacoreta Oriental

ETYMOLOGY: Described by Dr. Theodore Edvard Cantor, a Danish amateur zoologist who worked for the Bengal Medical Service of the East India Company and collected fishes in Penang for his Catalog of Malayan Fishes (Cantor 1849). It is believed he used the word *affinis*, meaning closely related, in naming *Thynnus affinis* because he stated that it had the general outline of the Skipjack Tuna *Katsuwonus pelamis*.

SYNONYMS: *Thynnus affinis* Cantor, 1849; *Euthynnus yaito* Kishinouye, 1915; *Wanderer wallisi* Whitley, 1937; *Euthynnus affinis yaito* Kishinouye, 1915; *Euthynnus alletteratus affinis* (Cantor, 1849)

TAXONOMIC NOTE: Some authors have considered the Kawakawa to be a subspecies of a worldwide *Euthynnus alletteratus*, others (Fraser-Brunner 1949) thought that the eastern Pacific *E. lineatus* was a subspecies of *E. affinis*, but all recent authors recognize *E. affinis* as one of three species in the genus.

FIELD MARKS:
1 All three species of the genus *Euthynnus* lack prominent stripes on the belly.
2 Vomerine teeth are present in the Kawakawa but absent in the Little Tunny in the Atlantic.
3 There are only 29–33 gill rakers, fewer than the 33–39 present in the Black Skipjack in the Eastern Pacific.

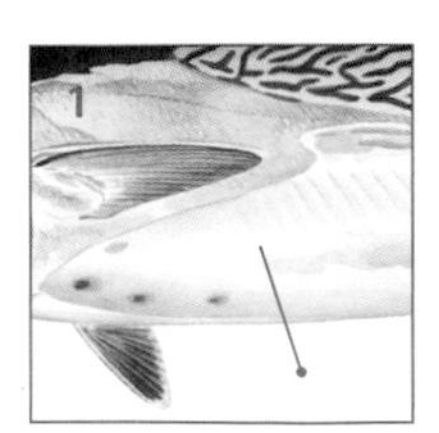

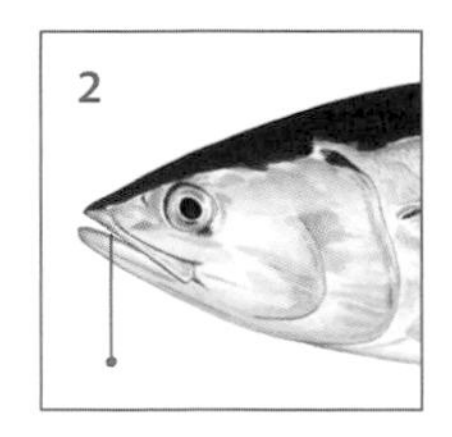

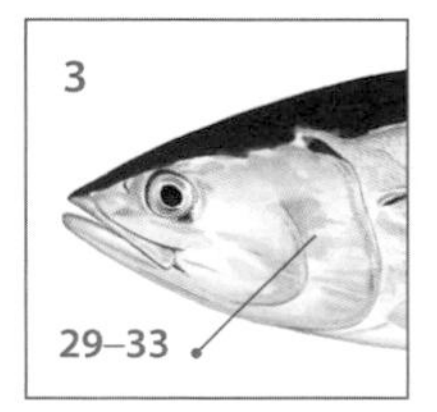

UNIQUE FEATURE: Kawakawa lack the protuberances on the posterior vertebrae that are present in the other two species of the genus.

DIAGNOSTIC FEATURES: Body robust, elongate, and rounded. The teeth are small and conical and in a single series on each jaw. The palatine and vomer have teeth. **Gill rakers:** Gill rakers on the first gill arch number 29–33, and there are 28 or 29 so-called gill teeth on the posterior surface of the gill arches. **Fins:** There are two dorsal fins, the first with 14–16 spines, contiguous or nearly contiguous with the second dorsal fin. The second dorsal fin has 10 or 11 fin rays and is followed by 8 finlets. The anal fin has 13 or 14 fin rays and is followed by 7 finlets. The pectoral fins are short. The interpelvic process is divided and is shorter than the pelvic fins. **Caudal peduncle:** A strong central keel is present on each side of the caudal-fin base between two smaller keels. **Scales:** The body is naked except for the anterior corselet. **Vertebrae:** Vertebrae number 20 precaudal plus 19 caudal for a total of 39. The 33rd and 34th vertebrae lack any swellings or protuberances. The osteology and soft anatomy were described and illustrated by Godsil (1954) and summarized by Yoshida (1979). **Color:** The dorsal surface varies from dark blue to yellowish green. The upper sides of the body are bluish, turning to deep purple or almost black on the head. There are two black spots on the head, one at the upper margin of the operculum and a second on the postero-ventral margin of the orbit. There are 15 or more narrow oblique to nearly horizontal dark wavy lines in the naked area above the lateral line. The belly is silvery white, with 1–11 black spots scattered between the pectoral and pelvic fins. These spots seem to appear and become particularly intense during feeding. The pectoral and pelvic fins are purplish, with their inner sides black.

GEOGRAPHIC RANGE: Kawakawa is an Indo-West Pacific species, but two stray individuals have been reported from the eastern tropical Pacific. Kawakawa are found in warm coastal waters, including around oceanic islands and archipelagos.

SIZE: Kawakawa reach at least 100 cm FL, about 15 kg. The IGFA all-tackle game fish weight record is of a 33 lb, 3 oz (15.05 kg) fish caught off Molokai, Hawaii in June 2014. The IGFA all-tackle game fish length record is of a 72 cm fish taken off Bribie Island, Australia, in February 2014.

HABITAT AND ECOLOGY: Kawakawa live in open waters but remain close to shore. They are found to depths of 50 m. They are a warm-water species, as shown by their travels to Taiwan from the northern Philippines with the warm Kuroshio Current from September to May and their movement away in June. They show increased abundance in Taiwan between October and April, while the catch is poorest during the coldest months. The young may enter bays and harbors. They form multi-species schools by size with other scombrid species, comprising from 100 to over 5,000 individuals. Kawakawa are preyed on by Skipjack and Yellowfin Tuna and billfishes such as Sailfish.

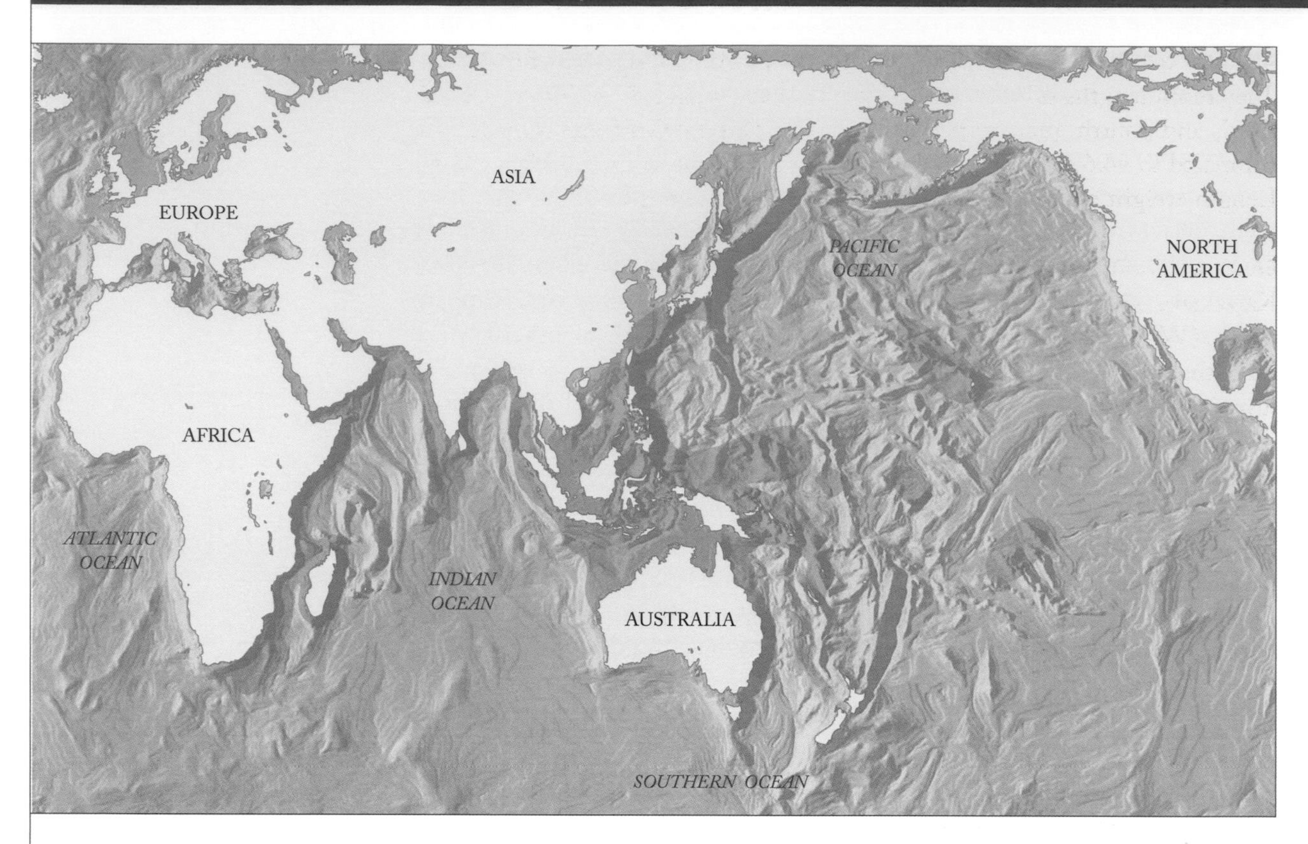

Above: Kawakawa range map.

FOOD: Kawakawa feed on small fishes, cephalopods, and crustaceans, with the relative proportions of each varying by location. In Indian waters, cephalopods were dominant (representing 56.5% of the diet by volume), while in Taiwan, peninsular Malaysia, the Gulf of Aqaba, and eastern Australia, fishes (including clupeoids, silversides, and lanternfishes) were dominant. In Indian waters, juvenile Kawakawa 41–150 mm long fed largely on fishes and switched to cephalopods at larger sizes, 151–660 mm, while in Taiwan smaller Kawakawa fed on lanternfishes and larval anchovies and switched to larger fishes such as jacks and mackerels with growth. The main food of juveniles from the Gulf of Mannar was the Commerson's Anchovy, *Stolephorus commersonii.*

REPRODUCTION: Kawakawa spawn extensively, both geographically and temporally, throughout their range, primarily in peripheral areas and around islands. Spawning occurs throughout the year in the Arabian Sea. Kawakawa appear to spawn in the waters of the northern Philippines and then migrate into Taiwan waters for feeding. A study conducted in Taiwan found the age at first maturity to be 2 years, while length at 50% maturity off India was reported to range from 37.7 cm to 49.7 cm. The larger size at first maturity (females at 49 cm, males at 49.7 cm) on the east coast of India may be the result of reduced fishing pressure. Average annual fecundity has been estimated at 210,000–680,000. The life cycle of Kawakawa was recently completed in land-based tanks, raising the possibility that this species might be a novel aquaculture species.

AGE AND GROWTH: Kawakawa are fast growing, with a longevity of 6–8 years. They reach lengths of 49, 69, 79, and 85 cm at the end of their first, second, third, and fourth year, respectively, in the Persian Gulf and Oman Sea, compared to 44.6, 64.9, and 77.5 cm for the first three years in Indian waters. Length-weight relationships were plotted for both regions.

EARLY LIFE HISTORY: There are descriptions and illustrations of four juvenile Kawakawa 33.5–156 mm FL and five larvae 3.7–10.5 mm from Philippine waters (Wade, 1950, 1951), four larvae 4.6–9.6 mm total length selected from more than 100 from east Indian waters (Matsumoto, 1958), and five larvae and juveniles 24.5–222 mm SL from India. At very early stages, larval Kawakawa may be confused with Oriental Bonito but usually have melanophores just anterior to the anus, which are absent in Oriental Bonito.

STOCK STRUCTURE: The stock structure of Kawakawa throughout its range is poorly known, and there is little information on stock structure in the Indian Ocean. Molecular analyses suggest that Kawakawa in the Philippines and southeast Asia are genetically homogeneous, as are those along the Indian coast.

FISHERIES INTEREST: Kawakawa are abundant in many parts of their range and are important in commercial fisheries. They are caught in multi-species fisheries, by surface trolling, with gill nets and purse seines, and as bycatch in industrial purse seines. They are marketed canned or frozen but are also utilized dried, salted, smoked, and fresh. Heavy metal levels were assessed for Kawakawa marketed at the Karachi Fish Harbor and found to be lower than the maximum permissible limit of international standards. Reported worldwide landings have gradually increased from 20,400 mt in 1950 to 385,923 mt in 2014. Within the Indian Ocean catches increased from 3,000 to over 150,000 mt between 1958 and 2015. The countries with the largest landings are Indonesia, India, the Philippines, Iran, and Thailand. In the Philippines and Indonesia, the catch includes many small individuals.

THREATS: Status of the Kawakawa populations varies by regions. In the northern Persian Gulf and Sea of Oman, Kawakawa has been highly exploited, leading to a recommendation that fishing regulations be applied to the population either by gradually increasing the mesh size of the gill nets or by restricting the fishery to certain seasons. The Kawakawa stock in the northeast Indian Ocean region may have entered the overfishing state due to the high fishing pressure on neritic fishes after 2008, when piracy activities intensified, forcing the fishery to move inshore. The stock along the Veraval coast of India was considered underexploited, but the exploitation rate off the coast of Sumatra indicated that Kawakawa were fully exploited. The estimated annual instantaneous fishing mortality in Tanzania was considerably larger than the target reference point, indicating that Kawakawa in the area are experiencing overfishing. It is listed as Least Concern on the IUCN Red List.

CONSERVATION: Kawakawa is listed as a highly migratory species in Annex I of the 1982 Convention on the Law of the Sea. It is widespread in the Indian and western Pacific oceans. However, there is little information on population trends. There are only sub-regional stock assessments for this species, generally based on short time series. For example, in Sri Lanka this species was considered to not be fully exploited. Based on a length-structured assessment for 2003–6 in Veravel, India, it seemed that catches were well below maximum sustainable yield, and in the Indian Ocean as a whole the stock is considered fully fished. There are no known conservation measures for this species, although some have been suggested. Some populations may be overexploited, so a comprehensive stock assessment is needed, especially as it seems that many catches are not being reported.

REFERENCES: Ahmed and Bat 2015; Al-Kiyumi et al. 2013; Al-Zibdah and Odat 2007; Bachok et al. 2004; Cantor 1849; Chiou and Cheng 1995; Chiou and Lee 2004; Chiou et al. 2004; Collette and Nauen 1983; Collette et al. 2011; Darvishi et al. 2003; Dayaratne and Silva 1991; Fitch 1953; Fraser-Brunner 1949; Ghosh et al. 2010; Godsil 1954; Griffiths et al. 2009; Iswarya Deepti and Sujatha 2012; Johnson and Tamatamah 2013; Jones 1960; Kaymaram and Darvishi 2012; Khan 2004; Kumar et al. 2012; Kumaran 1964; Matsumoto 1958; Muthias 1985; Nakamura and Magnuson 1965; Nishikawa 1988; Nishikawa et al. 1985; Rohit et al. 2012; Sabha Nissar et al. 2015a, 2015b; Santos et al. 2010; Schaefer 1987; Siraimeetan 1985; Sulistyaningsih et al. 2014; Taghavi Motlagh et al. 2010; Thomas and Kumaran 1963; Varghese et al. 2013; Wade 1950, 1951; Yazawa et al. 2015; Yesaki 1994; Yoshida 1979.

Little Tunny

Euthynnus alletteratus (Rafinesque, 1810)

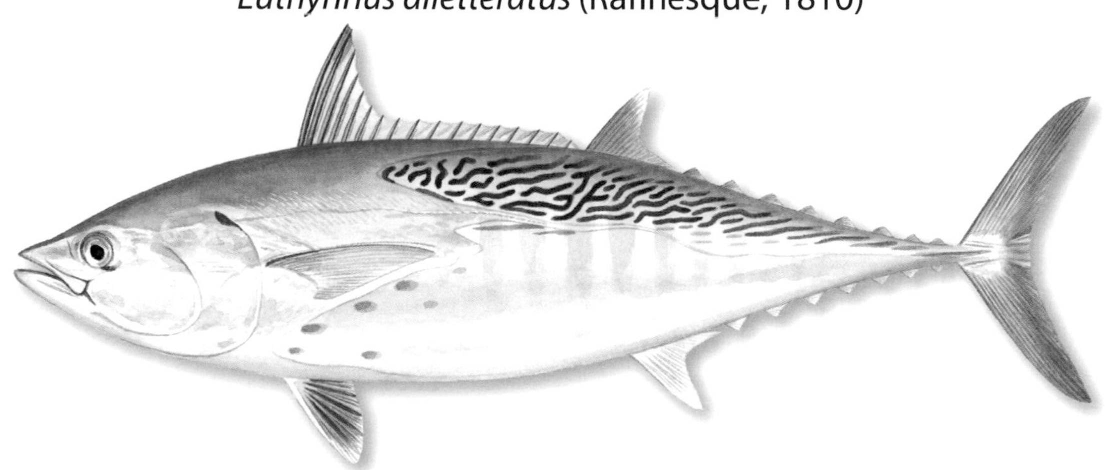

COMMON NAMES: English – Little Tunny
French – Thonine Commune
Spanish – Bacoreta

ETYMOLOGY: Described as *Scomber alletteratus* based on the Sicilian name "Alletteratu" by Professor Constantine Samuel Rafinesque-Schmaltz while in Sicily as secretary to the US Consul.

SYNONYMS: *Scomber Alletteratus* Rafinesque, 1810; *Scomber quadripunctatus* Geoffroy Saint-Hilaire, 1817; *Thynnus leachianus* Risso, 1827; *Thynnus thunina* Cuvier, 1829; *Thynnus brasiliensis* Cuvier, 1832; *Thynnus brevipinnis* Cuvier, 1832; *Euthynnus alletteratus aurolitoralis* Fraser-Brunner, 1949

TAXONOMIC NOTE: Some authors have considered that there were only two species in the genus *Euthynnus*, the Indo-West Pacific Kawakawa and the Atlantic Little Tunny, but all recent authors recognize three allopatric species in the genus and do not support recognition of the Gulf of Guinea population as a separate subspecies.

FIELD MARKS:

1 All three species of the genus *Euthynnus* resemble Skipjack but lack the prominent dark longitudinal stripes present on the Skipjack belly.
2 Vomerine teeth are absent in the Atlantic Little Tunny but are present in both the Kawakawa and Black Skipjack.
3 There are 37–40 gill rakers on the first gill arch of the Little Tunny, usually more than in either of the other two species, which have 29–39.

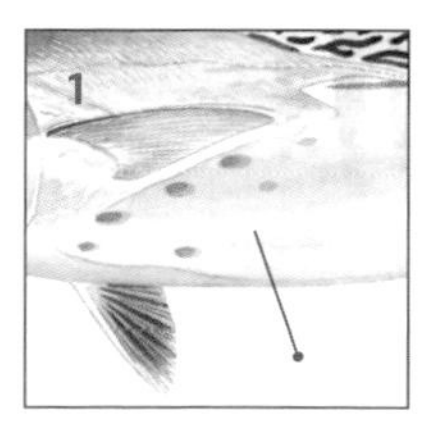

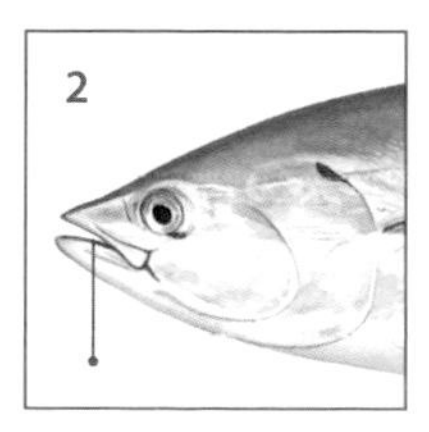

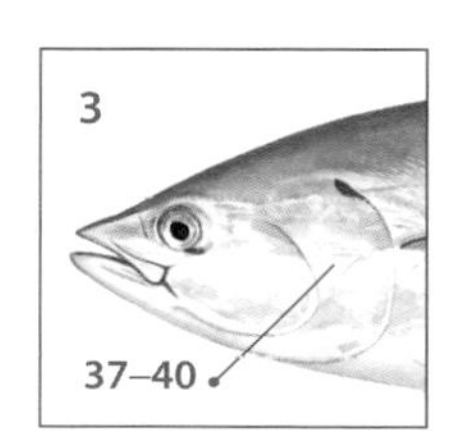

UNIQUE FEATURE: The Little Tunny is the only member of the genus that lives in the Atlantic Ocean, and it is the only member of the genus that lacks teeth on the vomer.

DIAGNOSTIC FEATURES: Body robust, elongate, and rounded. The teeth are small and conical and in a single series on each jaw, about 25–35 on each side of the lower jaw. The palatine has sharp and strong teeth but the vomer lacks teeth. **Gill rakers:** Gill rakers on the first gill arch number 37–45, and there are about 31 or 32 so-called gill teeth on the posterior surface of the first gill arch. **Fins:** There are two dorsal fins, the first with 14 or 15 spines, contiguous or nearly contiguous with the second dorsal fin, not broadly separated as in frigate tunas. The second dorsal fin has 12 or 13 fin rays and is followed by 8 finlets. The anal fin has 12 to 14 fin rays and is followed by 7 finlets. The pectoral fins are short. The interpelvic process is divided and is shorter than the pelvic fins. **Caudal peduncle:** A central keel is present on each side of the caudal-fin base between two smaller keels. **Scales:** The body is naked except for the anterior corselet. **Vertebrae:** Vertebrae number 20 precaudal plus 16–18 (usually 17) caudal, for a total of 37, rarely 36 or 38. The 31st and 32nd vertebrae have four conspicuous swellings or protuberances. The osteology and soft anatomy have been thoroughly described and illustrated by Godsil (1954). **Color:** The dorsal surface varies from dark blue to yellowish green. The upper sides of the body are bluish, turning to deep purple or almost black on the head. There are two black spots on the head, one at the upper margin of the operculum and a second on the postero-ventral margin of the orbit. There are broken, longitudinal black or dark bars running irregularly from the corselet toward the caudal region. The belly is dusky or silvery white, with an irregular number of black spots (usually 2–6) scattered between the pectoral and pelvic fins, but without the stripes present on the belly of the Skipjack. The pectoral and pelvic fins are purple, with their inner sides black.

GEOGRAPHIC RANGE: Little Tunny are endemic to the Atlantic Ocean in tropical and subtropical waters, including the Mediterranean Sea, Black Sea, Caribbean Sea, Gulf of Mexico, and the Gulf of Guinea. In the western Atlantic, they are found from the Gulf of Maine, Bermuda, and the Caribbean islands south to the border of Brazil and Argentina.

SIZE: 100 cm FL. The IGFA all-tackle game fish record is a 36 lb (16.32 kg) fish taken in Washington Canyon, New Jersey, in 2006. The longest game fish was 85 cm and caught at Satellite Beach, Florida, in June 2013.

HABITAT AND ECOLOGY: Little Tunny are reef associated and oceanodromous, occurring in neritic waters close inshore. They live in surface waters, mainly on the continental and insular shelves. They swim in fast-moving compact schools and are less migratory than Skipjack and other tunas. They are usually found in coastal areas with swift currents, near shoals and offshore islands. Along the Atlantic coast of the United States, most Little Tunny are caught in "green" inshore waters, not in the "blue" waters further offshore. In the Mediterranean they also occur far offshore.

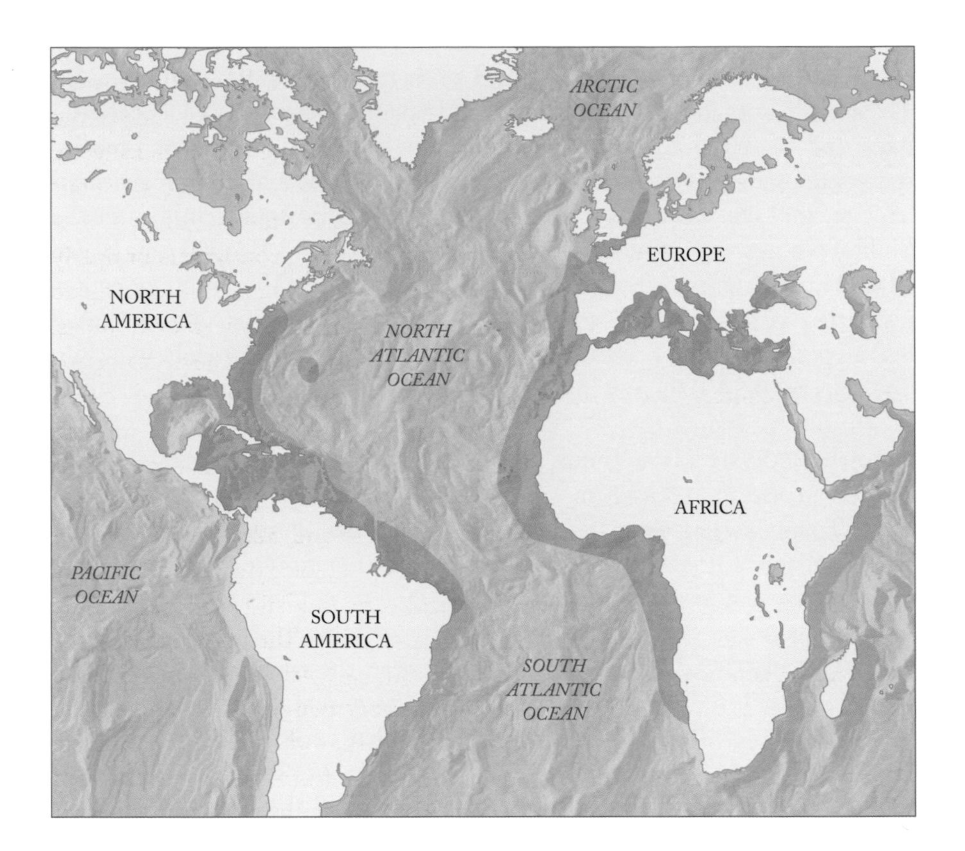

Left: Little Tunny range map.

FOOD: Little Tunny are opportunistic predators that feed primarily on fishes but also on crustaceans, squids, hyperiid amphipods, heteropods, and tunicates. Fishes were the dominant prey items in the Mediterranean, particularly pearlsides (*Maurolicus*) and larval fishes. Hyperiid amphipods were also important. A wide variety of fishes were consumed around the Cape Verde Islands. Fishes were the primary prey in the Gulf of Guinea, mainly due to the presence of two species, Atlantic Bigeye (*Priacanthus arenatus*) and Largehead Hairtail (*Trichiurus lepturus*). Fishes were also the most important prey in US waters, mainly adult sardines, anchovies, and jacks but also squids and crustaceans. As Little Tunny grow, crustaceans become less important in their diet and fishes more important. A phalanx-like group of up to nine individuals was observed attacking a ball-like school of dwarf herring that had formed under a night-light in the Caribbean Grenadine Islands. In Brazil, the three main fish families found in stomachs were herrings, jacks, and halfbeaks. Stomatopods and squids were also eaten. More than 70 species of fishes have been reported from Little Tunny stomachs.

REPRODUCTION: Spawning distributions of all three species of *Euthynnus* have been reported to be restricted primarily to peripheral areas and around islands within their respective ocean basins. Little Tunny spawn extensively, both geographically and temporally, throughout their range. Eggs are shed in several batches when the water is warmest. In Tunisia the sex ratio is 57.77% females. The gonado-somatic index indicated that spawning generally occurs between May and September in the Mediterranean Sea, with the most intensive spawning period between July and August. Size at first maturity ranged from 39.7 to 43.0 cm FL in several different regions. Absolute fecundity in the Gulf of Guinea was estimated to range from 342,000 to 2,127,000 eggs (mean 718,024) in females measuring between 42 and 76 cm FL, with egg counts increasing with increasing size. Off the southeastern coast of the United States, peak spawning occurs from April through August, particularly in July. The eggs are 0.84–0.94 mm in diameter and have one oil globule.

AGE AND GROWTH: Little Tunny have been aged by examining annual rings formed in the first dorsal-fin spines, 33rd vertebrae, and sagittal otoliths. Little Tunny have an estimated longevity of between 8 and 10 years, with an estimated age of first maturity of 2 or 3 years. Generation length is estimated to be approximately 4 years.

EARLY LIFE HISTORY: Juvenile Little Tunny were the most abundant young tunas in a 1959 study in the Gulf of Mexico. Three were illustrated, 4 mm, 6.5 mm, and 58 mm. Illustrations of a more complete series of eight larvae and juveniles, 5.5 mm notochord length to 58 mm SL, was included in Richards (2006). The incubation time is 48 hours at 26°C, and they hatch at 2.5 mm. Larval Little Tunny differ from the larvae of Skipjack and Bullet Tuna in having the first dorsal fin well pigmented. Growth of larvae in the Gulf of Mexico was rapid as in other scombrids, 1.07 mm/day.

STOCK STRUCTURE: There may be as many as four populations of Little Tunny: western Atlantic, Mediterranean, and two populations in the eastern Atlantic, central-eastern Atlantic from the western Sahara coast to Mauritania and Liberia and south-eastern from the Gulf of Guinea south to the Angola-Namibia border.

FISHERIES INTEREST: Little Tunny are a commercial species often taken in multi-species fisheries. In open waters the vast majority are taken by purse seines, and some on trolled lines; juveniles are also taken with beach seines. Specialized traps (madragues) are used in Tunisia and Morocco. Little Tunny are caught in the artisanal gill net fishery in northeast Brazil, and they are an important resource in Venezuela, where they are caught with beach seines and hook and line. Because of their abundance in inshore waters, Little Tunny are also a popular sport fish on light tackle, commonly taken by trolling jigs, spoons, or strip baits. Little Tunny are among the most commonly landed sport fish in southern Florida. They are also popular and effective as live

Sailfish bait. Little Tunny are caught throughout their range. In many instances, landings of Little Tunny are reported with other species as "small tuna," and several nations that are known to have catches do not report. There is also underreporting of Little Tunny discards from the purse seine fishery. As a result, reported landings are an underrepresentation of the true catch. Reported worldwide landings increased from 3,442 mt in 1950 to nearly 30,000 mt in 1989 and 1990 (FAO 2017). Over the past 20 years, reported landings have fluctuated between 13,000 and 24,000 mt without trend.

In the Caribbean, landings are aggregated with other species as "small tuna." This species is caught in small amounts in Brazil by several artisanal fisheries in northeast Brazil. In northeast Brazil, this species comprised 59.4% of total catch in a survey in Ceara state, 16.4% in Piaui state, and 15.6% in the north of Bahia state.

THREATS: Since 1991, the use of fish aggregating devices (FADs) by tropical purse seiners may have led to an increase in fishing mortality of Little Tunny and other small tropical tuna species, especially in the Gulf of Guinea. There is a general lack of information on the amount of discards and the mortality of these species as bycatch. Non-reporting continues to be a problem, and in many cases landings are not reported by species. The stock of *E. alletteratus* from the eastern coast of Alexandria, Egypt, is heavily exploited, and the fishing pressure exerted in the region is high. Based on available data, Little Tunny is listed as Least Concern on the IUCN Red List. However, close monitoring of catches should continue.

CONSERVATION: Little Tunny are a highly migratory species listed under Annex I of the 1982 Convention on the Law of the Sea. In Turkey there is a minimum landing size of 45 cm. High mortality from fishing along the eastern coast of Alexandria, Egypt, has led to a recommendation to reduce fishing mortality and to increase the minimum size limit to 44 cm FL to allow fish to spawn at least once.

SELECTED REFERENCES: Adams and Kerstetter 2014; Allman and Grimes 1998; Bahou et al. 2007, 2016; Bullis and Juhl 1967; Carlson 1952; Cayré and Diouf 1983; Collette and Nauen 1983; de Sylva and Rathjen 1961; Dragovich 1969; El-Haweet et al. 2013; Falautano et al. 1997; Fraser-Brunner 1949; Gaykov and Bokhanov 2008; Godsil 1954; González et al. 2008; Hajjej et al. 2009, 2010, 2011; Hattour 2009; IGFA 2018; Johnson 1983; Kahraman and Alicli 2007; Kahraman et al. 2008; Klawe and Shimada 1959; Landau 1965; Manooch et al. 1985; Matsumoto 1959; Mele et al. 2016; Menezes and Aragão 1977; Mota Alves and Aragao 1977; Oray and Karakulak 2005; Postel 1955; Ramírez-Arredondo 1990; Randall 1967, 1968; Richards 2006; Rodríguez-Roda 1966; Schaefer 2001; Valeiras et al. 2008; Yoshida 1979.

Black Skipjack

Euthynnus lineatus Kishinouye, 1920

COMMON NAMES: English – Black Skipjack
French – Barrilete Negro
Spanish – Thonine Noire

ETYMOLOGY: Named *lineatus* by the Japanese ichthyologist Kamakichi Kishinouye, presumably with reference to the 3–5 broad black stripes running along the upper sides of the fish.

SYNONYMS: This is one of the few species of tuna that has no synonyms.

TAXONOMIC NOTE: Some authors thought the eastern Pacific *E. lineatus* was a subspecies of *E. affinis*, but all recent authors recognize *E. lineatus* as one of three species in the genus.

FIELD MARKS:

1 The other two species of the genus *Euthynnus* resemble Skipjack Tuna but lack the prominent dark longitudinal stripes present on the Skipjack Tuna belly.
2 Vomerine teeth are present in both the Kawakawa and Black Skipjack but are absent in the Little Tunny.
3 There are 33–39 gill rakers on the first gill arch of the Black Skipjack, more than the 29–33 present in the Kawakawa but fewer than the 37–40 present in the Little Tunny.

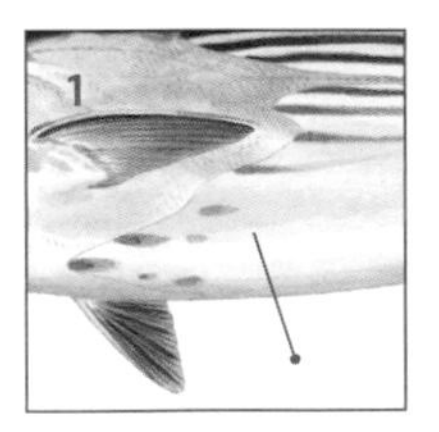

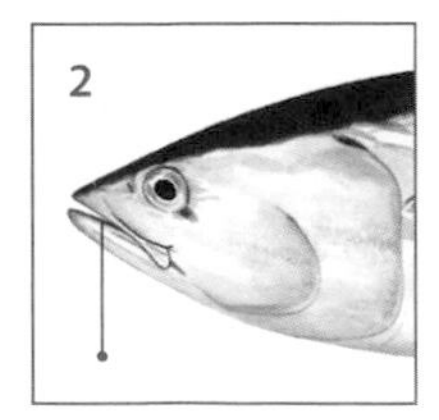

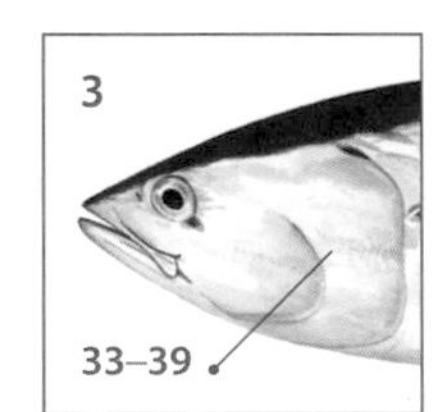

UNIQUE FEATURE: The Black Skipjack is the only species of the genus *Euthynnus* in the eastern Pacific.

DIAGNOSTIC FEATURES: Body robust, elongate, and rounded. The teeth are small and conical and in a single series on each jaw, about 25–30 on each side of the lower jaw. The palatine and vomer have teeth. **Gill rakers:** Gill rakers on the first gill arch number 32–41 and there are about 30 so-called gill teeth on the posterior surface of the first gill arch. **Fins:** There are two dorsal fins, the first with 14–15 spines, contiguous or nearly contiguous with the second dorsal fin. The second dorsal fin has 11 or 12 fin rays and is followed by 8 finlets. The anal fin has 11 to 13 fin rays and is followed by 7 finlets. The pectoral fins are short. The interpelvic process is divided and is shorter than the pelvic fins. **Caudal peduncle:** A central keel is present on each side of the caudal-fin base between two smaller keels. **Scales:** The body is naked except for the anterior corselet. **Vertebrae:** Vertebrae number 20 precaudal plus 16–18, usually 17, caudal, for a total of 37, rarely 36 or 38. The 31st and 32nd vertebrae have four conspicuous swellings or protuberances. The osteology and soft anatomy have been thoroughly described and illustrated by Godsil (1954). **Color:** The dorsal surface varies from dark blue to yellowish green. The upper sides of the body are bluish, turning to deep purple or almost black on the head. There are two black spots on the head, one at the upper margin of the operculum and a second on the postero-ventral margin of the orbit. There are 3–5 broad continuous black stripes running horizontally from the corselet to the caudal fin. The belly is dusky or silvery white, with an irregular number of black spots, (usually 2–6) spots, or streaks scattered between the pectoral and pelvic fins. The pectoral and pelvic fins are purple, with their inner sides black.

Below: Belly pattern variation.

GEOGRAPHIC RANGE: Black Skipjack are endemic to the eastern tropical Pacific and are found from San Simeon, California, and the lower half of the Gulf of California south to northern Peru, including all the offshore islands and the Galapagos. Two stray specimens (vagrants) were caught in the Hawaiian Islands.

SIZE: 84 cm FL. The IGFA all-tackle game fish record is of a 26 lb (11.79 kg) fish caught on Thetis Bank, Baja California, in 1991. The length-weight relationship has been plotted for 109 specimens that ranged in length from 365 to 667 mm TL and in weight from 2 to 13 lb, 13 oz.

HABITAT AND ECOLOGY: Black Skipjack are pelagic and oceanodromous and rarely occur where surface temperatures fall below 23°C. They can form multi-species schools with Yellowfin and Skipjack tunas. In captivity, Black Skipjack 46–55 cm FL swam average sustained speeds of 1.4–1.7 body lengths per second, similar to the swimming speeds of Kawakawa. As with other tunas, there is a warm central core in the Black Skipjack. The warmest body temperatures occur along the vertebral column between the first and second dorsal fins. The average temperature in this region was 34.2°C, while the surface temperature of the water where the fish were caught averaged 29.2°C. Deep red muscle temperatures of captives were 3.0–3.5°C above the temperature in the bait-wells in which they were tested. Black Skipjack have been found in the stomachs of Yellowfin Tuna and are undoubtedly also eaten by large billfishes.

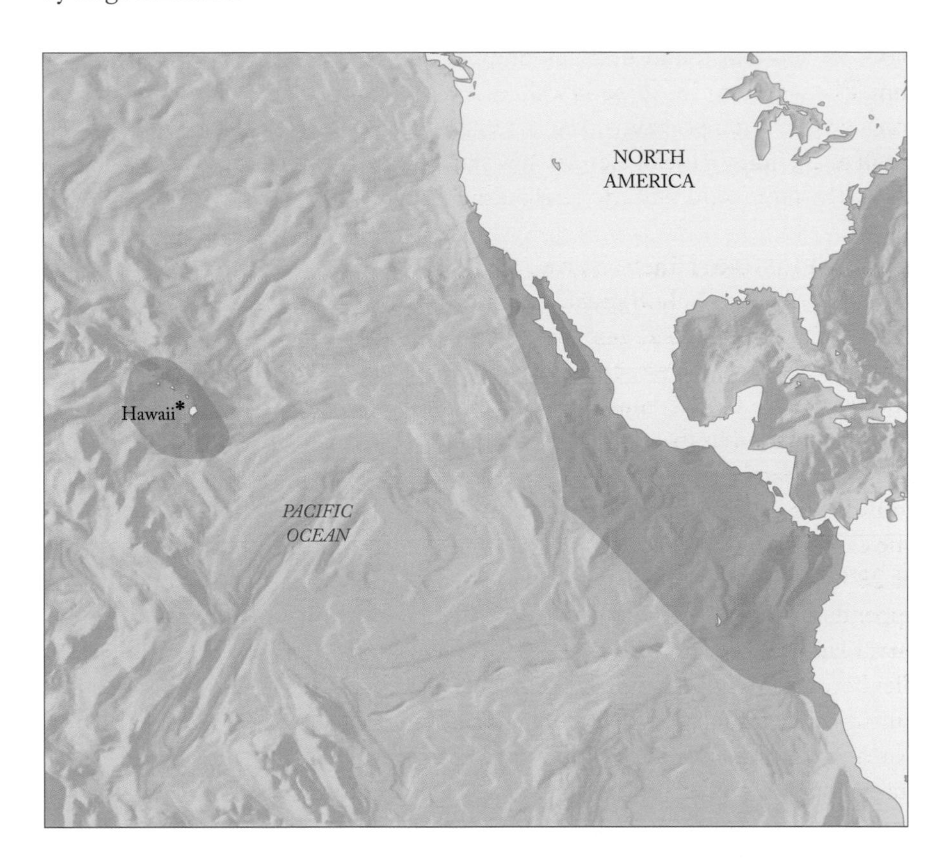

Right: Black Skipjack range map.

FOOD: There are few food studies of Black Skipjack, but an earlier investigator considered them feeders of "voracious habits and insatiable appetite." Some sampled stomachs contained squids, pelagic crabs, and small fishes such as Frigate Tuna and Sierra Spanish Mackerel. Gastric evacuation of force-fed small fishes was approximately 6 hours longer than for Skipjack. Black Skipjack are opportunistic predators that share their feeding patterns with other tunas and probably compete for food with species such as Yellowfin Tuna, Dolphinfish, and Oriental Bonito.

REPRODUCTION: The length at 50% maturity for Black Skipjack is estimated to be 47 cm FL. They spawn extensively, both geographically and temporally, throughout their range. Although the spawning areas of all three *Euthynnus* species have been reported to be restricted primarily to peripheral areas and around islands within their respective ocean basins, spawning in the eastern tropical Pacific has been shown to be widely distributed from coastal to oceanic waters. The batch fecundity of this species has been shown to increase with latitude in the eastern Pacific.

AGE AND GROWTH: Captive larval Black Skipjack grew at 1–4.8 mm/day for the first 15 days in captivity, but after 15 days growth increased to 3.4–4.8 mm/day. Greater growth rates, up to 7.86 mm/day, were found in a subsequent captive study, and this high growth rate exceeded the growth rate for most other pelagic fishes. Evidence of the development of central and lateral rete blood vessels was found in Black Skipjack as small as 95.9 mm FL and 125 mm FL, respectively. These blood vessels are what allow tunas to maintain warmer body temperatures than the surrounding water. Black Skipjack as small as 207 mm FL can elevate their red muscle temperature by at least 3°C above the temperature of the surrounding water.

EARLY LIFE HISTORY: Nineteen juvenile Black Skipjack, 48–86 mm total length, were described and photographed from off Central America. Two postlarval Black Skipjack, 14 and 22 mm long, were illustrated from among 27 specimens 7.5–23.5 mm that were caught off the Pacific coast of Central America. Seven larvae 5.0–21.0 mm were described and illustrated from among 135 specimens taken in the waters off the Pacific coast of Central America by the Danish research vessel *Dana* in 1920–22 and 1928–30. Postlarvae are mostly confined to surface waters within about 240 miles of the mainland except for one caught near Malpelo Island. Postlarvae were encountered at temperatures of 28°C to 32°C in the Gulf of California. Their diet was dominated by the appendicularian *Oikopleura dioica* and copepod nauplii and, to a lesser extent, larval lanternfishes and herring-like fishes. In larvae longer than 5 mm, the diet became almost exclusively *Oikopleura* and fish larvae. Feeding was mostly during daylight hours and was relatively low; 42% had food in their stomachs. Nutritional estimates in the Panama Bight indicated that the probability of starvation for preflexion larvae was quite high (41–43%) during the high-precipitation, reduced-upwelling season, while it was low for postflexion larvae and juveniles.

FISHERIES INTEREST: There is no specific fishery for Black Skipjack, but they are taken incidentally by tuna purse seines, live-bait pole-and-line gear, trolling, and sport fishing gear. FAO-reported landings fluctuated greatly from the 1970s to 2000, ranging from 0.5 mt in 1971 to a high of 3,299 mt in 1980. Since 2007 reported landings have been greater than 3,000 mt. Purse seine landings for this species reported by IATTC range from 3,000 to 6,000 mt since 2000, with no observable trend in these data. The main method of harvesting Black Skipjack is using purse seines, although they are also caught with pole and line. Black Skipjack is an important commercial fish in Golfo de Montijo, Panama. It is also targeted in Ecuador. Historically it was targeted and canned in Costa Rica. It was important previously as a sport fish in Gorgona, Colombia. Black Skipjack is commonly used as strip bait in sport fisheries for Sailfish and other billfishes throughout the eastern tropical Pacific. Black Skipjack is widespread in the eastern Pacific, appears to be fairly common, and has little directed fishing effort at present.

THREATS: Black Skipjack is listed as Least Concern on the IUCN Red List.

CONSERVATION: There are no known conservation measures for Black Skipjack. However, their distribution includes a number of marine protected areas in the eastern tropical Pacific region, especially the Cocos, Galapagos, and Malpelo marine protected areas. There have been previous IATTC area-wide closures in the Eastern Pacific for all tuna species (including Black Skipjack), such as the six-week periods from August 1 to September 11 and from November 15 to December 31.

SELECTED REFERENCES: Calkins and Klawe 1963; Collette and Nauen 1983; Collette et al. 2011; Dickson et al. 1994, 2000; Fraser-Brunner 1949; Godsil 1954; Graham 1973; Klawe and Calkins 1965; Matsumoto 1959, 1976; Matsumoto and Kang 1967; Mead 1951; Olson and Scholey 1990; Sánchez-Velasco et al. 1999; K. Schaefer 1984, 1987; M. Schaefer and Marr 1948; Yoshida 1979.

Butterfly Kingfish

Gasterochisma melampus Richardson, 1845

COMMON NAMES: English – Butterfly Kingfish
French – Thon Papillon
Spanish – Atún Argentino

ETYMOLOGY: Sir John Richardson was a Scottish naval surgeon, naturalist, and Arctic explorer who derived the name of the genus from the Greek words *gaster* and *chaisma*, meaning "stomach" and "cross-shaped configuration."

SYNONYMS: *Gasterochisma melampus* Richardson, 1845; *Lepidothynnus huttoni* Günther, 1889; *Chenogaster holmbergi* Lahille, 1903; *Gastrochisma boulengeri* Lahille, 1913

TAXONOMIC NOTE: Different-sized individuals of this species have been described four times in three different genera, from smallest to largest: *Gasterochisma melampus* Richardson, 1845, based on an 18 cm juvenile; *Gastrochisma* (sic) *boulengeri* Lahille, 1913, based on a 72.5 cm specimen; *Chenogaster holmbergi* Lahille, 1903, based on a 132 cm adult; and as *Lepidothynnus huttoni* Günther, 1889, based on a 167.6 cm adult.

FIELD MARKS:

1 The large body is covered with large cycloid scales.
2 The pelvic fins are enormous in juveniles, proportionally much smaller in adults, and fit into a deep ventral groove at all sizes.
3 The caudal peduncle has only two small keels on each side, but no lateral keel is present.

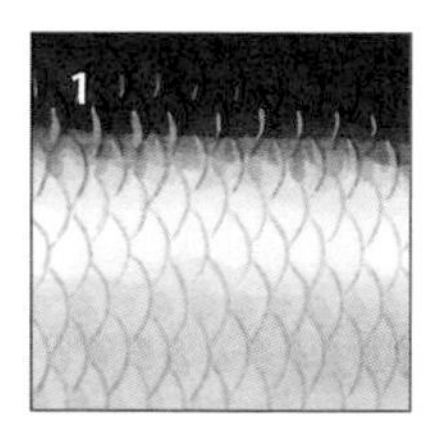

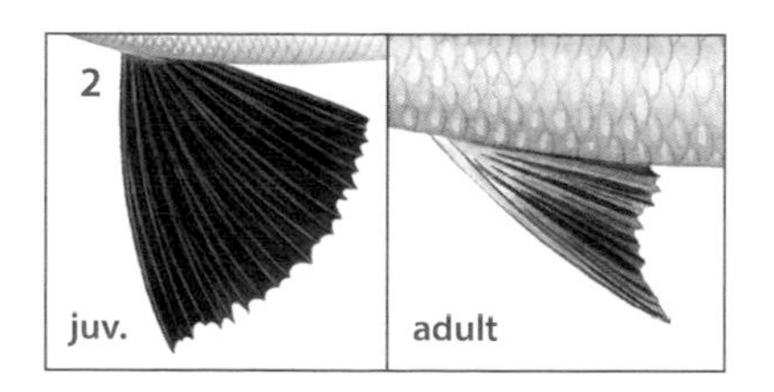

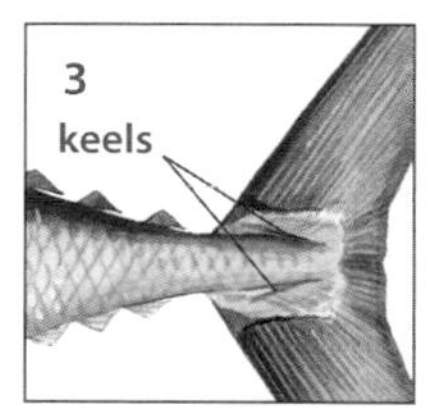

UNIQUE FEATURE: The scombrid with the largest scales.

DIAGNOSTIC FEATURES: The body is laterally compressed, its depth about 3.5 times in the fork length. The teeth are small and conical and in a single series. **Fins:** The two dorsal fins are separated by a wide interspace in adults. The first dorsal has 17–18 spines (rarely 15); the second dorsal has 10 or 11 rays and is followed by 6 or 7 finlets. The anal fin has 11 or 12 rays and is also followed by 6 or 7 finlets. The pectoral fins are short, with 19–22 rays. The pelvic fins are enormous in juveniles, longer than head length, becoming of more normal proportions for scombrids in adults. The pelvic fins fit into a deep ventral groove at all sizes. The interpelvic process is tiny and bifid. **Caudal peduncle:** The caudal peduncle has only two small keels on each side; no lateral keel is present. **Scales:** The body is covered with large cycloid scales, and no anterior corselet is present. **Swim bladder:** A swim bladder is present, with two anterior projections that extend into the back of the skull. **Inner ear:** Butterfly Tuna have no endolymphatic sac in their inner ear and therefore neither sagitta nor asteriscus otoliths are present. **Vertebrae:** Vertebrae number 21 precaudal plus 23 caudal, for a total of 44. Osteology was treated and illustrated by Kohno (1984). **Color:** Body deep bluish above, silvery below, without spots, stripes, or other prominent markings. The first dorsal fin is pale to translucent and the pelvic fins are blackish.

Below: Juvenile Butterfly Kingfish, approx. 22 cm.

GEOGRAPHIC RANGE: Butterfly Kingfish are circumglobal in southern temperate waters, mostly between 35° and 50° S, but there are records from off Argentina, Uruguay, and southern Brazil in the Atlantic north to 23°09′ N and 23°49′ N. There are two records from outside of southern waters in the Pacific, one from Ecuador and one from Hawaii, both of which are likely vagrants.

SIZE: A total of 1,708 specimens, 74 to 164 cm body length, were taken from throughout their range by the Japanese longline fishery for Southern Bluefin Tuna. Kohno (1984) based his osteological description on several specimens, one of which was 195 cm, apparently the longest recorded specimen. The IGFA all-tackle game fish record is of a 41.35 kg fish caught off Portland, Victoria, Australia, in May 2012.

HABITAT AND ECOLOGY: This enigmatic species occurs over deep oceanic waters, from the surface to below 656 ft (200 m). Their biology is poorly known, but they have a brain-heater organ derived from the lateral rectus eye muscle, not the superior rectus muscle as in billfishes. Butterfly Kingfish heater organs display exceptionally high aerobic capacities as indicated by the very high citrate synthase activity, which is higher than that found in any other vertebrate

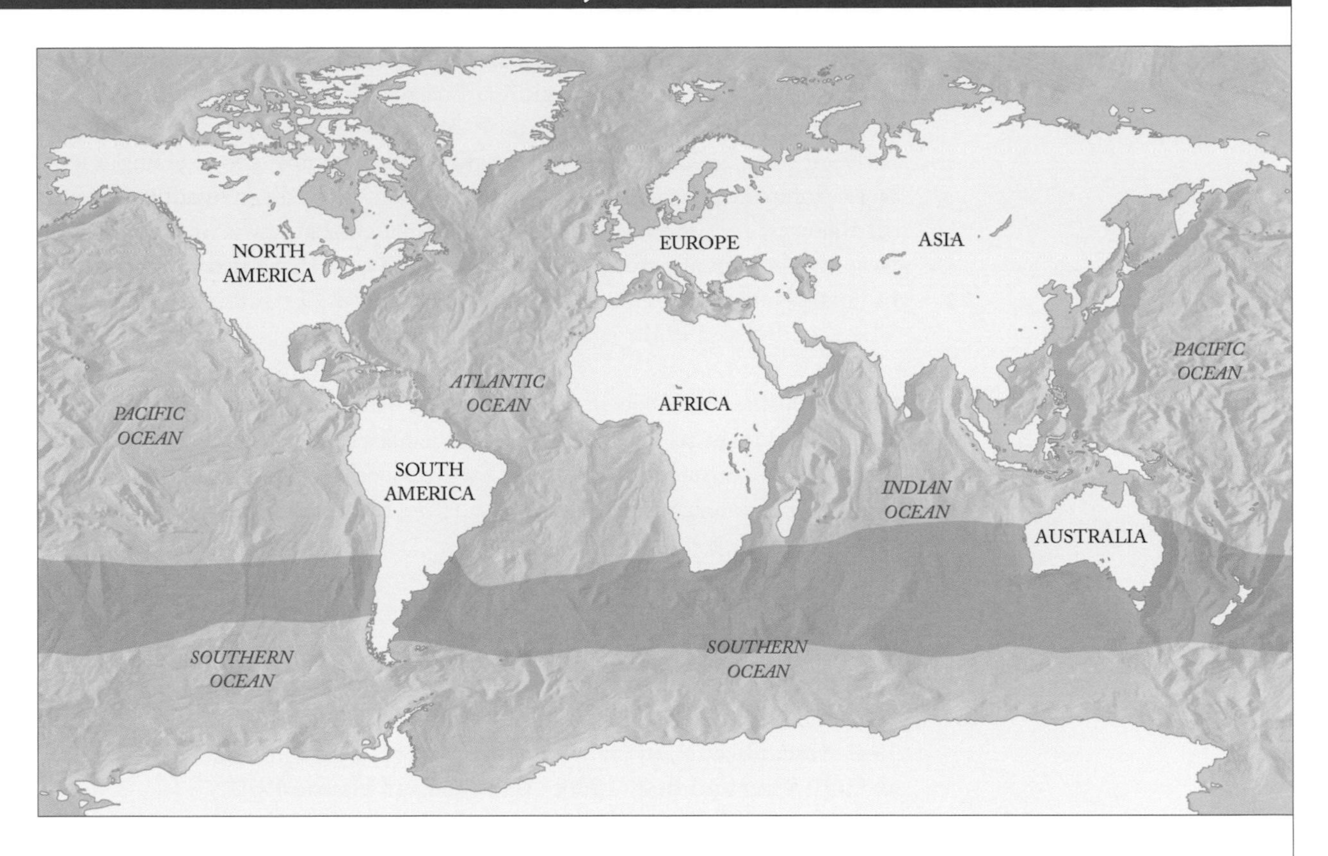

Above: Butterfly Kingfish range map.

tissue. The heater organ may allow Butterfly Kingfish to maintain an adequate cranial–water temperature gradient in their cold-water habitat.

FOOD: Squids constituted about half of the stomach contents of Butterfly Kingfish caught around the South Island of New Zealand. One-third of the stomach contents were fishes, and salps and crustaceans were of lesser importance. There was some indication of a shift from small mesopelagic species to larger mesopelagic species as the fish grew. Stomach contents of the specimen from north of the Hawaiian Archipelago included onychoteuthid and ommastrephid squids, vertebrae and fin rays from an unidentified fish, bird feathers, and parasitic nematodes.

REPRODUCTION: A recent study based on 25,564 individuals collected from longline operations from 1987 to 1996 has produced the only information on reproduction of the Butterfly Kingfish. Analysis of the gonosomatic index, oocyte sizes, and the presence of hydrated eggs concluded that the spawning area is between 85° and 130° W and 28° and 41° S in the southeast Pacific Ocean. The spawning season extends from mid-April to mid-July and is mostly in mid-May. Butterfly Kingfish have very large hydrated eggs, 1.6 mm in diameter, larger than almost all other scombrids. Spawning takes place at sea surface temperatures as low as 14–18°C, much lower than the usual condition in scombrids, 24°C or warmer. There is still no information on larvae.

FISHERIES INTEREST: Butterfly Kingfish are not of important commercial interest but are taken as bycatch by pelagic longliners targeting Southern Bluefin Tuna. Reporting of bycatch landings is fairly recent, with a reported global catch of 6–47 mt since 2000. More than 60 individuals weighing a total of about 4 mt were captured by Japanese longliners using sauries as bait in less than 2 weeks off the coast of Uruguay in 1963. It is estimated that approximately 900 mt were caught annually from 1990 to the early 2000s. They are sometimes eaten in Japan. Nutritional data on specimens from New Zealand was published by Vlieg and Body (1993).

CONSERVATION: This species is widespread and common in the Southern Ocean. It is taken as bycatch in longline fisheries for Southern Bluefin Tuna. There are no conservation measures for this species at this time. It is listed as Least Concern on the IUCN Red List. More research is needed on its biology and population trends.

REFERENCES: Block 1991; Coelho et al. 1990; Collette 2010; Collette and Nauen 1983; Collette et al. 2011; Cousseau and Figueroa 1989; Gauldie and Radtke 1990; Günther 1889; Horn et al. 2013; IGFA 2018; Ito et al. 1994; Itoh and Sawadaishi 2017; Kohno 1984; Lahille 1903, 1913; Richardson 1845; Rotundo et al. 2015; Santos and Nunan 2015; Tominaga 1967; Tullis et al. 1991; Vlieg and Body 1993; Warashina and Hisada 1972.

Shark Mackerel

Grammatorcynus bicarinatus (Quoy & Gaimard, 1825)

COMMON NAMES: English – Shark Mackerel
French – Tazard à Larges écailles
Spanish – Carite Cazón

ETYMOLOGY: Described by the French ichthyologists Jean Rene Constant Quoy and Jules Dumont d'Urville (1825, p. 357) based on specimens from their expeditions around the world on two French vessels *Uranie* and *Physicienne*. The species name *bicarinatus* refers to the caudal keels, although they also mention the two lateral lines early in their description.

SYNONYMS: *Thynnus bicarinatus* Quoy and Gaimard, 1825; *Grammatorcynus bicarinatus* (Quoy and Gaimard, 1825); *Grammatorcynus bilineatus* (non Rüppell, 1836)

TAXONOMIC NOTE: The Shark Mackerel was confused with the Double-lined Mackerel (*Grammatorcynus bilineatus*) until both species were separated (Collette 1983). References to *G. bicarinatus* from outside Australia all refer to the Double-lined Mackerel (*Grammatorcynus bilineatus*).

FIELD MARKS:

1. Both species of *Grammatorcynus* have two lateral lines, an upper one like other mackerels plus a lower one. However, the Shark Mackerel has a much smaller eye than the Double-lined Mackerel, only 3–4% of fork length compared to 7–9%.
2. There are only 12–15 gill rakers on the first arch instead of 18–24.
3. Most Shark Mackerel have a series of small dark spots along the ventral surface of the body, markings that are absent in the Double-lined Mackerel.

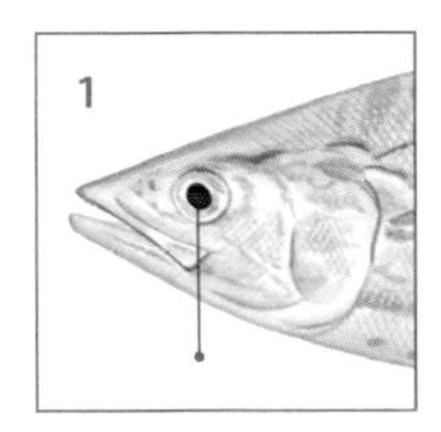

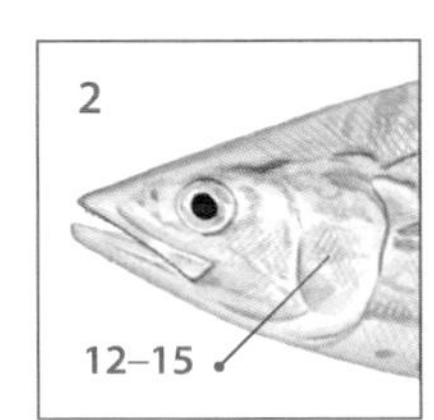

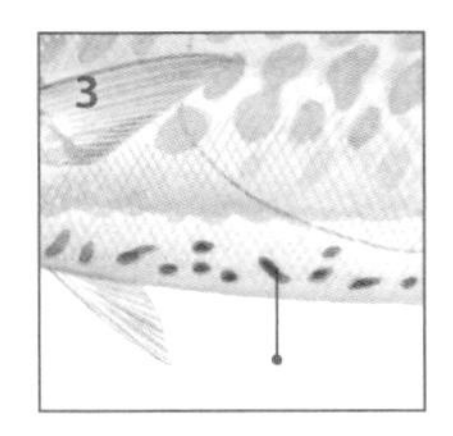

UNIQUE FEATURE: The larger of the two double-lined mackerels that is restricted to Australia and Papua New Guinea.

DIAGNOSTIC FEATURES: Gill rakers: There are relatively few gill rakers on the first arch, 12–15, usually 14 or 15. **Fins:** The first dorsal fin has 11–13 spines. The second dorsal fin is slightly higher and has 10–12 rays, followed by 6 or 7 finlets. The anal fin has 11–13 rays, also followed by 6 or 7 finlets. The pectoral fins are stout, with 22–26 rays. The interpelvic process is short and single. **Caudal peduncle:** The caudal peduncle is slender, with a well-developed lateral keel between the two smaller ones on each side. **Lateral line:** There are two lateral lines, the uppermost extending from the opercle to the lateral caudal keel, as is typical of most mackerels. The second lateral line branches off from the first under the third spine of the first dorsal fin, descends below the level of the pectoral fin, and runs posteriorly to join the upper lateral line at about the level of the last dorsal finlet. **Scales:** The body is covered with small scales, and there is no anterior corselet of enlarged scales. **Swim bladder:** A swim bladder is present. **Vertebrae:** The vertebrae number 14 precaudal plus 17 caudal, for a total of 31. Descriptions and illustrations of the osteology and soft anatomy of the Shark Mackerel are included in the revision of the genus by Collette and Gillis (1992). **Color:** The back and upper sides are mostly olive green, fading to a silvery white belly, frequently with small dark spots along the ventral margin of the body. Some become golden to brownish. When viewed by divers, Shark Mackerel display a prominent dark band along the side of the body following the lower lateral line.

Right: Golden color variation.

GEOGRAPHIC RANGE: Positively known only from the northwest and northeast coasts of Australia, occurring on the west coast south to Rottnest Island, off Perth, Western Australia, and on the east coast of Queensland south to northern New South Wales but apparently not through Torres Strait or around southern Papua New Guinea. Their greatest concentration seems to be in waters of the central section of the Great Barrier Reef.

SIZE: Shark Mackerel grow much larger than Double-lined Mackerel, reaching 28.5 lb (13 kg). The IGFA all-tackle angling record is of a 27 lb, 1 oz (12.3 kg) fish taken off Bribie Island, Queensland, Australia, in March 1989.

REPRODUCTION: Little is known about the biology of Shark Mackerel, and most of the references to "*Grammatorcynus bicarinatus*," including all those from outside Australia, actually refer to the Double-lined Mackerel, *Grammatorcynus bilineatus*. Gonads mature in early southern spring.

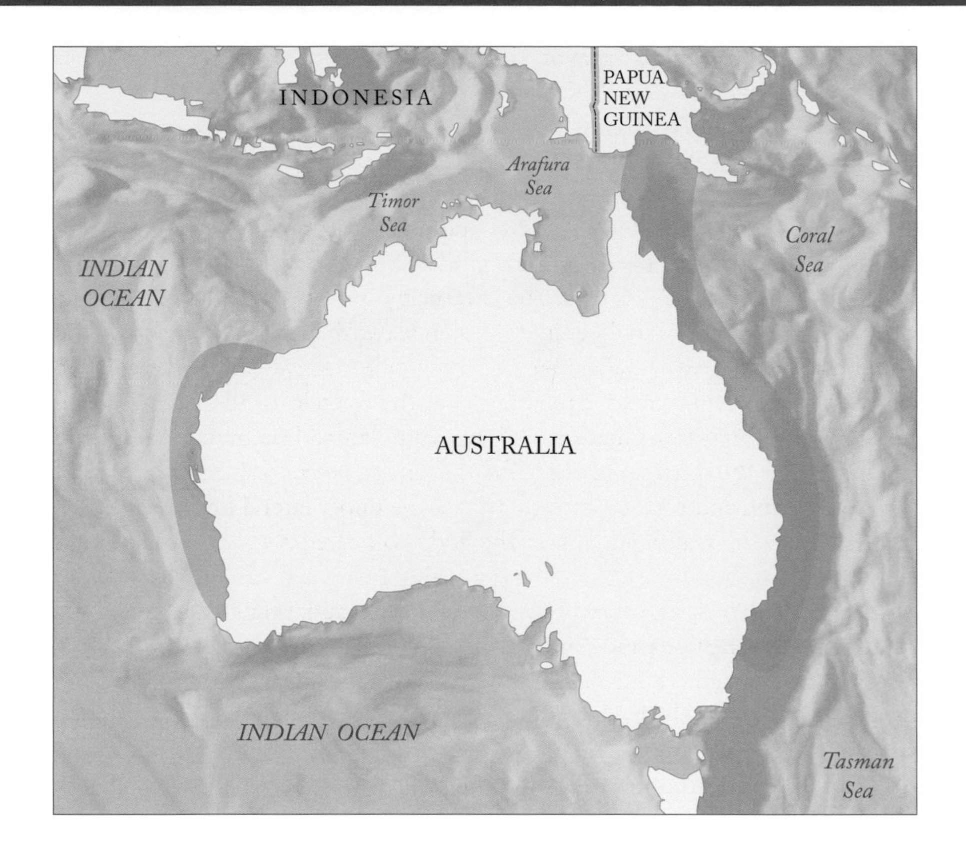

Left: Shark Mackerel range map.

FISHERIES INTEREST: There is a directed commercial fishery for this species in northeast Australia, with landings from the Great Barrier Reef of 43.2 mt in 2003. This is a target species for light-tackle recreational fishermen on the Great Barrier Reef and the Gascoyne and Pilbara bioregions of Western Australia. The common name Shark Mackerel refers to the shark-like ammonia odor of the flesh, which can be eliminated by brushing with lemon juice prior to grilling.

THREATS: The recreational fisheries on both the western and eastern coasts of Australia do not seem to have adversely affected populations. More research is needed on this species' biology and population trends, especially as there is a directed fishery for this species in Queensland. It is listed as Least Concern in the IUCN Red List.

CONSERVATION: The recreational fishery is managed with bag limits in Western Australia and a combination of bag limits and a minimum size in Queensland. At least a portion of this species' range overlaps the Great Barrier Reef marine protected area.

REFERENCES: Collette 1983, 2001; Collette and Gillis 1992; Collette and Nauen 1983; Collette et al. 2011; Grant 1987; Hutchins 1979; IGFA 2018; Kingsford and Welch 2007; McPherson 1984, 1988; Pepperell 2010; Quoy and Gaimard 1825; Randall et al. 1990; Sumner et al. 2002; Williamson et al. 2006.

Double-lined Mackerel

Grammatorcynus bilineatus (Rüppell, 1836)

COMMON NAMES: English – Double-lined Mackerel
French – Thazard-kusara
Spanish – Carite-cazón Pintado

ETYMOLOGY: Described by the German naturalist and explorer Wilhelm Peter Eduard Simon Rüppell (1836, pp. 39–40) based on specimens he collected and named *bilineatus*, referring to the two lateral lines characteristic of both the Double-lined Mackerel and its sister species, the Shark Mackerel.

SYNONYMS: *Thynnus bilineatus* Rüppell, 1836; *Nesogrammus piersoni* Evermann and Seale, 1907; *Grammatorcynus bilineatus* (Rüppell, 1836)

TAXONOMIC NOTE: Double-lined Mackerel were often confused with Shark Mackerel (*Grammatorcynus bicarinatus*) until Lewis (1981) recognized two species with electrophoretic data in his doctoral dissertation and provided specimens to Collette (1983), who clarified the taxonomy of the genus. This confusion has continued (Auster 2008), but any references to *Grammatorcynus bicarinatus* from outside Australia refer to Double-lined Mackerel, not Shark Mackerel.

FIELD MARKS:

1 Both species of *Grammatorcynus* have two lateral lines, an upper one like other mackerels plus a lower one. However, the Double-lined Mackerel has a larger eye, 4.1–6.0% of fork length versus 3.1–4.6%.
2 There are 18–24 gill rakers on the first arch instead of 12–15.
3 A series of small dark spots along the ventral surface of the body are absent.

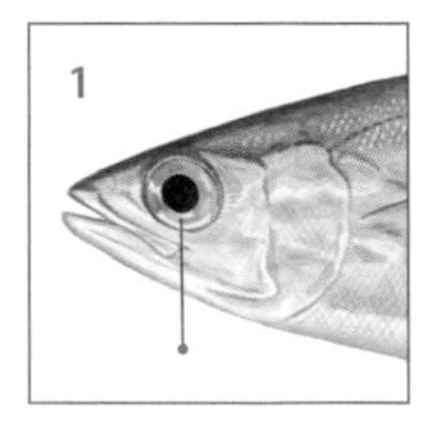

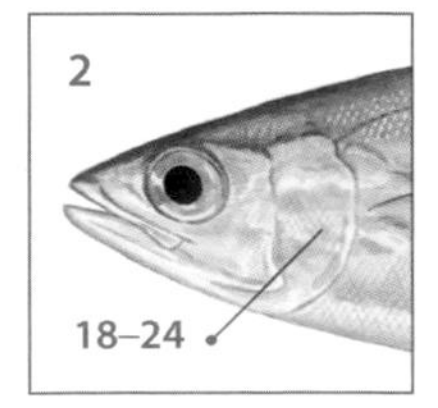

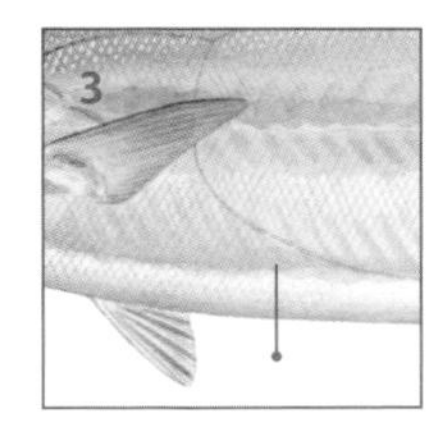

UNIQUE FEATURE: The smaller of the two double-lined mackerels and the one with the larger distribution.

DIAGNOSTIC FEATURES: Gill rakers: There are 19–24 gill rakers on the first arch. **Fins:** The first dorsal fin has 11–13 spines, usually 12; the second dorsal fin is slightly higher and has 10–12 rays, followed by 6 or 7 finlets. The anal fin has 11–13 rays, also followed by 6 or 7 finlets, rarely only 5. The pectoral fins are stout, with 22–26 rays. The interpelvic process is short and single. **Caudal peduncle:** The caudal peduncle is slender, with a well-developed lateral keel between the two smaller ones on each side. **Lateral line:** There are two lateral lines, the uppermost extending from the opercle to the lateral caudal keel, as is typical of most mackerels. The second lateral line branches off from the first under the third spine of the first dorsal fin, descends below the level of the pectoral fin, and runs posteriorly to join the upper lateral line at about the level of the last dorsal finlet. **Scales:** The body is covered with small scales, and there is no anterior corselet of enlarged scales. **Swim bladder:** A swim bladder is present. **Vertebrae:** The vertebrae number 12 precaudal plus 19 caudal, for a total of 31. Descriptions and illustrations of the osteology and soft anatomy of the Double-lined Mackerel are included in Silas (1963) and the revision of the genus by Collette and Gillis (1992). **Color:** In life, the back is bright blue to pale green, fading to a silvery white belly, without the small dark spots along the ventral margin of the body that are characteristic of the Shark Mackerel.

Below: Pale green color variation.

GEOGRAPHIC RANGE: Double-lined Mackerel are widespread in the Indo-Pacific, from the Red Sea eastward to the Andaman Sea and from the Ryukyu Islands to the northern coasts of Australia and out into the central Pacific, including the Marshall Islands and Fiji and Tokelau. It is not clear whether the distribution is continuous around the Indian subcontinent. Recent records document its occurrence in the Gulf of Aqaba, Kagoshima Province, in southern Japan and Orchid Island, Taiwan.

SIZE: Double-lined Mackerel reach at least 63 cm FL and 3.3 kg in weight. The IGFA all-tackle angling record is of a 6 lb, 10 oz (3 kg) fish caught off Willis Island, Queensland, Australia, in November 2006.

HABITAT AND ECOLOGY: Double-lined Mackerel are reef associated and oceanodromous. They inhabit open water but are often seen swimming near outer reef walls or deep clear-water slopes. They are found mostly in shallow reef waters, where they form large schools. In Palau, schools containing up to several hundred fish appeared only on incoming tides.

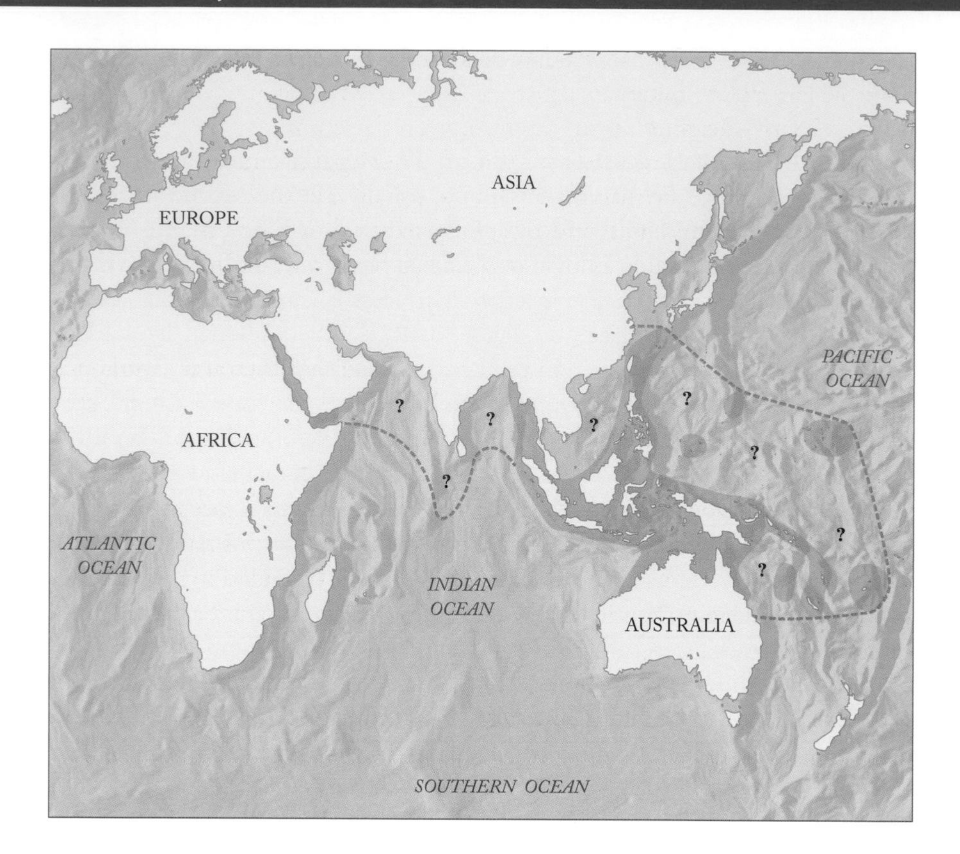

Right: Double-lined Mackerel range map.

FOOD: In the Andaman Islands, Double-lined Mackerel feed on crustaceans, herring-like fishes (*Sardinella* and *Thrissocles*), and other fishes, including barracudas (*Sphyraena*) and triggerfishes (*Balistes*). Three search-and-attack strategies on Scissortail Fusiliers (*Caesio caerulaurea*) were observed on the Great Barrier Reef: linear search by single or groups of fish; ambush from a stationary position on the seafloor; and ambush from resting schools of co-occurring predators.

REPRODUCTION: Size at first maturity in Fiji seems to be attained at 40–43 cm FL, and the spawning season may be from October to March. Females mature by about 42 cm, males by 40 cm, in the Andaman Islands. In Palau, fish with mature gonads occurred from December to July.

EARLY LIFE HISTORY: Eight larval Double-lined Mackerel 8.5–17.5 mm from Philippine waters were described, and the smallest and largest of these were illustrated. Nishikawa provides a detailed account of development, including illustration of four individuals, based on 62 postlarval and juvenile specimens (4.8–56.9 mm SL), mostly collected from Papua New Guinea waters. Double-lined Mackerel differ from the morphologically similar mackerels of the genera *Scomber* and *Rastrelliger* in having preopercular spines and characteristic pigment blotches on the body.

FISHERIES INTEREST: In the Andaman Islands, Fiji, and Tokelau, Double-lined Mackerel are taken in artisanal fisheries. One of its Palauan names is *biturchturch*, meaning "urine," referring to the ammonia smell given off by these fish if they are cooked without removing the kidney tissue along the backbone. The smell disappears if the fish is filleted and the backbone discarded. The flesh is mild and pleasantly flavored. Double-lined Mackerel are probably most valuable as prime trolled bait for Black Marlin off the Great Barrier Reef.

THREATS: There is no present evidence that artisanal fisheries and capture for bait have adversely impacted Double-lined Mackerel populations. It is listed as Least Concern on the IUCN Red List.

CONSERVATION: Double-lined Mackerel are abundant and widespread in the Indo-West Pacific. They are caught in minor artisanal fisheries and by trolling in at least some portions of their range, but there is no population information available. There are no species-specific conservation measures for this species.

REFERENCES: Auster 2008; Collette 1983, 2001; Collette and Gillis 1992; Collette and Nauen 1983; Collette et al. 2011; Hata et al. 2011; IGFA 2018; Johannes 1981; Khalaf 2005; Lewis 1981, 1988; Lewis et al. 1983; McPherson 1984; Nishikawa 1979; Pepperell 2010; Preston et al. 1987; Rüppell 1836; Silas 1963; Wade 1951; Zylich et al. 2011.

Dogtooth Tuna

Gymnosarda unicolor (Rüppell, 1836)

COMMON NAMES: English – Dogtooth Tuna
French – Thon à gros yeux
Spanish – Casarte ojón

ETYMOLOGY: The generic name *Gymnosarda* means "naked bonito," referring to the mostly naked body, and the species name *unicolor* was used by the German naturalist and explorer Wilhelm Peter Eduard Rüppell because of the distinctive lack of markings on the body.

SYNONYMS: *Thynnus (Pelamis) unicolor* Rüppell, 1836; *Pelamys nuda* Günther, 1860; *Scomber vau* Curtiss, 1938; *Gymnosarda nuda* (Günther, 1860)

TAXONOMIC NOTE: Recognized as a monotypic genus by almost all authors but was thought to include *Cybiosarda elegans* by Fraser-Brunner (1950).

FIELD MARKS:

1 As the common name implies, Dogtooth Tuna have very large conical teeth in both the upper and lower jaws and two patches of teeth on their tongue.
2 Their eyes are very large.
3 Unlike many species of mackerels and tunas, they lack spots, lines, and other markings on the body.

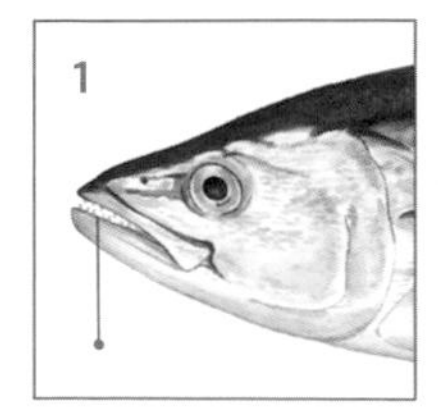

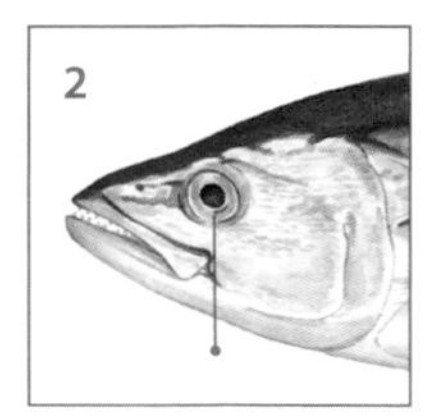

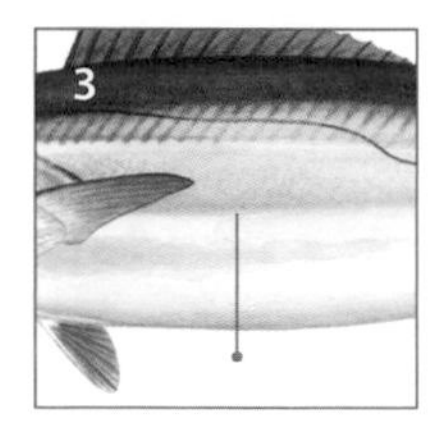

UNIQUE FEATURE: The Dogtooth Tuna has the largest teeth of the scombrids.

DIAGNOSTIC FEATURES: The body is elongate and moderately compressed. The mouth is fairly large, with the upper jaw extending back to the middle of the eye. Dogtooth Tuna have a differently shaped head than do other bonitos: the interorbital distance is much wider, the eyes are larger, and the postorbital distance is shorter. **Teeth:** There are 14–31 large conical teeth in the upper jaw and 10–24 in the lower jaw. These large and conspicuous teeth have led to the common name of Dogtooth Tuna. There are two patches of fine teeth on the tongue. **Gill rakers:** There are 11–14 gill rakers on the first gill arch. **Olfactory:** The olfactory apparatus is among the best developed of any of the Scombridae, with 48–56 laminae in the nasal rosette, more than in any other scombrid, which average 20–39. **Fins:** The dorsal fins are close together, the first with 13–15 spines and with a straight margin, the second followed by 6 or 7 finlets. The anal fin has 12 or 13 rays followed by 6 finlets. There are 25–28 rays in the pectoral fins. The interpelvic process is large and single. **Caudal peduncle:** The caudal peduncle is slender, with a well-developed lateral keel between two smaller keels. **Scales:** The body is naked posterior to the corselet except for the lateral line scales, the scaly dorsal-fin base, and the scaly caudal keel. **Swim bladder:** Unlike other bonitos, Dogtooth Tuna have a large swim bladder. **Lateral line:** The lateral line strongly undulates. **Vertebrae:** It is the only bonito with 19 precaudal and 19 caudal vertebrae. Descriptions and illustrations of the osteology and soft anatomy of the Dogtooth Tuna are included in the revision of the bonitos (Collette and Chao 1975). Drawings of the viscera were also presented by Blanc and Postel (1958) and Silas (1963). **Color:** The back and upper sides are a brilliant blue-black, the lower sides and belly are silvery, and there are no spots, lines, or other markings on the body. The anterior edge of the first dorsal fin is darker than the rest of the fin; the other fins are grayish. The second dorsal and anal fins are usually tipped with white.

GEOGRAPHIC RANGE: This Indo-Pacific species is found from the Red Sea and East Africa to French Polynesia, north to Japan, south to Australia. Its distribution is disjunct, as this species is found primarily around reefs. A map of specimens examined and valid literature records is presented in Collette and Chao. No evidence of genetic differentiation was found between populations of *Gymnosarda unicolor* distributed across a 7500 km range in the Indo-West Pacific from the Cocos Islands (Indian Ocean) to Marion Reef (Pacific Ocean). There are several recent records of Dogtooth Tuna from Taiwan caught as bycatch on a pelagic longline.

SIZE: Maximum length 8.1 feet (247 cm) FL. The IGFA all-tackle angling record is a virtual tie between two fish from Mauritius, a 230 lb, (104.32 kg) fish caught off LeMorne in 1993 and a 230 lb, 6 oz (104.5 kg) fish caught off Rodriguez Island in 2007. While jigging off Latham Island, Tanzania, in November 2013, an angler landed an enormous Dogtooth Tuna that weighed 226 lb, 10 oz (102.8 kg), only 4 lb lighter than the IGFA record.

One maximum age estimate was 12 years, but a 165 lb (75 kg) individual was aged by its otoliths and estimated to be 19 years old.

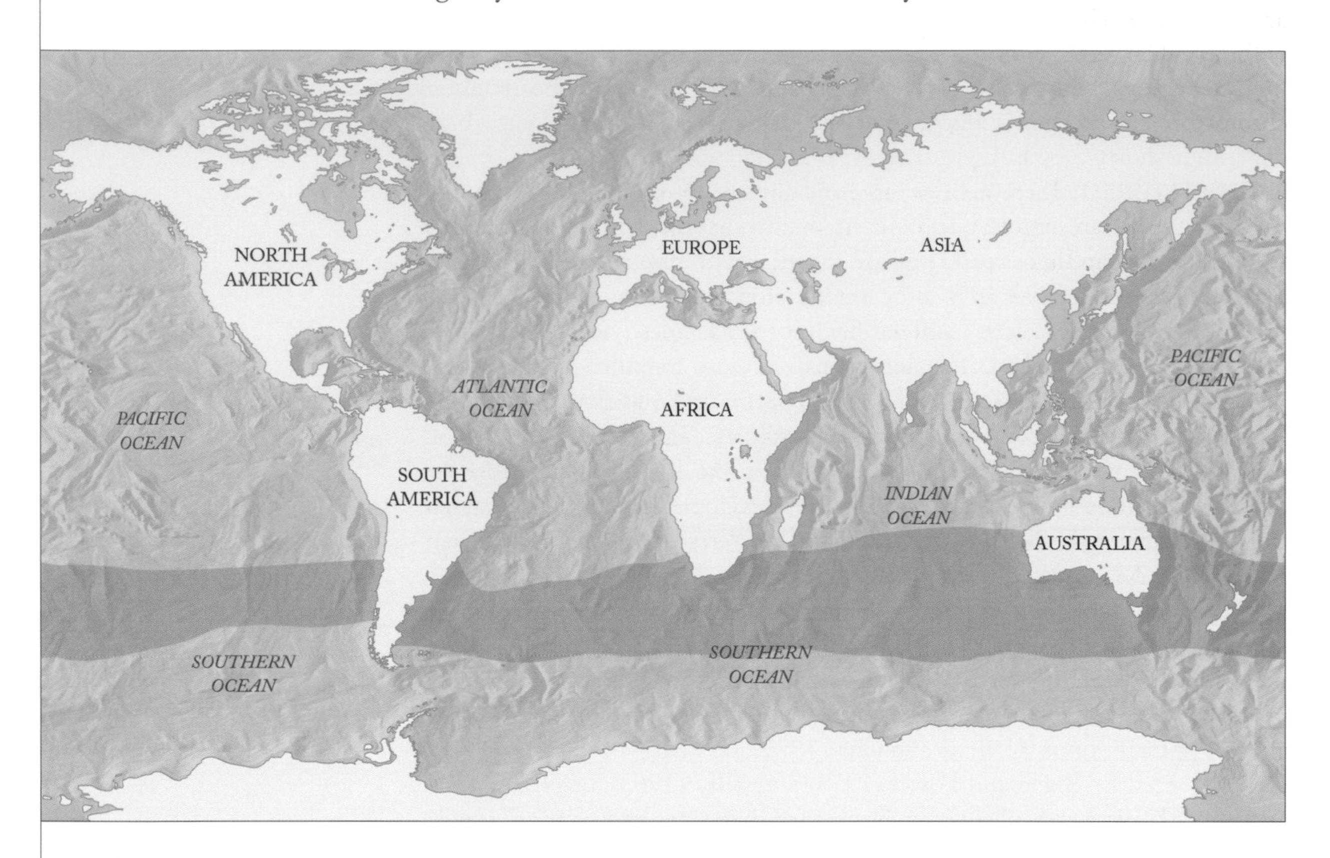

Above: Dogtooth Tuna range map.

HABITIAT AND ECOLOGY: Dogtooth Tuna are reef associated and oceanodromous, found offshore mainly around coral reefs, and may occur to depths of at least 300 m. They are generally solitary or occur in small schools of six or fewer.

FOOD: They prey on small schooling fishes such as scads (*Decapterus*), unicornfishes (*Naso*), wrasses (*Cirrhilabrus)*, and fusiliers (*Caesio* and *Pterocaesio*).

REPRODUCTION: Dogtooth Tuna spawn throughout the year, with peak spawning from August to March off India and March, August, and December along the coast of Samarai, Papua. Length at first maturity was estimated at 65 cm FL in Fiji and 69 cm off India.

EARLY LIFE HISTORY: A total of 24 larvae, 2.28–9.22 mm SL, and a single juvenile, 41 mm SL, were described from a wide area of the tropical and subtropical western Pacific between approximately 10° N and 20° S, with clear concentrations near the shallow seas along islands such as the Caroline, Solomon, and New Hebrides. Juveniles at Tsushima Island, Japan, grew from 24.0–27.5 cm FL and 200–300 g weight in early January to 36.0–39.5 cm FL and 640–910 g in late November to 41.0–45.0 cm FL and 1,115–11,170 g by the middle of the next January. Larvae have been recorded off the eastern coast of Taiwan. Five larvae, 2.48–9.06 SL, were illustrated by Okiyama and Ueyanagi, showing development of the elongate snout, wide mouth with fang-like teeth, and spiny armature of the preopercle, supraorbital, and pterotic regions. Unlike closely related species, there is a complete absence of chromatophores in the tail region throughout the larval phase.

FISHERIES INTEREST: There are few fisheries directed specifically at Dogtooth Tuna, but they are regularly caught in small numbers, mostly in artisanal fisheries with handlines, pole, and trolling during certain seasons in many parts of their range. They are also prized by sport fishermen for their fighting prowess. A summary of recreational fishing for "doggies" was presented by Steynberg. Dogtooth Tuna are taken regularly in small numbers by handlines from July to September at night in reef areas off the Indian Lakshadweep Islands, and they are sometimes taken by offshore mackerel trollers off the Cairns and Townsville districts of eastern Australia. Worldwide reported landings fluctuate but in general show a gradual increase in reported catch from negligible in the 1960s to 100–500 mt in the 1970s and 1980s, to between 1,200 and 2,300 mt in the 2000s. They are marketed canned and frozen. They have been reported to sometimes be ciguatoxic, but this is not common. More than half of the Dogtooth Tuna examined from the Seychelles exceeded the provisional maximum permissible level of 1.0 mg of total mercury established by the US Food and Drug Administration, with two fish having levels of 3.3 and 4.4 ppm.

THREATS: This is a minor commercial species that is caught mainly by pole and line and is also caught in sport fisheries. It is regularly caught in small numbers, mostly in artisanal fisheries, with handlines, pole, and trolling during certain seasons in many parts of its range. Initial high catches are usually not maintained, perhaps because it is a solitary species that does not school. It is listed as Least Concern on the IUCN Red List.

CONSERVATION: There are no current species-specific conservation measures in place, although Dogtooth Tuna occur in some protected areas within their range, such as the Great Barrier Reef. They are widespread in the Indo-West Pacific. No population information is available, but they are caught in artisanal and recreational fisheries throughout their range. There may be localized depletions given that it is a solitary species, but it is unlikely that there are currently widespread population declines.

REFERENCES: Bentley et al. 2014; Blanc and Postel 1958; Brouard and Grandperrin 1984; Chiu and Chen 1993; Collette 2001; Collette and Chao 1975; Collette and Nauen 1983; Fukusho and Fujita 1972; Grant 1982, 1987; IGFA 2013, 2016; Joshi et al. 2012; Lewis et al. 1983; Matthews 1983; Okiyama and Ueyanagi 1977. Pepperell 2013; Randall 1980; Ronquillo 1964; Silas 1963; Sivadas and Anasukoya 2005; Steynberg 2014.

Skipjack Tuna

Katsuwonus pelamis (Linnaeus, 1758)

COMMON NAMES: English – Skipjack Tuna
French – Bonite à Ventre Rayé
Spanish – Listado
Japanese – Katsuo

ETYMOLOGY: The Swedish naturalist Carl Linnaeus (1758:148), who devised the binomial nomenclature system we use, selected the Latin name *pelamis*, which means "a young tuna." The generic name *Katsuwonus* was constructed by the Japanese biologist Kamakichi Kishinouye in 1915 from the Japanese name for the Skipjack Tuna, Katsuo.

SYNONYMS: *Scomber pelamis* Linnaeus, 1758; *Scomber pelamides* Lacepède, 1801; *Thynnus vagans* Lesson, 1829; *Katsuwonus pelamis* (Linnaeus, 1758); *Euthynnus pelamis* (Linnaeus, 1758)

TAXONOMIC NOTE: Some authors have included the Skipjack Tuna in the genus *Euthynnus* but recent authors agree with its placement in the monotypic genus *Katsuwonus*.

FIELD MARKS:
1 Skipjack have more gill rakers (50–63) than other tropical species of tunas.
2 The four to six conspicuous longitudinal dark bands on the silvery belly distinguish Skipjack from all other members of the family Scombridae.

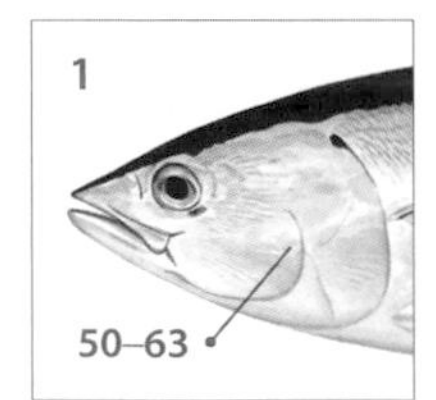

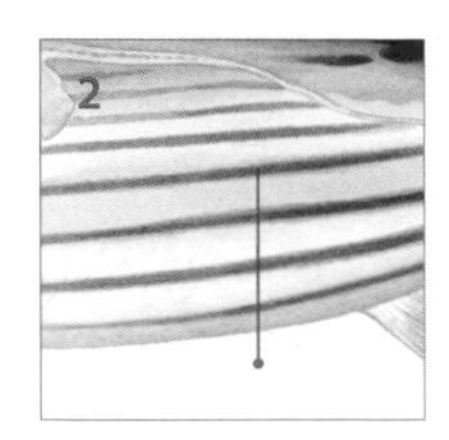

UNIQUE FEATURE: The 4 to 6 conspicuous relatively broad, longitudinal dark bands along the lower half of the body distinguish Skipjack from all other members of the family Scombridae.

DIAGNOSTIC FEATURES: The body is fusiform, elongate, and rounded. **Teeth:** The dentition is weak and confined to a single row of teeth on the upper and lower jaws. No teeth are present on the vomer or palatine bones in the roof of the mouth. The tongue is smooth, with vertical cartilaginous ridges on each side. **Gill rakers:** Gill rakers are numerous, 50 to 63 on the first gill arch, with a very small gap between the gill rakers. **Fins:** The two dorsal fins are separated by a small interspace, not longer than the diameter of the eye. The first dorsal fin has 14 to 17 spines; the second has 13 to 16 rays and is followed by 7 to 10 finlets. The anal fin has 13 to 17 rays followed by 6 to 8 finlets. The pectoral fins are short, with 24 to 32 rays. The interpelvic process is small and divided. **Scales:** The body is naked except for the anterior corselet and small scales along the lateral line. **Swim bladder:** A swim bladder is absent. **Vertebrae:** There are typically 20 precaudal vertebrae plus 21 caudal vertebrae, for a total of 41. The soft anatomy, circulatory system, and skeleton were well described and illustrated by Godsil and Byers (1944). **Color:** The back is dark purplish blue; the lower sides and belly are silvery, with 4 to 6 very conspicuous longitudinal dark bands on the silvery belly, which in live individuals may appear as discontinuous lines of dark blotches. Dark vertical bands may appear on the dorsolateral surfaces when feeding.

GEOGRAPHIC RANGE: Skipjack Tuna are circumglobal in seas warmer than 15°–18°C. They are found throughout the warm Atlantic, including the Gulf of Mexico and Caribbean Sea. They are common in the eastern Atlantic, including the Gulf of Guinea and the western Mediterranean, but are absent from the eastern Mediterranean Sea and Black Sea. They are widespread throughout the Indo-West Pacific. In the eastern Pacific, they occur from British Columbia to northern Chile, including waters of the oceanic islands.

SIZE: Maximum size recorded is 111 cm FL and 34.5 kg. The IGFA all-tackle game fish record is a 45 lb, 4 oz (20.54 kg) Skipjack caught over Flathead Bank in Baja California, Mexico, in November 1996.

HABITAT AND ECOLOGY: Skipjack Tuna are pelagic and oceanodromous in offshore waters to depths of 330 m, with one record down to 596 m. Skipjack Tuna exhibit a strong tendency to school in surface waters, often in association with floating objects, sharks, and whales. They often form mixed schools with small Yellowfin Tuna and Bigeye Tuna when associated with floating objects and in the western Atlantic are commonly found in schools with Blackfin Tuna. As with other higher tunas, Skipjack maintain elevated body temperatures, 4–10°C above the temperature of the water around them. Lacking a swim bladder, Skipjack rely on lift from their pectoral fins to help keep them up in the water, constantly swimming at about 69 cm/sec, or 1.6 body lengths/sec.

Their lower temperature limit appears to be about 15–18°C but the upper temperature limit apparently varies with size, from 30°C or more for small fish to 25°C for the largest fish. Skipjack are also limited to water with unusually high concentrations of dissolved oxygen, at least 3.0–3.5 ml/l (4–5 ppm) for long-term survival. They are preyed upon by large pelagic fishes, particularly billfishes. In one study 6,867 stomachs from six species of Pacific billfishes contained 1,742 Skipjack.

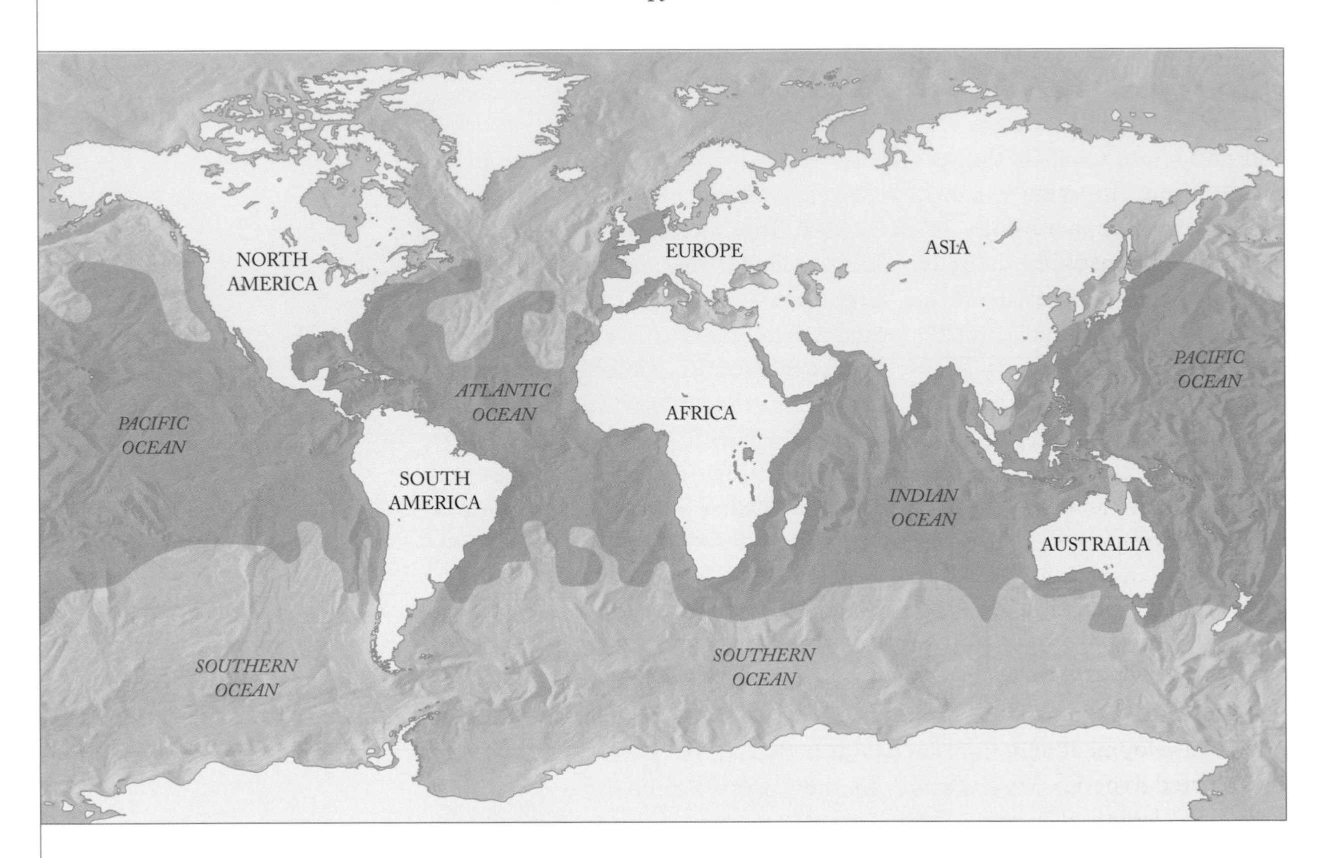

Above: Skipjack Tuna range map.

MOVEMENTS: Results from extensive conventional tagging programs of Skipjack Tuna in the Pacific, Atlantic, and Indian oceans demonstrate a strong relationship between time at large and mean displacements, with mean net movements increasing over a two-year time period to approximately 1,500–2,000 km. However, with median lifetime displacement ranges of less than 500 nautical miles, Skipjack is not as highly migratory as the bluefin tunas or Albacore. A limited number of movements across the Atlantic Ocean and from the eastern Pacific to the central or western Pacific Ocean have been reported. Skipjack Tuna movement patterns are strongly influenced by seasonal and interannual environmental variation, and general movement patterns fit well with advection-diffusion models. Acoustic tagging studies have demonstrated that Skipjack have very short residence times (typically less than a few days) with moored FADs.

FOOD: Skipjack Tuna feed on fishes, squids, and crustaceans. At least 48 different species of fishes have been recorded from stomachs of Skipjack Tuna in the Atlantic Ocean. Fishes constituted 75% of the volume of stomach contents around Cuba, squids 23%, and crustaceans only 2%. In the western South Atlantic Ocean, the lightfish *Maurolicus* and a euphausiid were the principal food items. Reef-inhabiting fishes such as lizardfishes, jacks, snappers, and sea basses were important components of the food of Skipjack from around the Marquesas and Tuamotu islands in the central Pacific. In the eastern Pacific Ocean, 88% of the volume of food consisted of two families of crustaceans (Euphausiidae and Galatheidae) and five families of fishes (lightfishes, flyingfishes, lanternfishes, cutlassfishes, and anchovies). Mackerels and juvenile tunas, including frigate tunas, Little Tunny, Blackfin Tuna, and Skipjack, are important components of the diet in some regions. Larger Skipjack eat comparatively more fishes and fewer crustaceans and squids. The very small gap between the numerous gill rakers enables smaller Skipjack to efficiently feed on smaller food items such as crustaceans.

Skipjack feeding peaks in the early morning and late afternoon. Skipjack also feed at night, particularly on deep-scattering prey organisms, including meso-pelagic fishes, squids, and crustaceans. In captivity, the average weight of food eaten by a Skipjack was 136 g, equivalent to 8.6% of its body weight. The cumulative amount eaten in a day is even higher—about 15% of its body weight. Shrimp exoskeletons passed through the digestive tract of captive Skipjack in an hour and a half. Gut evacuation rates for Skipjack are much shorter than for other piscivorous fishes, apparently in response to their higher metabolic rate.

REPRODUCTION: Skipjack Tuna are batch spawners that have asynchronous oocyte development and a protracted spawning season. Age of first maturity is estimated to be 1.5 years at lengths of 40–55 cm FL, depending on the area. Estimated length at 50% maturity for females is 42–43 cm in the Atlantic and the Indian oceans. Sex ratio is about 1:1 but fisheries that rely on young immature fish are dominated by females, while those that capture older fish are mostly male. After reaching sexual maturity this species spawns repeatedly, provided sufficient energy reserves and sea surface temperatures of 24–29 °C. It spawns throughout the year in the Caribbean and other equatorial waters and from spring to early fall in subtropical waters, with the spawning season becoming shorter as distance from the equator increases. In tropical waters, reproductively active female Skipjack Tuna are capable of spawning near daily. Batch fecundity increases with body length but is highly variable. Individual batch fecundity ranges from 80,000 eggs for a 44 cm female from Madagascar to 1.25 million eggs for a large (75 cm) female from the Seychelles Islands.

AGE AND GROWTH: Skipjack Tuna are very fast growing, relatively short lived, and extremely fecund. Longevity, as calculated from growth rings on the first dorsal spine or sagittal otolith, is estimated to be at least 7 years. Average fish lengths in the Gulf of Guinea were 29.2 cm at age 1, 42.9 cm at age 2, 55.5 cm at age 3, and 69.1 cm at age 4. Average fish lengths from North Carolina at ages 1–4 were 40.6, 49.3, 56.9, and 63.8 cm, respectively, differing from the Gulf of Guinea mostly in the estimated length at age 1. Back-calculated lengths at age from otoliths of fish from the western and central Pacific Ocean were 36.2, 43.3, 48.3, 52.6, 56.5, 60.8, and 63.2 cm FL for ages 1–7.

EARLY LIFE HISTORY: There are illustrations of 7 Skipjack larvae, 3.7–14.5 mm TL, from the central Pacific, and a series of 9 larval and juvenile Skipjack Tuna, 3.7 mm notochord length to 47 mm SL, are included in Richards (2006). They hatch at 3.0 mm notochord length. Larval Skipjack Tuna differ from the larvae of other scombrids in pigmentation and myomere number. They have a distinct chromatophore on the ventral margin of the caudal region and there are no chromatophores where the pectoral fins meet. They occur from the lower portion of the mixed layer to the upper portion of the thermocline in a temperature range of 20–25°C and a salinity range of 33.6–35.5 ppt. Model simulations of the effect of climate change predict the distribution of larval Pacific Ocean Skipjack will shift toward the eastern Pacific and to higher latitudes. Juveniles actively feed during daytime from morning to sunset but do not feed at night. Fish larvae are the principal food of juvenile Skipjack. Three stanzas of growth were found for Skipjack Tuna in the central Pacific Ocean using growth increments on their otoliths: 1.6 mm/day for fish up to about 27.0 cm, 0.8 mm/day for fish between 27.0 and 71.4 cm, and 0.3 cm/day for fish between 71.4 and 80.3 cm.

STOCK STRUCTURE: No significant genetic differentiation has been found between Skipjack Tuna from the Atlantic and Pacific oceans, suggesting that there is sufficient gene flow within and between oceans to prevent the accumulation of genetic stock structure. However, a high level of genetic diversity was found between populations in India and Japan, and phylogenetic analyses provide support for the presence of two divergent mtDNA clades in the northwestern region of the Indian Ocean.

FISHERIES INTEREST: In 2014 global landings of Skipjack Tuna were greater than 3 million mt, the third highest fisheries catch, exceeded only by Alaska Pollack and Peruvian Anchovy. Skipjack comprise about 60% of the commercial tuna catch worldwide and are mostly used for canning. They are captured near the surface mostly with purse seine gear and increasingly in association with FADs (fish aggregating devices). They are also caught on pole-and-line gear, in gill nets, drift nets, and longlines, and by recreational fishers.

Global catches of Skipjack Tuna increased from 165,000 mt in the 1950s to over 1 million mt in the late 1980s to over 2 million mt in the 2000s and are currently in excess of 3 million mt. Skipjack stocks are assessed and managed by four recognized tuna RFMOs.

Atlantic Ocean: Skipjack are managed as two stocks in the Atlantic Ocean, with the eastern stock contributing about 85% of the landings. In total, Skipjack landings in the Atlantic gradually increased from over 100,000 mt in the 1970s to over 250,000 mt in the 2010s, with considerable year-to-year variation. Catch has increased in the Mediterranean in recent years, but statistics from this region are incomplete. Purse seine catches account for more than 80% of Atlantic Skipjack landings overall, with bait boat catches representing a little less than 20%.

Eastern Pacific Ocean: Purse seine catches account for the vast majority of Skipjack Tuna landings in the eastern Pacific. Catches were less than 50,000 mt until the late 1970s, increased to 50,000–100,000 mt in the 1990s, and currently fluctuate between 200,000 to over 300,000 mt. Reported landings in 2016 were 342,000 mt. Skipjack recruitment in the eastern Pacific region is highly variable and is thought to be the reason for the large variations in annual catches and apparent stock size.

Western and Central Pacific Ocean: The western and central Pacific Ocean is responsible for the greatest production of Skipjack Tuna from throughout the world's oceans. Landings, primarily by purse seine vessels, increased from under 350,000 mt in the 1960s to over 1 million mt in the 1990s, reaching 2 million mt in the 2010s.

Indian Ocean: In the Indian Ocean, Skipjack are captured in a variety of fisheries, with the purse seine, gill net, and pole-and-line fisheries each contributing between 20 and 35% of the landings, which were 394,000 mt in 2015. Skipjack landings in the Indian Ocean reached 100,000 mt in the mid-1980s, exceeded 400,000 mt in the 2000s, and peaked at 600,000 mt in the mid-2000s.

THREATS: Continued increases in purse seine fleet capacities and targeting Skipjack within mixed-species tuna aggregations associated with FADs has caused increased catches of small Yellowfin and Bigeye tunas, resulting in growth overfishing of both species in some management areas and raising concerns about the sustainability of Skipjack catches. It is listed as Least Concern on the IUCN Red list worldwide and in each region. However, in some regions there may be signs of overfishing and uncertainty in estimating population size and trends. More research and monitoring is needed for this species to develop better population models and to ensure that current fishing mortality does not exceed estimated MSY.

CONSERVATION: The Skipjack Tuna is listed as a highly migratory species in Annex I of the 1982 Convention on the Law of the Sea. Although it is heavily fished, it is considered relatively abundant and is fast growing, short lived, and very fecund.

Indian Ocean: The IOTC assessed Skipjack Tuna in 2014. The stock was not overfished and overfishing was not occurring. Catches over the past several years have been well below the estimated MSY.

Western and Central Pacific Ocean: Skipjack Tuna were assessed in the western and central Pacific Ocean in 2016. The stock was not overfished and overfishing was not occurring. Although fishing mortality has been gradually increasing, it is currently estimated to be less than two-thirds of that necessary to support MSY. In the Western and Central Pacific there is an area-wide temporal closure on the use of FADs but the numbers of FADs has been increasing.

Eastern Pacific Ocean: The 2016 assessment, using stock status indicators from a simple model for Skipjack Tuna in the eastern Pacific Ocean, concluded that the stock is probably not overfished or experiencing overfishing. However, there is a significant trend in decreasing average weight and an increasing trend in apparent recruitment of Skipjack Tuna observed in recent years, but it is not known if those are the result of continued high levels of recruitment or recruitment overfishing. There are no specific conservation and management measures for Skipjack Tuna, but the purse seine fishery has been managed through area-wide temporal closures, as well as a spatio-temporal closure, implemented to reduce fishing effort on Bigeye and Yellowfin tunas.

Atlantic Ocean: The 2014 assessment of Skipjack Tuna indicated that neither the eastern stock nor western stock is likely overfished or experiencing overfishing. The ICCAT Standing Committee on Research and Statistics has recommended that catch and effort not exceed the 2012–13 levels in the east Atlantic and that catches in the western Atlantic not exceed the estimated MSY of 30,000–32,000 mt. While not directed at Skipjack Tuna, the purse seine and bait boat fisheries in the eastern Atlantic have been subject to a variety of time/area closures on FAD fishing in the Gulf of Guinea to reduce fishing mortality on Yellowfin and Bigeye tunas. The current moratorium, which entered into force in 2016, includes the area from 4° S and 5° N latitude and from the African coast to 20° W longitude during the months of January and February.

SELECTED REFERENCES: Ankenbrandt 1985; Batts 1972a, 1972b, 1972c; Dammannagoda et al. 2011; Dragovich 1969; Ely et al. 2005; Gilman et al. 2016; Godsil and Byers 1944; Graves et al. 1984; Magnuson 1969; Matsumoto et al. 1984; Nakamura 1965; Said Koya et al. 2012; Schaefer 2001b; Schaefer and Fuller 2007; Tanabe 2002; Yoshida 1971.

Plain Bonito

Orcynopsis unicolor (Geoffroy Saint-Hilaire, 1817)

COMMON NAMES: English – Plain Bonito
French – Palomette
Spanish – Tassarte

ETYMOLOGY: The French naturalist Étienne Geoffroy Saint-Hilaire (1817) named the Plain Bonito *unicolor* because of its lack of any prominent stripes or bars.

SYNONYMS: *Scomber unicolor* Geoffroy Saint-Hilaire, 1817; *Cybium bonapartii* Verany, 1847; *Cybium altipinne* Guichenot, 1861; *Thynnus peregrinus* Collett, 1879; *Pelamys unicolor* (Geoffroy Saint-Hilaire, 1817); *Pelamichthys unicolor* (Geoffroy Saint-Hilaire, 1817); *Sarda unicolor* (Geoffroy Saint-Hilaire, 1817); *Cybium veranyi* (non Döderlein, 1872)

TAXONOMIC NOTE: The Plain Bonito has been placed in seven different genera—*Scomber*, *Cybium*, *Pelamys*, *Orcynopsis*, *Thynnus*, *Pelamichthys*, and *Gymnosarda*—over its more than 250-year history but is morphologically distinct enough to have its own monotypic genus, *Orcynopsis* (Postel 1950, Collette and Russo 1975).

FIELD MARKS:

1 They differ from all other Atlantic scombrids by having two distinct patches of teeth on their tongues instead of only one. This character is shared with two Indo-Pacific species, the Leaping Bonito and the Dogtooth Tuna.

2 As noted under etymology, the Plain Bonito differs from most other scombrids in lacking stripes or bars. They have short and deep bodies.

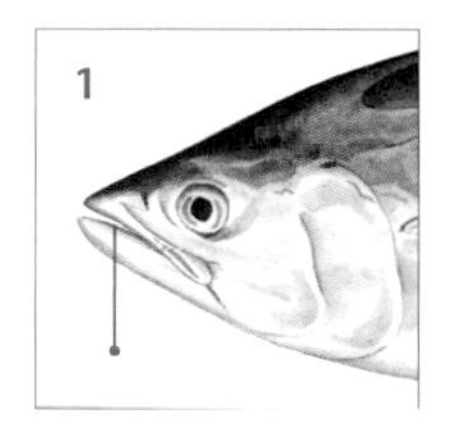

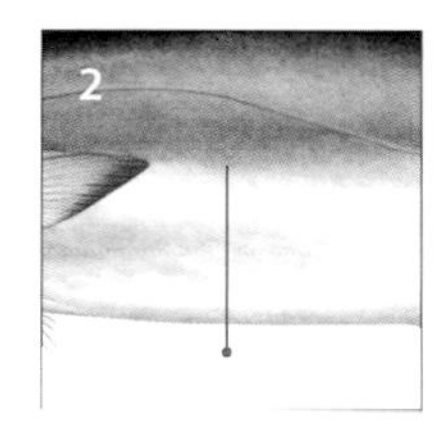

UNIQUE FEATURE: The rarest east Atlantic scombrid.

DIAGNOSTIC FEATURES: The body is relatively short and deep and strongly compressed. **Mouth and teeth:** The mouth is relatively large, with the upper jaw reaching to the posterior margin of the eye. There are two tooth patches on the upper surface of the tongue. There are 13–29 large conical teeth on each side of the upper jaw and 10–23 on the lower jaw. **Gill rakers:** There are 12–17 (usually 14–16) gill rakers on the first gill arch. **Olfactory:** The olfactory rosette has 25–28 lamellae. **interorbital:** The interorbital width is 23.0–31% of head length. **Fins:** The dorsal fins are close together, with the first short and high with 12–14 spines and an almost straight outline. The second dorsal has 12–15 rays and is followed by 7–9 finlets. The anal fin has 14–16 rays followed by 6–8 finlets. The pectoral fins are short, with 21–23 rays. The interpelvic process is small and bifid. **Caudal peduncle:** The caudal peduncle is slender, with a well-developed lateral keel between two smaller keels on each side. **Scales:** The body is naked behind the well-developed corselet except for a band of scales along the bases of the dorsal fins and patches of scales around the bases of the pectoral, pelvic, and anal fins. **Swim bladder:** No swim bladder is present. **Liver:** The liver has an elongate right lobe and a short left lobe that tends to fuse with the middle lobe. Ventral views of the viscera were presented by Postel (1950, 1956) and Collette and Russo (1975). **Vertebrae:** There are 17 or 18 precaudal vertebrae plus 19–21 caudal vertebrae, for a total of 37–39. The osteology and soft anatomy were described and illustrated by Collette and Chao (1975). **Color:** The back is blue-black with or without a faint mottled pattern laterally but without any prominent stripes or spots. The lower sides are silvery. The anterior three-quarters of the first dorsal fin are black. The second dorsal fin and dorsal finlets are dark. There is some yellow pigment on the anal fin.

Above: Faint mottled pattern.

GEOGRAPHIC RANGE: The Plain Bonito is an eastern Atlantic endemic species. Their range is centered in the southern Mediterranean Sea, with scattered records from France, Italy, Israel, and Syria, and extends south along the coast of West Africa to Dakar, Senegal, where it is found year-round. There are only three records from northern European waters, all from southern Scandinavia. Robert Collett described one individual collected near Oslo, Norway, in 1879 as a new species, *Thynnus peregrinus*, but this name is a junior synonym of *O. unicolor*. In response to seasonal temperature, the southern distribution boundary shifts northward from Dakar (14° N) to southern Morocco (22° N) during the summer, and in the winter the center of its distribution shifts southward from the southern coast of Portugal to the coast of Morocco.

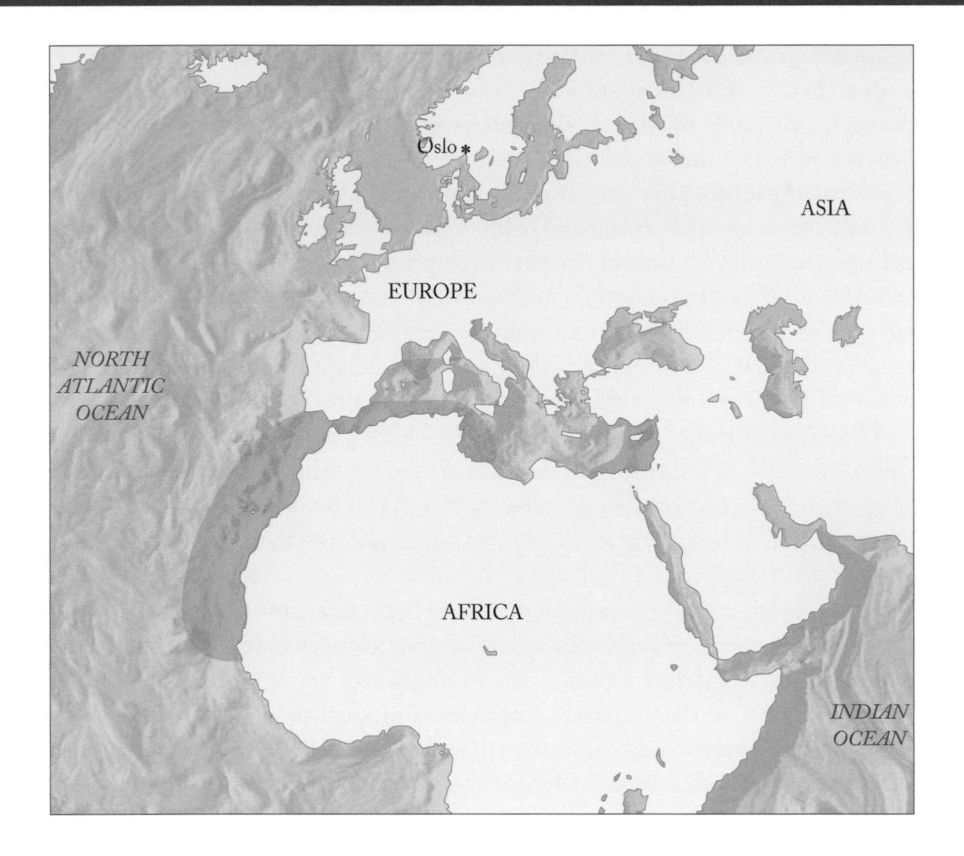

Left: Plain Bonito range map.

SIZE: Maximum length is 130 cm FL, commonly to 90 cm. The maximum weight is 29 lb (13.1 kg), commonly to 11 lb (5 kg). Females grow larger than males. Maturity is reached at about 70–80 cm in FL.

HABITAT AND ECOLOGY: This is a pelagic, oceanodromous, and neritic species, confined primarily to temperate waters, but juveniles may be encountered in waters of up to 30°C. They form small schools at the surface and are also frequently associated with birds. The best sources of biological information on the Plain Bonito are Postel and more recently, a doctoral thesis on small tunas in Tunisia by Hattour and its subsequent publication as a book.

FOOD: Plain Bonito feed on small fishes, especially sardines (42% frequency), jacks (23%), anchovies (21%), mackerel (12%), porgies (9%), and cephalopods off Tunisia; and sardines, anchovies, Chub Mackerel, and other fishes off Cape Verde.

REPRODUCTION: Length at first maturity is 43.5cm FL for females and 45 cm FL for males. In Tunisian waters, both sexes mature at 2 years and spawning occurs from May to September. A female weighing 11 to 13 lb (5 or 6 kg) may carry 500,000 to 600,000 eggs, which are spawned in batches.

FISHERIES INTEREST: Plain Bonito are of minor commercial importance. It seems there is no fishery directly targeting this species. Some catches are taken incidentally in Morocco, Algeria, and Tunisia. They are taken with beach seines off Ghana and Togo and as bycatch in tuna traps in Tunisia and southern Portugal. They are marketed canned or frozen. Smoked, they are reported to be similar to smoked salmon. Reported landings range from 100 mt reported in 1950 to over 3,100 mt reported in the 1960s. Since that time reported catches have generally fluctuated between 200 and 1,000 mt without trend. The estimated Mediterranean catch from 1965 (first declared catch) to 2005 fluctuated between 1 and 252 mt. The Mediterranean countries that reported landings of this species were Algeria (with highest landing values) and Tunisia. In the Mediterranean and Black Sea, 28% of the total reported catch in 1980–2007 comprised small tuna species, however, several countries from this region do not report catches to ICCAT. It is commonly believed that catches of small tunas are unreported or underreported data in all areas.

THREATS: Small tunas are exploited mainly by coastal fisheries and often by artisanal fisheries, although substantial catches are also made, either as target species or as bycatch, by purse seiners, midwater trawlers, handlines, troll lines, drift nets, surface-drifting longlines, and small scale gill nets. Several recreational fisheries also target small tunas. Since 1991, the increased use of FADs by tropical purse seiners may have led to an increase in fishing mortality of small tropical tuna species. There is a general lack of information on the mortality of these species as bycatch, exacerbated by the confusion regarding species identification. There is no fishery directed at this species and no known specific threats. This species was listed as Least Concern on the IUCN Red List, but recent information on the apparent reduction in catch in the Mediterranean led to ranking the species as Vulnerable on the European Red List.

CONSERVATION: There are no species-specific conservation measures for Plain Bonito in place at this time. This species is sporadically caught and reported, with fluctuating catch data. It is generally taken only incidentally.

REFERENCES: Aggrey-Fynn and Sackey-Mensah 2012; Belemlih et al. 1990; Belhabib et al. 2015; Ben-Tuvia 1971; Collette and Chao 1975; Collette and Nauen 1983; Collette et al. 2011; Di Natale et al. 2009; dos Santos et al. 2002; Hattour 2000, 2005, 2009; Hattour et al. 2005; ICCAT 2009; Lam et al. 2008; Louisy 2015; Nieto et al. 2015; A. Postel 1955; E. Postel 1950, 1954, 1955; 1956; Quigley 2012; Saad 2005; Seret and Opic 1981; STEFC 2009.

Short Mackerel

Rastrelliger brachysoma (Bleeker, 1851)

COMMON NAMES: English – Short Mackerel
French – Maquereau Trapu
Spanish – Caballa Rechoncha

ETYMOLOGY: Named *brachysoma*, a combination of the Greek words meaning "short" and "body," by the prolific Dutch ichthyologist and army surgeon Pieter Bleeker (1851:356).

SYNONYMS: *Scomber brachysoma* Bleeker, 1851; *Rastrelliger brachysoma* (Bleeker, 1851)

TAXONOMIC NOTE: The Short Mackerel is distinct enough that it has only been described as new one time, unlike most scombrids.

FIELD MARKS:
1 The Short Mackerel has the deepest body of the three species of *Rastrelliger*, its depth at the margin of the gill cover 23–27% fork length.
2 Black blotch behind pectoral-fin base absent.
3 Three to six spots present along the dorsal-fin base that may be retained in older specimens.

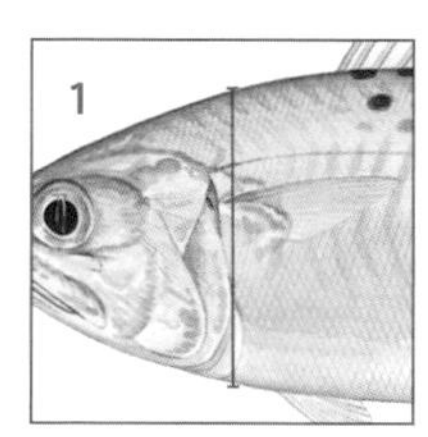

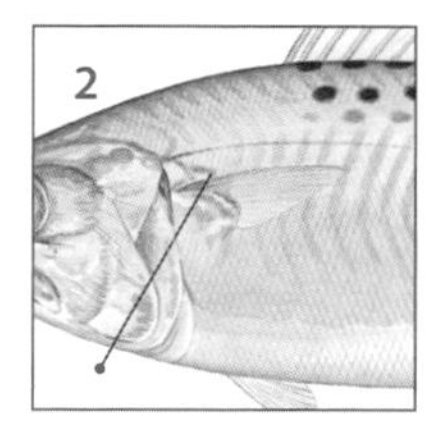

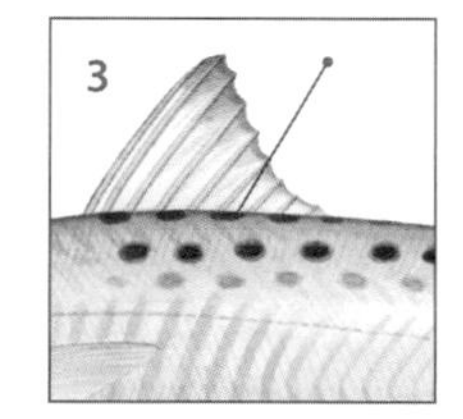

UNIQUE FEATURE: The Short Mackerel has the deepest body of the three species of *Rastrelliger*.

DIAGNOSTIC FEATURES: Body compressed and very deep, its depth at the margin of the gill cover 23–27% fork length. The head length is equal to or less than the body depth. An adipose eyelid is present. **Mouth and teeth:** The vomer and palatines are toothless. The premaxillary and mandible each have a row of small, nearly conical, slightly curved teeth with smooth edges. **Gill rakers:** The gill rakers are very long, 16–24 on the upper limb, 1 in the middle, and 30–48 on the lower limb. **Fins:** The first dorsal fin has 8 to 10 spines, usually 9. The second dorsal has 11–13 (usually 12) rays, followed by 4–6 finlets, usually 5. A rudimentary, skin-covered spine is present anterior to the origin of the anal fin. The anal fin has 10–13 rays, followed by 5 finlets. There are 18–21 pectoral-fin rays. The second dorsal fin and the anal fin are covered with a thick layer of scales. **Scales:** The entire body is covered with thin deciduous scales. **Swim bladder:** A swim bladder is present. **Lateral line:** The cephalic lateral line system has many more and much finer branches on the top of the head than are present in the Indian Mackerel. **Vertebrae:** Vertebrae number 13 precaudal plus 18 caudal, for a total of 31. There are 11 interneural bones under the first dorsal fin, 12 under the second dorsal fin, plus 6 under the dorsal finlets. **Intestine:** The length of the intestine is 3.2–3.6 times fork length. The digestive tract was described by Senerat et al. (2015). **Genome sequence:** The complete mitochondrial genome sequence was published by Jondeung and Karinthanyakit (2010). Morphological and molecular characters that distinguish Short Mackerel from the other two species of *Rastrelliger* were summarized by Muto et al. (2016). **Color:** In life, bluish green above and immaculate silvery below. Three to six dark spots present along the base of the dorsal fin may be retained in older specimens. The dorsal fins are dusky. The pectoral, pelvic, and anal fins are yellowish hyaline. The dorsal and caudal fins have dusky margins. The caudal fin is yellowish.

GEOGRAPHIC RANGE: Short Mackerel are found in the central Indo-West Pacific from the Andaman Sea east to Thailand, Indonesia, Papua New Guinea, the Philippines, Solomon Islands, and Fiji.

SIZE: Short Mackerel are reported to reach a maximum length of 34.5 cm FL and are commonly found at 15–20 cm FL. Maximum weight is 7.5 oz (213 g).

HABITAT AND ECOLOGY: Short Mackerel are pelagic and are also found in estuarine habitats with slightly reduced salinities and in areas where surface temperature ranges between 20°C and 30°C. It forms schools of equally sized individuals.

FOOD: Short Mackerel feed primarily on phytoplankton, especially diatoms such as *Chaetoceros* spp. and *Coscinodiscus* spp., but they also consume a smaller proportion of zooplankton.

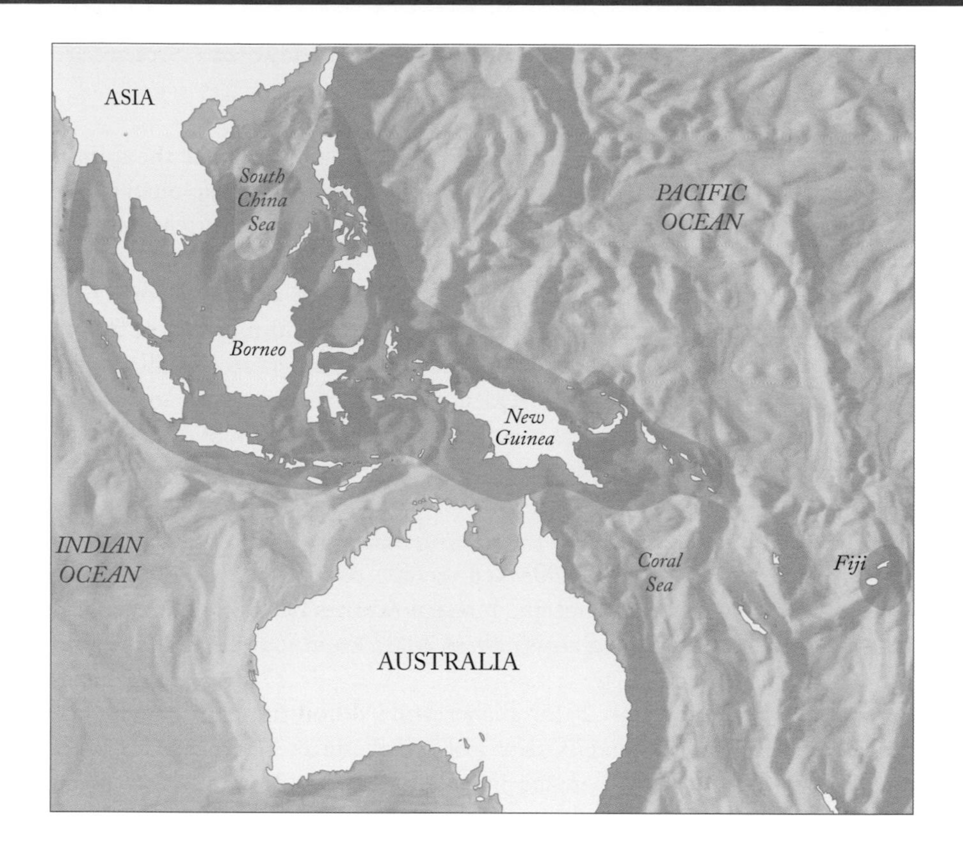

Left: Short Mackerel range map.

REPRODUCTION: The spawning season for Short Mackerel on the west coast of Thailand (Andaman Sea) seems to occur from February to April in the northern part of the area and in August in the southern part of the area. Length at sexual maturity occurs at about 17.3 cm TL and at an age of 7.5 months.

EARLY LIFE HISTORY: Most larval Short and Indian Mackerels were found in seas shallower than 90 m, indicating that the nursery area of their larvae is coastal. Four juvenile Short Mackerel from the Philippines, 54, 72, 98, and 110 mm FL, were illustrated by Manacop. At these sizes, all display about two rows of round or quadrangular spots on their backs between the lateral line and dorsal fins.

FISHERIES INTEREST: Short Mackerel are targeted by commercial and artisanal fisheries throughout their range, but landings are often reported in combination with mixed *Rastrelliger* spp. They are first recruited at the young age of 3 months. Estimated landings of Short Mackerel have steadily increased from less than 50,000 mt in the 1960s to over 300,000 mt in 2015. In 1958, Short Mackerel was the most important commercial species in the Philippine chub mackerel fishery, which also included Indian Mackerel and Pacific Chub Mackerel. The mackerel fishery in the Andaman and Nicobar islands is supported by Short and Indian mackerels, with total landings of

806–1,955 mt. The best fishing seasons are March–June and September–October using gill nets and boat seines.

THREATS: There is no consolidated information on abundance or the impact of fisheries on Short Mackerel. They were considered as overexploited in the Java Sea. Reported worldwide landings for *Rastrelliger* species have steadily increased since 1950 to over 800,000 mt in 2015, but no effort information is available. Given that effort is assumed to be increasing, it is not known how populations of Short Mackerel are affected by current and historical fishing pressure. Given the absence of an international management body, further monitoring of Short Mackerel is needed on the national level, in addition to species-specific data on landings, effort, and population status. Short Mackerel are listed as Data Deficient on the IUCN Red List.

CONSERVATION: Reduction in abundance of Short Mackerel in the Gulf of Thailand over 56 years (1953–2008) led the Thai Department of Fisheries to issue thirteen Notifications (specific management measures) relating to gear type and to closures of fishing areas in the Gulf to avoid spawning populations.

REFERENCES: Basheer et al. 2015; Blaber and Milton 1990; Bleeker 1851; Collette 2001; Collette and Nauen 1983; Collette et al. 2011; Hariati et al. 2017; Indaryanto et al. 2015, Jamaluddin et al. 2010; Jondeung and Karinthanyakit 2010, 2015; Jones and Rosa 1965, 1967; Jones and Silas 1964; Lavapie-Gonzalez et al. 1997; Madhu et al. 2002; Manacop 1958; Matsui 1967, 1970; Muto et al. 2016, 2017; Saikliang 2014; Sudjastani 1974; Sutthakorn and Saranakomkul 1987.

Island Mackerel

Rastrelliger faughni Matsui, 1967

COMMON NAMES: English – Island Mackerel
French – Maquereau des îles
Spanish – Caballa Isleña

ETYMOLOGY: The Island Mackerel was named by US ichthyologist Tetsuo Matsui (1967) in honor of James L. Faughn, who led the "Naga" Expedition.

SYNONYMS: *Pneumatophorus australasicus* (non Cuvier, 1832); *Scomber australasicus* (non Cuvier, 1832); *Rastrelliger faughni* Matsui, 1967

TAXONOMIC NOTE: The first good study of the Island Mackerel was as the misidentified *Pneumatophorus australasicus* from the Philippines (Manacop 1958). Three out of four molecular studies indicate that the Island Mackerel is the basal species of the three species of *Rastrelliger*.

FIELD MARKS:

1 Island Mackerel are easily distinguished from the other two species of the genus *Rastrelliger* by having the shortest and fewest gill rakers on the lower arch, only 20–25 in fish greater than 50 mm SL compared to 30–46 in the other two species.
2 Black spot present behind pectoral-fin base.
3 Two to six large spots present under first dorsal-fin base.
4 Two rows of black spots present along the back.

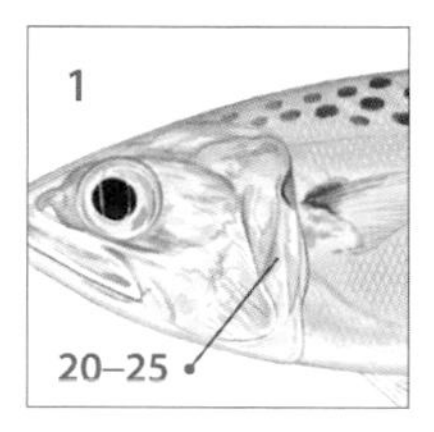

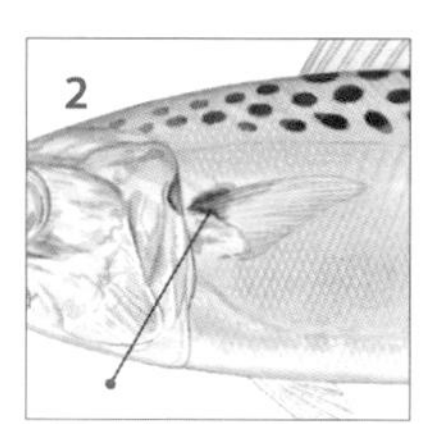

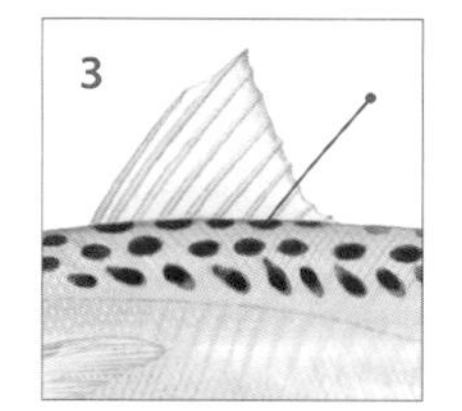

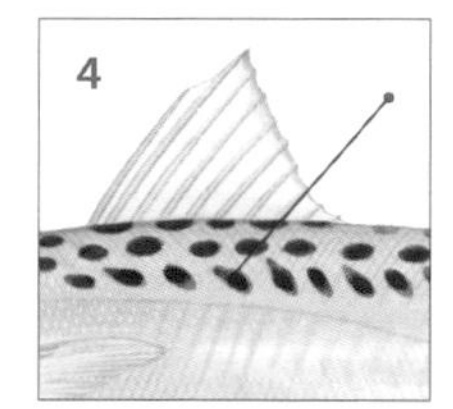

UNIQUE FEATURE: The Island Mackerel combines the compressed body shape of *Rastrelliger* with the few, short gill rakers of the species of *Scomber*.

DIAGNOSTIC FEATURES: The body is fusiform and somewhat laterally compressed, with the body depth less than head length. The snout is longer than the eye diameter. An adipose eyelid is present. **Mouth and teeth:** The vomer and palatines are toothless. The premaxilla and dentary each have a row of small, nearly conical, slightly curved teeth with smooth edges. **Gill rakers:** There are relatively few gill rakers on the first arch, 11–14 on the upper limb, 1 in the middle, and 22–25 on the lower limb, for a total of 34–40. The gill rakers are short (3.8–5.6% SL), similar to the species of mackerels in the genus *Scomber*, much shorter than in the other two species of *Rastrelliger*. **Fins:** The first dorsal fin has 9 or 10 spines, usually 10. The second dorsal has 11–13 rays, usually 12, followed by 4–6 finlets, usually 5. A rudimentary, skin-covered spine is present anterior to the origin of the anal fin. The anal fin has 12 or 13 rays, followed by 5 finlets. There are 18–21 pectoral fin rays. **Scales:** The entire body is covered with thin deciduous scales. **Swim bladder:** A swim bladder is present. **Vertebrae:** Vertebrae number 13 precaudal plus 18 caudal, for a total of 31. There are 11 interneural bones under the first dorsal fin, 12 under the second dorsal fin, plus 6 under the dorsal finlets. **Intestine:** The length of the intestine is equal to or less than the fork length. **Molecular characters:** Morphological and molecular characters that distinguish Island Mackerel from the other two species of *Rastrelliger* were summarized by Muto et al. (2016). **Color:** The dorsal surface is dark, the ventral surface silvery to yellowish silver. There are two rows of black spots on the upper surface below the base of the dorsal fin from the origin of the first dorsal fin to the caudal peduncle. Two to six large spots are present at the base of the first dorsal fin.

GEOGRAPHIC RANGE: The Island Mackerel was originally described from specimens from Taiwan, the Philippines, and Indonesia, but the range was later extended to the coastal waters of Madras, India, and to the island of Groote Eylandt in the Gulf of Carpentaria, northern Australia.

SIZE: The largest individual mentioned in the original description was 204 mm SL. Island Mackerel up to 260 mm were taken in Madras, India, in February and March.

HABITAT AND ECOLOGY: Island Mackerel are epipelagic and neritic, occurring in waters where surface temperatures do not fall below 17°C. They form schools of equally sized individuals. Little life history information exists for this species.

FOOD: Island Mackerel feed on zooplankton, thus complementing the planktonic food spectrum of the other two *Rastrelliger* species, which feed on phytoplankton and zooplankton. Analyses of stomach contents during the spawning season off Madras, India, showed very low feeding intensity; only partly digested anchovies and larval invertebrates were found in a few stomachs.

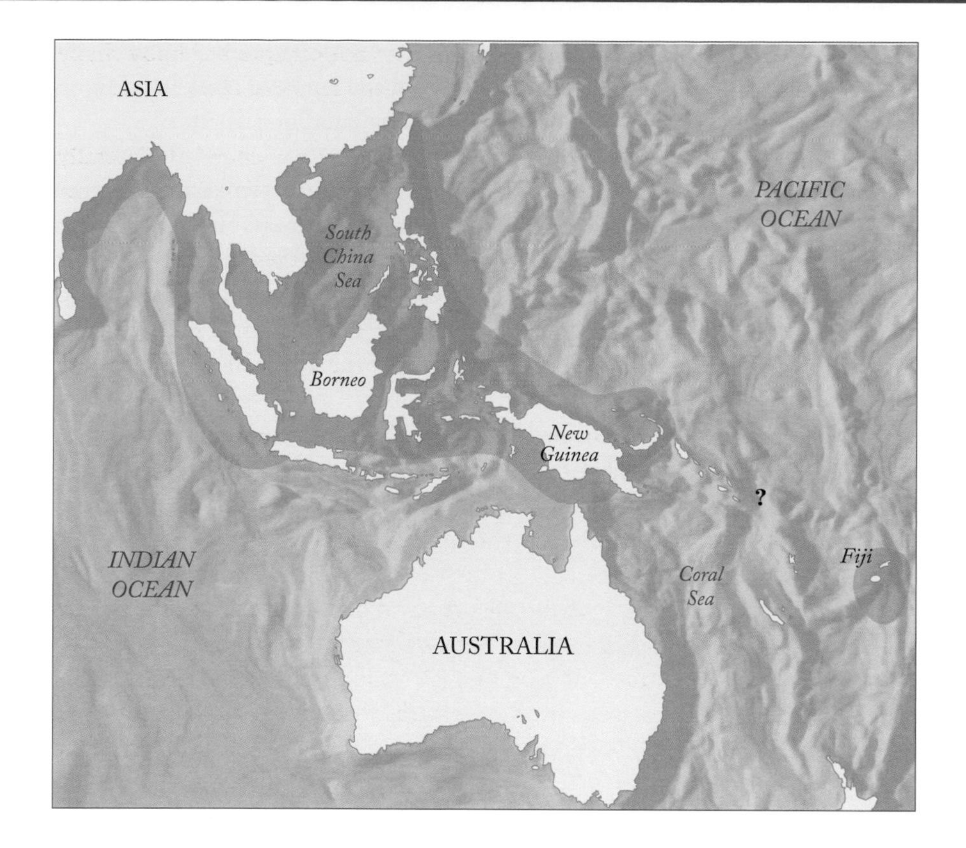

Left: Island Mackerel range map.

REPRODUCTION: Gonads of adult Island Mackerel were well developed during February and March off Madras, suggesting peak spawning at that time. Juveniles as small as 35–60 mm were taken in March and April.

AGE AND GROWTH: Longevity is estimated to be up to 3 years in the Philippines. Estimated length at first capture in the Philippines was 10.2 cm.

EARLY LIFE HISTORY: No information is available.

FISHERIES INTEREST: Island Mackerel are targeted commercially with purse seines, fish corrals, gill nets, cast nets, drift nets, and by dynamiting. They are marketed fresh, frozen, canned, dried, salted, and smoked. They are caught along with other species of *Rastrelliger* off Taiwan, the Philippines, Indonesia, Malaysia, and India. Appreciable quantities of Island Mackerel were reported as landed at Visakhapatnam and Kakinada on the east cost of India.

THREATS: There is no information on population status or general abundance of Island Mackerel. They are targeted in commercial and artisanal fisheries throughout their range, but landings are primarily reported as mixed *Rastrelliger* species. Reported worldwide landings for *Rastrelliger* species have steadily increased since 1950 to over 800,000 mt in 2015, but no effort information is available. Island Mackerel are listed on the IUCN Red List as Data Deficient. Given the absence of an international management body, further monitoring of this species is needed on the national level, in addition to species-specific data on landings, effort, and population structure.

CONSERVATION: Island Mackerel are widespread in southeastern Asia. Although effort is assumed to be increasing, it is not known how populations of this species population are affected by current and historical fishing pressure. Island Mackerel are listed on the IUCN Red List as Data Deficient.

REFERENCES: Aripin and Showers 2000; Basheer et al. 2015; Blaber et al. 1994; Collette 2001; Collette and Nauen 1983; Collette et al. 2011; Gnanamuttu 1971; Gnanamuttu and Girijavallabhan 1984; Jamaluddin et al. 2010; Jondeung and Karinthanyakit 2015; Lavapie-Gonzalez et al. 1997; Manacop 1958; Matsui 1967; Muto et al. 2016, 2017; Pillai 1993.

Indian Mackerel

Rastrelliger kanagurta (Cuvier, 1816)

COMMON NAMES: English – Indian Mackerel
French – Maquereau des Indes
Spanish – Caballa de la India

ETYMOLOGY: Described by the great French anatomist and ichthyologist Georges Cuvier (1816) as *kanagurta* based on plate 136 of Russell (1803:28), presumably using a local common name in Coromandel, India.

SYNONYMS: *Scomber kanagurta* Cuvier, 1816; *Scomber canagurta* Cuvier, 1829; *Scomber loo* Lesson, 1829; *Scomber delphinalis* Cuvier, 1832; *Scomber chrysozonus* Rüppell, 1836; *Scomber microlepidotus* Rüppell, 1836; *Scomber moluccensis* Bleeker, 1856; *Scomber uam* Montrouzier, 1857; *Scomber reani* Day, 1871; *Scomber lepturus* Agassiz, 1874; *Rastrelliger serventyi* Whitley, 1944; *Rastrelliger kanagurta* (Cuvier, 1816)

TAXONOMIC NOTE: The Indian Mackerel has been described as a new species ten times by eight different authors, while the other two species of the genus, the Island and Short mackerels, were only described once.

FIELD MARKS:

1 The Indian Mackerel body is deeper, 4.3–5.2 times in fork length, than in the Island Mackerel, but not as deep as in the Short Mackerel, 3.7–4.3 times in fork length.
2 At least one black spot is present near the lower pectoral-fin margin.
3 Golden longtitudinal bands run the length of the upper body.

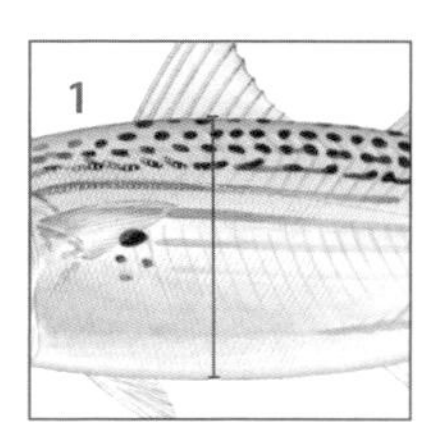

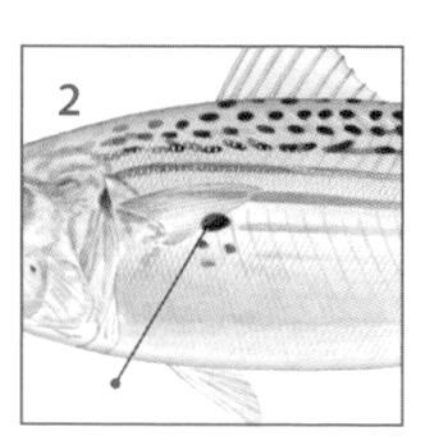

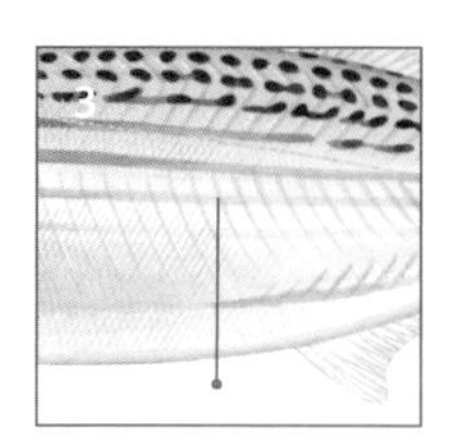

UNIQUE FEATURE: The Indian Mackerel is the most widespread member of the genus, with the other two species, the Island and Short mackerels, occurring in parts of the range of the Indian Mackerel.

DIAGNOSTIC FEATURES: Body compressed and moderately deep, its depth at the margin of the gill cover contained 4.3–5.2 times in fork length. The head is longer than the body depth. An adipose eyelid is present. **Mouth and teeth:** The vomer and palatines are toothless. The premaxillary and mandible each have a row of small, nearly conical, slightly curved teeth with smooth edges. **Gill rakers:** The gill rakers are very long, visible when the mouth is open, 30–46 on the lower limb. **Fins:** The first dorsal fin has 8 to 10 spines, usually 9. The second dorsal has 11–13 (usually 12) rays, followed by 4–6 finlets, usually 5. A rudimentary, skin-covered spine is present anterior to the origin of the anal fin. The anal fin has 10–13 rays, followed by 5 finlets. There are 18–21 pectoral-fin rays. The second dorsal fin and the anal fin are covered with a thick layer of scales. **Scales:** The entire body is covered with thin deciduous scales. **Swim bladder:** A swim bladder is present. **Vertebrae:** Vertebrae number 13 precaudal plus 18 caudal, for a total of 31. There are 11 interneural bones under the first dorsal fin, 12 under the second dorsal fin, plus 6 under the dorsal finlets. **Intestine:** The length of the intestine is 1.4–1.8 times fork length. **Osteology and genome sequence:** The osteology was described and illustrated by Gnanamuttu (1966). The complete mitochondrial genome sequence was published by Chen et al. (2013). Morphological and molecular characters that distinguish the Indian Mackerel from the other two species of *Rastrelliger* were summarized by Muto et al. (2016). **Color:** Narrow dark longitudinal bands (golden in fresh specimens) run the length of the upper part of the body. There is a prominent black spot on the body near the lower margin of the pectoral fin. The dorsal fins are yellowish with black tips. The caudal and pectoral fins are yellowish. The other fins are dusky.

GEOGRAPHIC RANGE: Indian Mackerel is an Indo-West Pacific epipelagic species found from the Red Sea and East Africa south to Natal, east to Indonesia, north to the Ryukyu Islands and China, south to Australia and Melanesia, and east to Samoa and the Marshall Islands. It entered the eastern Mediterranean Sea through the Suez Canal.

SIZE: Indian Mackerel reach at least 35 cm FL, commonly to 25 cm. A record size Indian Mackerel, 360 mm TL weighing 560 grams, was caught off Karwar, India. They reach 11–18 cm at the end of their first year, 21–24 cm at the end of their second, 25–28.5 cm at the end of their third, 27.6–30.2 cm at the end of their fourth, and 33.0 cm at the end of their fifth year.

HABITAT AND ECOLOGY: Indian Mackerel is a common, coastal pelagic species occurring in areas where surface temperatures are at least 17°C. They inhabit coastal bays, harbors, and deep lagoons, usually in turbid plankton-rich waters, and are often found in large schools of equally sized individuals.

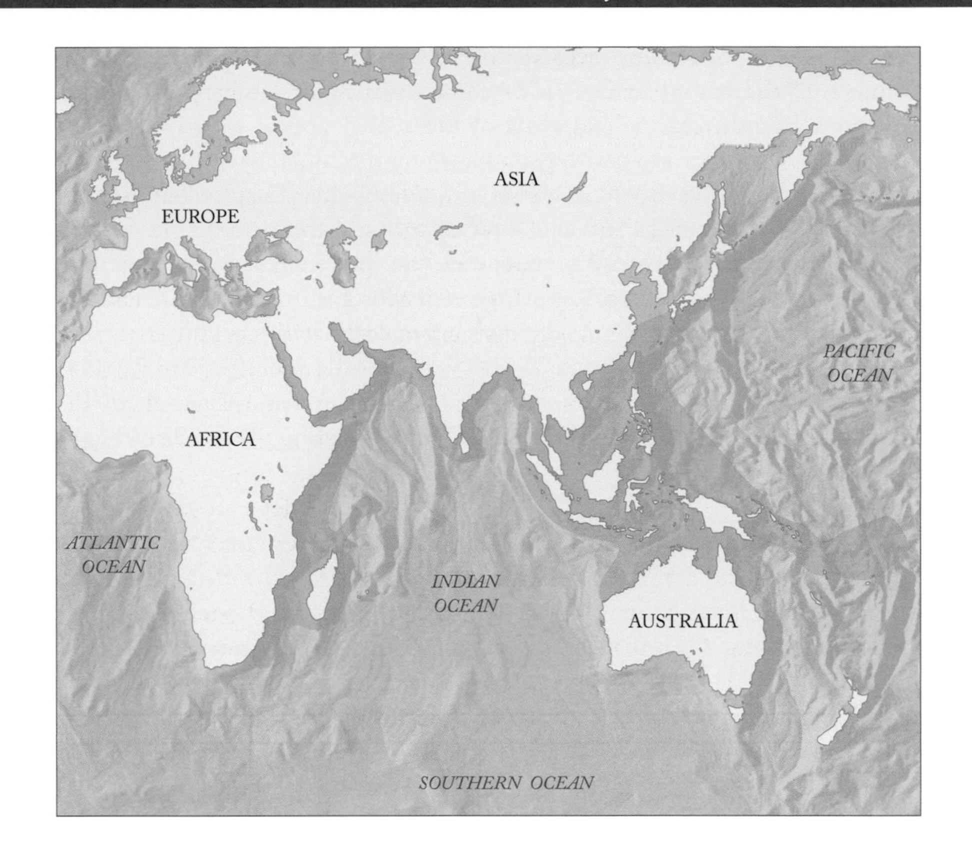

Left: Indian Mackerel range map.

FOOD: Copepods form the most important food items of adult Indian Mackerel on the west coast of India and the Andaman Islands (35–75% by volume), followed by pelagic tunicates, phytoplankton, stomatopod larvae, sergestids, decapod larvae, molluskan larvae, and amphipods. Feeding intensity was low during the peak spawning activity, October–December. During the pre-monsoon season in coastal waters of west Bengal, diatoms, copepods, rotifers, and green algae were dominant, while blue-green algae and dinoflagellates had the highest percentage of occurrence during the monsoon season. Juvenile Indian Mackerel feed on phytoplankton such as diatoms and small zooplankton such as cladocerans, ostracods, and larval polychaetes. With growth, they gradually change their diet, a process that is reflected in the relative shortening of their intestine.

Below: Indian Mackerel feeding position.

REPRODUCTION: Spawning activity of Indian Mackerel peaked during October–December off Kerala, southwestern India, and October–April in the Andaman Islands. On the east coast of India, they appear to spawn almost year-round (October–July), with peak spawning during January–April. Indian Mackerel have a long spawning season in Mozambique, March to April, with the peak in December or January. There appear to be two spawning seasons in the western waters of Aceh, Indonesia, the first from January to March and the second from August to October. Spawning is in several batches, and six batches of eggs were observed in some females. Batch fecundity has been estimated as between 55,000 and 315,000 ova in India and 29,000 to 520,000 in Aceh, Indonesia. Length at first maturity for males was 17.3–18.4 cm TL and for females, 17.4–18.8 cm TL in India and 19.6 cm TL to 20 cm FL in Indonesia.

EARLY LIFE HISTORY: Most larval Indian and Short Mackerels were found in seas shallower than 90 m, indicating that their larval nursery areas are coastal. Indian Mackerel have spherical pelagic eggs with a smooth chorion. The eggs measure 0.91–0.98 mm in diameter and contain a single large oil globule, 0.23–0.25 mm in diameter. Yolk-sac larvae measure 1.49–2.84 mm SL. Preflexion larvae measure between 1.8 and 3.3 mm SL; flexion larvae are 2.85–4.85 mm SL. Early development stages were photographed in Sree Renjima et al. Unlike most other scombrid larvae, *Rastrelliger* larvae lack preopercular spines. The number of gill rakers increases with size until a length of about 51 mm. Two juvenile Indian Mackerel from the Philippines, 26 and 48 mm FL, were illustrated by Manacop. Eight larvae and juveniles, 8.7–52.7 mm, were selected from among 210 specimens from the inshore waters of Vizhingam on the southwest coast of India for description and illustration by Balakrishnan and Rao.

STOCK STRUCTURE: No evidence for geographical structure was evident in a molecular study (mtDNA cytochrome *b*) of 23 populations of Indian Mackerel from the Indonesian-Malaysian Archipelago and the Persian Gulf, although the Persian Gulf was recognized as a distinct subpopulation. Morphometric and genetic analyses of Indian Mackerel from both sides of peninsular India (east: Mandapam, Bay of Bengal; west: Kochi and Karwar, Arabian Sea) indicated that the populations could all belong to the same stock. A microsatellite study of Indian Mackerel from four localities from the Andaman Sea indicated a single population with moderate levels of genetic variation. The most recent molecular study also concluded that samples from 10 localities on the Indian coast did not show significant differentiation; however, these samples did significantly differ from samples from Thailand.

FISHERIES INTEREST: Indian Mackerel is targeted by commercial and artisanal fisheries throughout its range using a wide variety of gear such as seines, gill nets, cast nets, and trawls. India produces 90% of the world production of Indian Mackerel, of which 77% is from the west coast and 23% from the east coast. However, landings are frequently reported as mixed *Rastrelliger* spp. in areas where more than one species occurs. Reported worldwide landings for *Rastrelliger* species have steadily increased since 1950 to over 800,000 mt, but little effort information is available.

THREATS: Indian Mackerel are widespread, attain full sexual maturity at a small size, spawn year-round, and show little geographic variation. Some populations appear to be stable; others are overexploited. Much of the catch are age 1 and may not have spawned, so gill net mesh size should be increased to catch only larger individuals. Catch data is frequently not given by species in regions where there is more than one species of the genus, so total catch of each of the three species of *Rastrelliger* is not known. No recent stock assessments were available when the three species were assessed, which led to a listing as Data Deficient on the 2011 IUCN Red List. Given the absence of an international management body, further monitoring of this species is needed on the national level, in addition to species-specific data on landings, effort, and population status.

CONSERVATION: Environmental factors together with high exploitation rates of Indian Mackerel seem to have caused a decline in catch along the Mangalore-Malpe coast. The exploitation rate in the Gulf of Suez, Egypt, was higher than the predicted value, indicating that this population is overexploited. However, exploitation rates over the years along the Tuticorin coast indicated that this stock of Indian Mackerel still remains underexploited. Any expansion should only be for the gill net fishery, which harvests larger and older fish, and not for the trawl fishery, which harvests mostly pre-adults, which may not have had the opportunity to spawn. The stocks in Marudu Bay, Sabah, the southeast coast of India, and Oman were also recently evaluated as not overexploited. Given that effort is assumed to be increasing and that there is some evidence of localized declines, it is not known whether other populations of Indian Mackerel are affected by current and historical fishing pressure.

REFERENCES: Abdussamad et al. 2006, 2010; Agassiz 1874; Akib et al. 2015; A. M. Amin et al. 2015; S.M.N. Amin et al. 2014; Arrafi et al. 2016; Balakrishnan and Narayana Rao 1971; Basheer et al. 2015; Bhendarkar et al. 2013; Chen et al. 2013; Collette 1970, 2001; Collette and Nauen 1983; Collette et al. 2011; Cuvier 1816; Das et al. 2016; Day 1871; Dhulkhed and Annigeri 1983; Gnanamuttu 1966; Hariati and Fauzi 2017; Jamaluddin et al. 2010; Jayabalan et al. 2014; Jayasankar et al. 2004; Jondeung and Karinthanyakit 2010, 2015; Jones and Rosa 1965, 1967; Luther 1973; Manacop 1958; Matsui 1967, 1970; Mehanna 2001; Montrouzier 1857; Munpholsri et al. 2013; Muto et al. 2016, 2017; Randall et al. 2005; Rao 1962; Rohit and Gupta 2004; Russell 1803; Seshappa 1969; Sivadas and Bhaskaran 2009; Sivadas et al. 2006, 2016; Sousa and Gislason 1985; Sree Renjima et al. 2016; Sudjastani 1974; Sukumaran et al. 2017; Yohannan and Abdurahiman 1998; Zaki et al. 2016.

Australian Bonito

Sarda australis (Macleay, 1881)

COMMON NAMES: English – Australian Bonito
French – Bonite Bagnard
Spanish – Bonito Austral

ETYMOLOGY: Named *australis* by the Australian ichthyologist William Macleay (1881:557) because of its occurrence in the Southern, or Austral, Hemisphere.

SYNONYMS: *Pelamys australis* Macleay, 1881; *Pelamys schlegeli* McCoy, 1888; *Sarda australis* (Macleay, 1881); *Sarda chiliensis australis* (Macleay, 1881)

TAXONOMIC NOTE: Some fishery biologists, such as Silas (1964), have considered *Sarda australis* to be a subspecies or synonym of the Oriental Bonito, *Sarda orientalis*, but Collette and Chao (1975) morphologically and Viñas et al. (2010) genetically clearly showed the validity of the species.

FIELD MARKS:

1 Bonitos of the genus *Sarda* all have 5–10 narrow dark stripes on the upper part of their bodies, but the 10 stripes are closer to horizontal in the Australian Bonito and tend to extend onto the belly in some individuals.

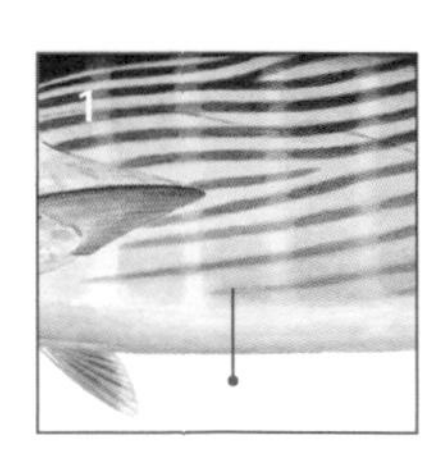

UNIQUE FEATURE: The bonito restricted to eastern Australia and New Zealand.

DIAGNOSTIC FEATURES: Teeth: There are 16–26 teeth on each side of the upper jaw, 11–20 on the lower jaw. Teeth are sometimes present on the vomer in the roof of the mouth. **Jaw:** The supramaxilla bone is intermediate in width compared to other species of bonitos. **Gill rakers:** There are 19–21 gill rakers on the first gill arch. **Fins:** There are 18 or 19 spines in the first dorsal fin and 15–17 rays in the second dorsal fin, usually followed by 7 dorsal finlets. There are 14–17 rays in the anal fin, usually followed by 6 anal finlets. The pectoral fin has 25–27 rays, modally 26. **Scales:** The body is covered with very small scales posterior to the corselet. **Vertebrae:** Vertebrae number 23 or 24 precaudal plus 21 or 22 caudal, for a total of 45 or 46. **Soft anantomy, osteology:** The soft anatomy and osteology were described and illustrated by Collette and Chao (1975). **Color:** The dorsal stripes are closer to being horizontal than in other species of *Sarda* and extend onto the belly in some individuals.

GEOGRAPHIC RANGE: Australian Bonito has been known from around southeastern Australia and Norfolk Island, and its known range was extended to northern New Zealand only relatively recently.

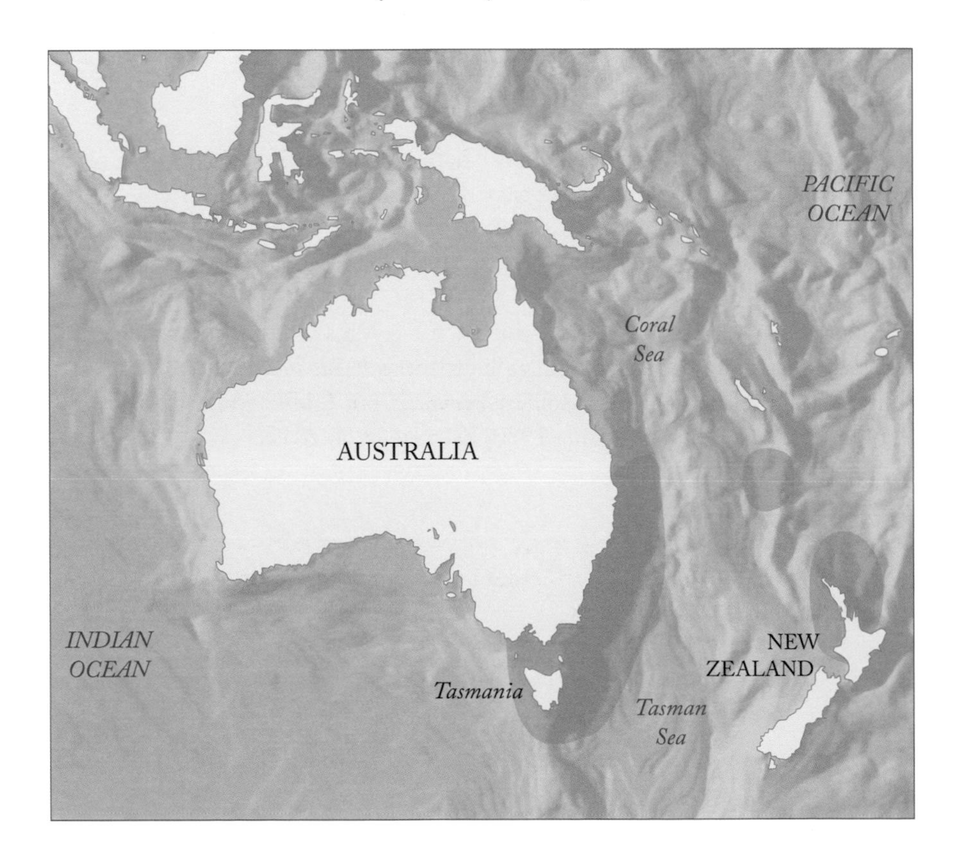

Left: Australian Bonito range map.

SIZE: Maximum size is 100 cm FL; it is commonly caught at 40–45 cm FL and 1.8–2.3 kg weight. The IGFA all-tackle angling record is a 20 lb, 11 oz (9.4 kg) fish caught off Montague Island, New South Wales, in April 1978.

HABITAT AND ECOLOGY: The biology of Australian Bonito is not well known. They are pelagic and oceanodromous. They school by size and mature from January through April. They are found in schools in the inshore coastal waters of Queensland and feed on various fishes.

REPRODUCTION: The gonads were at their largest relative to body weight during the austral spring/summer months off New South Wales, with the mean gonadal somatic index peaking during November. The size at sexual maturity was estimated to be about 36 cm FL, suggesting that this species matures at 1 year of age.

FISHERIES INTEREST: Australian Bonito are an important target of recreational fishers, and there is an established commercial line fishery along southeastern Australia. They provide first-rate sport to anglers using the lightest tackle. Australian Bonito are a secondary target species for the recreational land-based game fishery along the southeastern coast of Australia. They are also used as bait for billfishes, sharks, and tunas. Landings in the commercial line fishery fluctuate annually between 100 and 250 mt.

CONSERVATION: In New South Wales there is a recreational bag limit of 10 Australian Bonito per angler per day. It is listed as Least Concern on the IUCN Red List but more information is needed on this species' biology, harvest levels, and population trends.

REFERENCES: Collette 2001; Collette and Chao 1975; Collette and Nauen 1983; Collette et al. 2011; Grant 1982, 1987; Griffiths 2012; IGFA 2018; James and Habib 1979; Macleay 1881; Rowling et al. 2010; Silas 1964; Stewart et al. 2013; Viñas et al. 2010; Yoshida 1980; Zischke et al. 2012.

Eastern Pacific Bonito

Sarda chiliensis (Cuvier, 1832)

COMMON NAMES: English – Eastern Pacific Bonito
French – Bonite du Pacifique Oriental
Spanish – Bonito del Pacífico

ETYMOLOGY: The great French anatomist and ichthyologist Georges Cuvier (1832:163) named the species *chiliensis* for the source of his material, Valparaiso, Chile. The American (although originally French) ichthyologist Charles Frédérick Girard (1858:106-107) named the northern population *lineolatus* based on its prominent stripes.

SYNONYMS: *Sarda chiliensis* Cuvier, 1832; *Sarda chiliensis chiliensis* Cuvier, 1832; *Pelamys lineolata* Girard, 1858; *Sarda lineolata* (Girard, 1858); *Sarda chiliensis lineolata* (Girard, 1858)

TAXONOMIC NOTE: Two geographic subspecies are recognized: *S. c. lineolata* occurs from Alaska to the tip of Baja California, the southwest Gulf of California, and the Revillagigedo Islands; *S. c. chiliensis* occurs from Ecuador to Chile.

FIELD MARKS:

1 All four bonitos in the genus *Sarda* have 5–10 narrow dark stripes on the upper part of their bodies.
2 The Eastern Pacific Bonito has more gill rakers in the first gill arch (23–27) than the other three species (8–23).

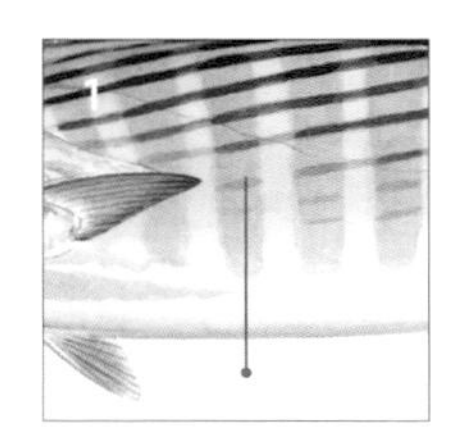

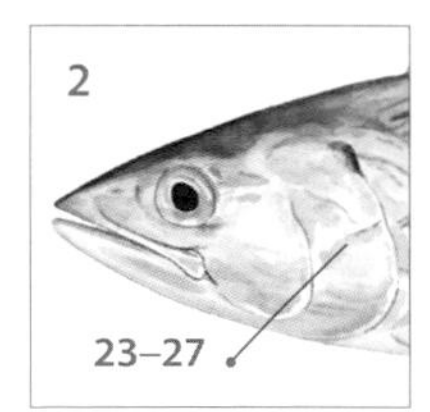

UNIQUE FEATURE: The Eastern Pacific Bonito is the only scombrid that has its geographical distribution separated into two parts by a different species of bonito—the Oriental Bonito.

DIAGNOSTIC FEATURES: Teeth: There are 18–30 teeth on each side of the upper jaw, 14–25 on the lower jaw. Vomerine teeth are always absent. **Jaw:** The supramaxilla is wider than in the other species of bonitos (Collette & Chao, 1975:fig. 32f). **Gill rakers:** There are 23–27 gill rakers on the first gill arch. The gap between the gill rakers is smaller, 1.8 mm, than in the Oriental Bonito, mean 2.3 mm. **Fins:** There are 17–19 spines in the first dorsal fin and the length of the dorsal fin base is 26.7–31.4% of fork length. There are 12–15 rays in the second dorsal fin, usually followed by 7 or 8 finlets. There are 12–15 rays in the anal fin, usually followed by 6 or 7 anal finlets. The pectoral fin has 23–26 rays, usually 24 or 25. **Scales:** Eastern Pacific Bonito, like other members of its genus, have the body completely covered with small scales. **Vertebrae:** Vertebrae number 22–24 precaudal plus 20–23 caudal, for a total of 42–46, usually 44 or 45. **Soft anantomy, osteology:** In a ventral view of the viscera, the spleen is prominent in the posterior part of the visceral cavity. The soft anatomy and osteology have been described and illustrated (Godsil 1954, 1955; Collette and Chao 1975). **Color:** The dorsal stripes are slightly oblique. Dark vertical bars frequently appear when the fish are feeding. Two individuals, 37.1 and 47.2 cm FL, caught off California lacked the characteristic bonito striping but had normal coloration in all other respects.

GEOGRAPHIC RANGE: The Eastern Pacific Bonito is restricted to the eastern Pacific and is divided into two geographically distinct subspecies separated by the Oriental Bonito. The northern subspecies, *Sarda chiliensis lineolata*, ranges from the Gulf of Alaska to the tip of Baja California, Mexico, and Socorro Island, one of the Revillagigedo Islands. This subspecies is centered between southern California and central Baja California, Mexico, and moves farther north in warm-water years. The southern subspecies, *Sarda chiliensis chiliensis*, is found off the western coast of South America from Mancora, Peru, just south of the Gulf of Guayaquil to Talcahuano and Valdivia, Chile.

SIZE: The maximum reported size of Eastern Pacific Bonito is at least 79 cm FL in the southern hemisphere and 102 cm FL in the northern hemisphere, where a fish may reach 11.3 kg. The IGFA all-tackle game fish record is of a 21 lb, 5 oz (9.67 kg) individual taken off California in October 2003.

HABITAT AND ECOLOGY: The Eastern Pacific Bonito is a coastal pelagic species that schools by size. In captivity at Marineland of the Pacific in Palos Verde, California, Eastern Pacific Bonito swam continuously against the current with pectoral fins extended, pelvic fins and first dorsal fin adpressed, and mouth agape at an angle of 10–15 degrees. The pectoral fins produce the lift necessary to overcome the weight of this species that lacks a swim bladder. Their undisturbed cruising speed in tanks at the Monterey Bay Aquarium was about

1.21 body lengths per second, with a tailbeat frequency similar to that of Yellowfin Tuna. Several studies have used the Eastern Pacific Bonito to represent bonitos (Sardini) as a phylogenetic intermediate between more primitive mackerels (Scombrini) and more advanced tunas (Thunnini) in comparing differences in red muscle development and how this effects swimming.

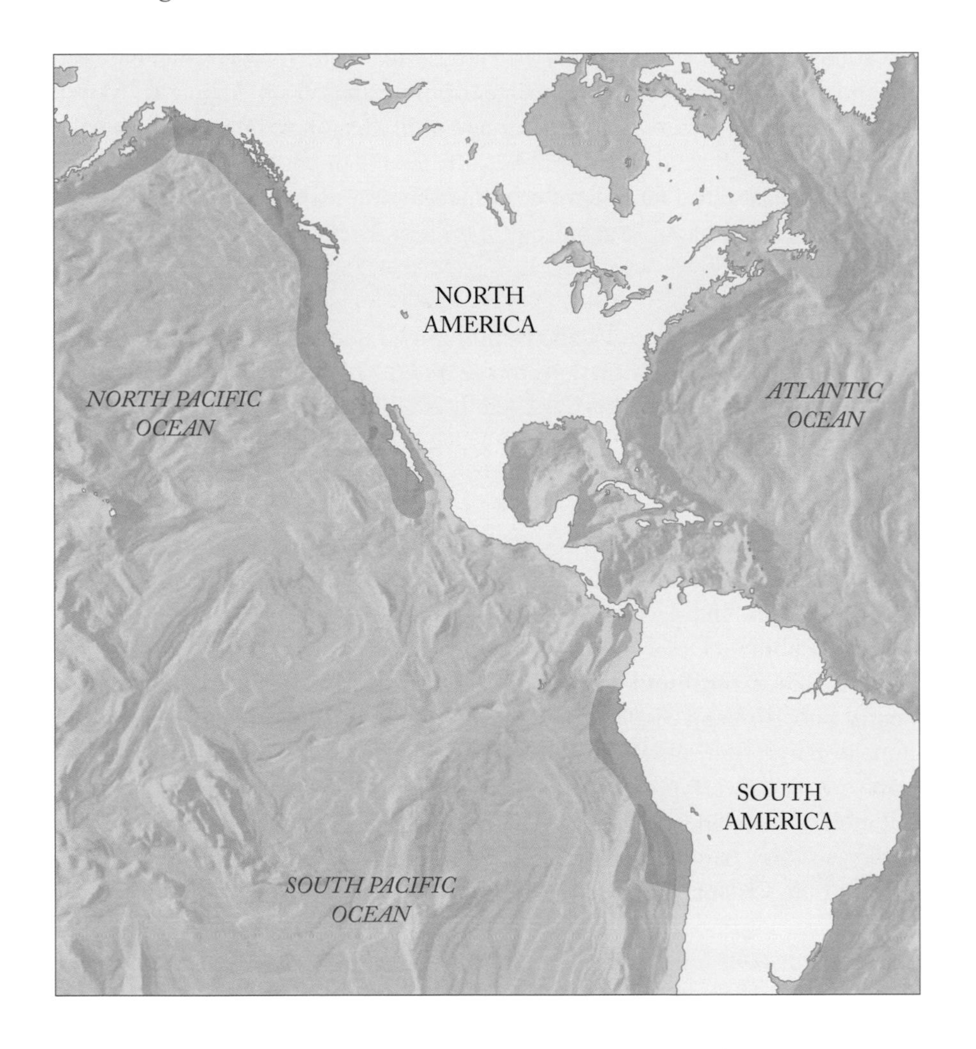

Left: Eastern Pacific Bonito range map.

FOOD: Eastern Pacific Bonito are primarily piscivorous predators, consuming about 6% of their body weight per day. Off southern California, 12 species of fishes were the principal contributions to their diet, providing 91.1% by numbers and 81.8% by volume of the diet. Northern Anchovy alone represented 75.5% of prey items by number and occurred in 56.3% of the individuals. Other species of fishes provided 15.6% by numbers, 5.9% by volume, and 30.1% by occurrence. Common squid contributed 8.1% by numbers and 18.0% by volume, and occurred in 25.1% of the stomachs containing food. In a later report, also from southern California, the diet was 71.4% fishes, 26.2% crustaceans (11.9% euphausiids), by frequency of occurrence. The crustacean most important to both the Eastern Pacific Bonito and the Skipjack was the euphausiid *Nyctiphanes simplex*.

REPRODUCTION: In California, Eastern Pacific Bonito males reach sexual maturity at 1 or 2 years of age and a length of about 50 cm, and females mature at age 2 and a length of 56 cm. Spawning occurs primarily off Baja California, beginning in February or early March in the south, progressing northward in the following four or five months. In Peru, sexual maturity is reached between 47 and 53 cm. Spawning occurs in nearshore waters from August to March but is concentrated between October and February. Older fish mature earlier in the season and tend to live further offshore than younger fish. Spawning is in batches, and the number of eggs shed in one season by a 600 mm, 3 kg fish has been estimated at about half a million. Fecundity increases with size of the female.

AGE AND GROWTH: Eastern Pacific Bonito grow rapidly during their first three years, with much slower growth from age 3 to 6. The mean lengths of each age group in the California fishery are: age I, 20.3 in (51.5 cm); II, 24.9 in (63.3 cm); III, 27.4 in (69.5 cm); IV, 28.7 in (72.9 cm); V, 29.4 in (74.8 cm); and VI, 29.8 in (75.7 cm). Off Peru, they reach an age of at least 6 years.

EARLY LIFE HISTORY: Fertilized eggs of Eastern Pacific Bonito were collected in shallow waters of 16–20°C off the coast of Baja California, Mexico, in the spring of 1966. The developing eggs were 1.4–1.8 mm in diameter and had 2–6 oil globules. Descriptions of postlarvae and juveniles from California (27, 16.7–54.5 mm) and Peru (8, 38.0–200 mm) with illustrations of two postlarvae (16.7 and 33.0 mm) were presented by Pinkas. Drawings of a 43 mm postlarva from off Baja California and of a 160 mm juvenile caught off northern Peru were presented by Klawe. Counts of gill rakers on the first gill arch of small individuals, 34–135 mm, were within the same range as for adults, so this character can be used to differentiate juvenile Eastern Pacific Bonito from Oriental Bonito.

STOCK STRUCTURE: The two geographically separated populations of Eastern Pacific Bonito are recognized as subspecies using available names: *Sarda chiliensis chiliensis* for the southern subspecies and *Sarda chiliensis lineolata* for the northern subspecies. There are only slight meristic differences between the two populations, with the northern subspecies having slightly more vertebrae and teeth in the upper and lower jaws than the southern population. No genetic differences have been found between the populations. However, as the populations are currently geographically isolated, with the widespread Oriental Bonito occurring in the waters between the two populations, and as the names of the subspecies have been used in the literature, there is some practical value in continuing to use the subspecific names for the two populations.

FISHERIES INTEREST: Eastern Pacific Bonito are targeted in commercial purse seine fisheries in the southern and northern eastern Pacific and in a popular recreational fishery in southern California. Overall landings have fluctuated greatly over time in the north and in the south, with the southern stock typically accounting for >90% of the reported landings. FAO-reported landings for *Sarda chiliensis* (both subspecies) increased from 35,000 mt in 1950 to over 113,000 mt in 1961. Landings remained between 65,000 and 95,000 mt through 1972, and then decreased to less than 6,000 mt by 1986. Since that time, landings have fluctuated between 1,000 and 40,000 mt and increased to almost 94,000 mt in 2015. Variation in landings of Eastern Pacific Bonito are believed to be tightly linked to variations in the abundance of anchovies, their principal prey item.

Peru is responsible for the vast majority of landings of Eastern Pacific Bonito in the southern hemisphere, with Chile representing less than 10% of the total. Reported landings from Peru show much greater variation than those from Chile, although there is some concern over the accuracy of reported catches from Chile, especially for the 1980s. Decreases in reported landings from Peru during the 1990s may have been affected by changes in effort, as some vessels targeted Jack Mackerel and Pacific Chub Mackerel.

Landings in California and Mexico have fluctuated greatly over the past 50 years, from less than 1,000 mt to nearly 14,000 mt in the early 1970s. Commercial landings have declined steadily since the mid-1980s but have increased moderately in recent years, from 291 mt in 1997 to 803 mt in 2008. The trend over the last 15 years seems to be low landings for most years, interspersed with high-yield years. Competition with higher-valued fisheries was likely responsible for part of the decline in landings observed during the 1980s and 1990s. Increased regulation, decreased stocks, and lower market demand contributed to the decline. In 1982, Mexico began restricting foreign vessel access to its nearshore fisheries. Prior to this closure, 50–90% of Eastern Pacific Bonito landed in the United States was caught off the coast of Baja California, Mexico. Now less than 10% originates in Mexican waters. In addition, Eastern Pacific Bonito support a robust recreational fishery off southern California, with anglers fishing from shore, jetties, piers, and boats. Estimated recreational landings have varied greatly over the past 60 years, peaking at close to 1.4 million fish in the mid-1960s.

THREATS: The Eastern Pacific Bonito is listed as Least Concern on the IUCN Red List. However, given the extreme fluctuations in catch data, this species should be carefully monitored.

CONSERVATION: There are no range-wide assessments for Eastern Pacific Bonito or regional assessments of the northern or southern populations. Based on catch data, Eastern Pacific Bonito SSB appears to have experienced dramatic fluctuations over the last 20 to 40 years. These changes are likely due to changes in fishing pressure and to the availability of anchovies, their primary food. Eastern Pacific Bonito populations fluctuate on a decadal scale in a similar manner to Northern Anchovy. These fluctuations are usually associated with warm- and cold-water periods of the Pacific Decadal Oscillation. There are few conservation measures for Eastern Pacific Bonito. In Peru, the minimum size for catch is 52 cm FL, and the maximum tolerance of juveniles is 10% of the catch, but there are no catch quotas. Recreational anglers in California are restricted to a bag limit of 10 Eastern Pacific Bonito, with a minimum size of 61 cm (24 in) FL, although there is a tolerance for 5 of the 10 fish to be below the minimum size. Much of landings data are combined for two bonito species, the Eastern Pacific and Western Pacific bonitos, complicating management.

REFERENCES: Altringham and Block 1997; Ancieta Calderón 1963, 1964; Bernard et al. 1985; Black 1978; Campbell and Collins 1975; Collette and Chao 1975; Collette and Nauen 1983; Collette et al. 2011; Collins and MacCall 1977; Cuvier 1829; Cuvier and Valenciennes 1832; Dowis et al. 2003; Ellerby et al. 2000; Girard 1858; Godsil 1954, 1955; Goldberg and Mussieet 1984; IGFA 2018; Klawe 1961; Kuo 1970, 1975; Magnuson and Heitz 1971; Magnuson and Prescott 1966; Matsumoto et al. 1969; McFarlane et al. 2000; Oliphant 1971; Pinkas 1961; Pinkas et al. 1968; Sepulveda et al. 2003; Sokolovskii 1971; Vildosa 1960, 1963a, 1963b; Viñas et al. 2010; Yoshida 1980.

Oriental Bonito

Sarda orientalis (Temminck & Schlegel, 1844)

COMMON NAMES: English – Oriental Bonito, Striped Bonito
French – Bonite Oriental
Spanish – Bonito Mono

ETYMOLOGY: Coenraad Jacob Temminck was a Dutch ornithologist, illustrator, and collector, and Hermann Schlegel was a German-born zoologist who spent most of his life in the Netherlands. Together they produced the Fauna Japonica, containing the description of *Pelamys orientalis* (1844:99) from Japan in the Orient, leading to its name.

SYNONYMS: *Pelamys orientalis* Temminck and Schlegel, 1844; *Sarda velox* Meek and Hildebrand, 1923; *Sarda orientalis serventyi* Whitley, 1945; *Sarda orientalis* (Temminck and Schlegel, 1844)

TAXONOMIC NOTE: Three geographic populations of Oriental Bonito were named as separate taxa: *orientalis* from Japan in 1844, *velox* from the Pacific coast of Panama in 1923, and *serventyi* from Western Australia in 1945, but all are currently considered part of a single Indo-Pacific species.

FIELD MARKS:
1 All bonitos have dorsal stripes but the stripes, are slightly oblique and do not extend onto the belly in the Oriental Bonito.
2 The Oriental Bonito has fewer gill rakers on the first gill arch, 8–13, than the other bonitos, 16–27.

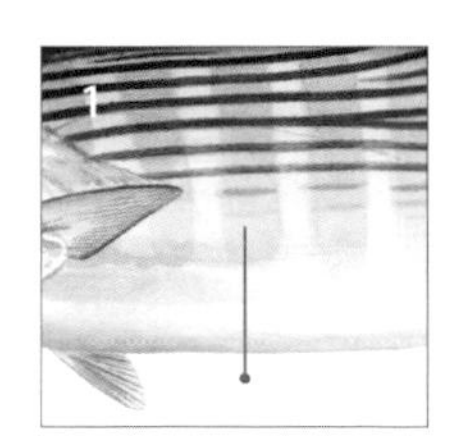

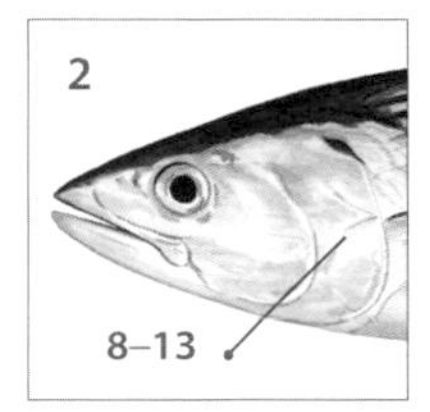

UNIQUE FEATURE: The Oriental Bonito is the most widespread species in the genus, occurring throughout the Indo-Pacific and even extending into the eastern tropical Pacific.

DIAGNOSTIC FEATURES: Teeth: There are 12–20 teeth on each side of the upper jaw, 10–17 on the lower jaw. Teeth are present on the palatines but are absent from the vomer. **Jaw:** The supramaxilla is narrower than in the other bonitos. **Gill rakers:** There are 8–13 gill rakers on the first gill arch, usually 10–12, fewer than in other bonitos. The gap between the gill rakers is larger, 2.3 mm, than in the Eastern Pacific Bonito, 1.8 mm. **Fins:** There are 17–19 spines in the first dorsal fin, usually 18, and the length of the dorsal-fin base is 28.2–32.7% of fork length. There are 14–17 rays in the second dorsal fin, usually followed by 8 finlets. There are 14–16 rays in the anal fin, usually followed by 6 anal finlets. The pectoral fin has 23–26 rays, usually 24 or 25. The interpelvic process is divided. **Scales:** Oriental Bonito, like the other members of its genus, have the body completely covered with small cycloid scales. **Vertebrae:** Vertebrae number 23–25 precaudal plus 20–22 caudal, typically for a total of 44 or 45. **Soft anatomy, osteology:** In a ventral view of the viscera, the spleen is prominent in the posterior half of the visceral cavity. The soft anatomy and osteology have been described and illustrated (Godsil 1954, 1955; Collette and Chao 1975). **Color:** Body dark blue above, silvery below, with 6 to 8 black longitudinal oblique stripes. Dark vertical bars frequently appear when the fish are feeding.

GEOGRAPHIC RANGE: Oriental Bonito are widespread throughout the Indo-Pacific, but there are many gaps in its known distribution. There is a population in the eastern Pacific between the tip of Baja California south to Ecuador (and the Galapagos Islands) between populations of the Eastern Pacific Bonito to the north and south. It occurs in Hawaii but is not common there. It occurs along both coasts of Honshu, Japan, and is recorded from the coasts of China and the Philippines. It is widespread in the Indian Ocean, including southwest Australia, India, the Seychelles Islands, and South Africa. Based on its known disjunct distribution, it is likely that there are separate populations.

SIZE: The maximum reported length of Oriental Bonito in the Indian Ocean is 102 cm FL. The IGFA all-tackle game fish record is of a 23 lb, 8 oz (10.65 kg) fish caught off Victoria, Mahe, Seychelles Islands, in February 1975. Oriental Bonito attain a length of 40 cm by the end of the first year and 50 cm by the end of the second year. Longevity was estimated at 4.4 years.

HABITAT AND ECOLOGY: Oriental Bonito are a coastal pelagic species that can be found schooling with small tunas to depths of 30 m and at temperatures of 13.5–23°C. They also live around islands such as Hawaii and the Seychelles.

FOOD: Oriental Bonito feed mainly on fishes, particularly clupeoids, but also eat squids and decapod crustaceans. Juveniles, 9.8–19.9 cm FL, from the Gulf of Mannar, India, fed largely on fishes, particularly anchovies and sardines, unlike juvenile frigate tunas and Kawakawa that eat larger quantities of zooplankton.

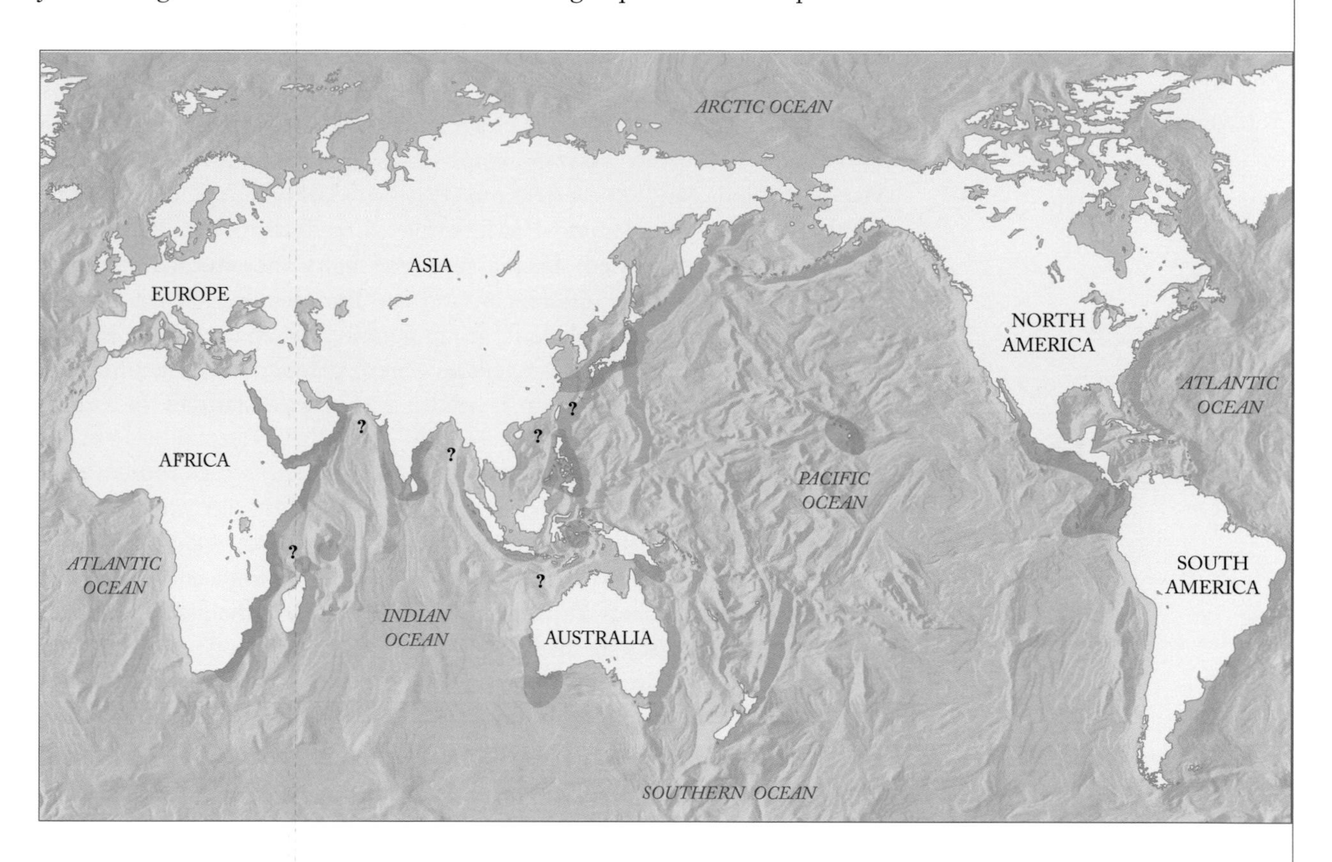

Above: Oriental Bonito range map.

REPRODUCTION: Spawning of Oriental Bonito presumably occurs in the coastal waters of the tropical zone in the Indo-Pacific. Size of first maturity along the Kerala coast of India was 42 cm FL. Relative fecundity varied from 293,793 to 696,512 eggs for fishes 39–52 cm FL.

EARLY LIFE HISTORY: Artificially fertilized eggs of Oriental Bonito are buoyant, spherical, and 1.32–1.45 mm in diameter. Eggs hatched in about 50 hours at water temperatures of 20–24°C. Newly hatched larvae measured 4.1–4.3 mm TL and initiated swimming and feeding on rotifers on day 2. Yolk and oil globules were completely absorbed by day 4. Larvae shifted from preflexion to flexion around day 7 at about 6 mm SL and to postflexion around day 11 at about 9 mm SL. Fed first on zooplankton, then on small fishes, larvae grew to 14 mm in 10 days, 74 mm in 20 days, 106 mm in 30 days, 219 mm in 42 days, and to 290 mm 99 days after hatching. This rapid growth rate is nearly equal to that of the Japanese Spanish Mackerel and is among the fastest growth rates among marine fishes larvae hatched from pelagic eggs. Seven larvae, 3.16–22.17 mm SL, from Japan were described and illustrated by Nishikawa, and two juveniles from India, 80 and 158 mm SL, were illustrated by Jones. Larval Oriental Bonito resemble other scombrid larvae, with their large head, large eyes, and head supination. Larvae larger than 5–6 mm SL differ from other scombrid larvae in having spines on the supraorbital ridge, distinct black pigment on the pelvic fins, and a large black pigment patch on the caudal-fin base. Photomicrographs of developing larvae were published by Kaji et al.

STOCK STRUCTURE: Although three geographic populations of Oriental Bonito were named as separate taxa: *orientalis* from Japan, *velox* from the Pacific coast of Panama, and *serventyi* from Western Australia, all are currently considered part of a single Indo-Pacific species. Comparisons of small samples from Japan and the eastern Pacific revealed slight anatomical differences and slight morphometric differences, and the lineages from Japan formed a strongly supported monophyletic group genetically separate from eastern Pacific specimens. A more comprehensive genetic comparison of the different populations throughout their range is needed to clarify the status of the various populations of the species.

FISHERIES INTEREST: Directed fisheries for Oriental Bonito are not well developed, and landings are generally included with other species or not reported. They are caught with other scombrids with troll lines, encircling nets, purse seines, and drift nets. Almost all reported catches (about 93%) of both species of Pacific bonitos are made by purse seines. Landings reported to FAO were below 1,000 mt through 2006, and then increased sharply to 6,055 mt in 2008. Since that time, they have fluctuated between 1,700 and 7,700 mt. Most of the reported landings have been from the Indian and western Pacific oceans.

In the Eastern Pacific, landings are often combined for Oriental and Eastern Pacific bonitos. Historical catch for these species in the eastern Pacific ranged from about 26 to 14,227 mt, peaking in 1990. Reported catches peaked again in 2007. It is not known what drives these fluctuations (abundance or purse seine vessels switching target species), and it is most likely that only a small proportion of these landings data are for Oriental Bonito, as the majority of the catch was from Chile and Peru, which are probably Eastern Pacific Bonito.

THREATS: A lack of reporting of landings to the species level and a general lack of reporting for Oriental Bonito are of concern. The limited distribution of some species of bonito, together with the growing demand for bonito for high-quality canned products, require close monitoring of this species. The Oriental Bonito is listed as Least Concern on the IUCN Red List. However, more information is needed on population status, landings, and biology especially as there has been a recent regional decline in the catch of bonitos in the eastern tropical Pacific.

CONSERVATION: There are no range-wide assessments for Oriental Bonito. The exploitation level in India is well below MSY level. There are no known conservation measures for Oriental Bonito. There is a need for robust fishery data to support the provision of management advice for both species of bonito in the eastern Pacific, and there is a need to collect data on catches from the Western Central Pacific Ocean and from artisanal fisheries throughout the Pacific.

REFERENCES: Collette and Chao 1975; Collette and Nauen 1983; Collette et al. 2011; Godsil 1954, 1955; Harada et al. 1974; IGFA 2018; Jones 1960; Kaji et al. 2002a, 2000b; Kikawa et al. 1963; Kumaran 1964; Magnuson and Heitz 1971; Meek and Hildebrand 1923; Nishikawa 1988; Silas 1963, 1964; Siraimeetan 1985; Sivadas et al. 2012; Sivasubramaniam 1969; Temminck and Schlegel 1844; Thomas and Kumaran 1963; Viñas et al. 2010; Whitley 1945; Yoshida 1980.

Atlantic Bonito

Sarda sarda (Bloch, 1793)

COMMON NAMES: English – Atlantic Bonito
French – Bonite à dos rayé
Spanish – Bonito atlántico

ETYMOLOGY: In his description of *Scomber sarda*, the German physician and naturalist Marcus Élieser Bloch (1793:44) referred to a figure of a juvenile bonito labelled Pelamyde Sarda in Rondelet (1554:248).

SYNONYMS: *Scomber sarda* Bloch, 1793; *Scomber mediterraneus* Bloch and Schneider, 1801; *Scomber palamitus* Rafinesque, 1810; *Scomber ponticus* Pallas, 1811; *Thynnus brachypterus* Cuvier, 1829; *Pelamys sarda* Cuvier in Cuvier and Valenciennes, 1831; *Palamita sarda* Bonaparte, 1831; *Pelamis sarda* Valenciennes, 1844; *Sarda pelamys* Gill, 1862; *Sarda mediterranea* Jordan and Gilbert, 1882

TAXONOMIC NOTE: There is some genetic differentiation between populations from the western North Atlantic and those from the Mediterranean, but they are not considered distinct species (Viñas et al. 2010).

FIELD MARKS: 1 Bonitos of the genus *Sarda* all have 5–10 narrow dark stripes on the upper part of their bodies, but the stripes are more oblique in the Atlantic Bonito.

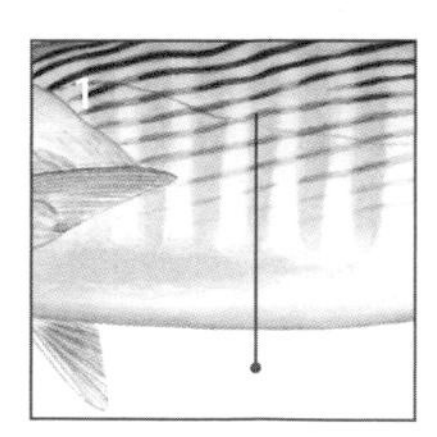

UNIQUE FEATURE: Atlantic Bonito have more spines in the first dorsal fin (20–23) than the three other species of bonitos (17–19) and more vertebrae (50–55 compared to 43–6).

DIAGNOSTIC FEATURES: Body tuna shaped, thick, and stout, one-fourth as deep as long, tapering anteriorly to a pointed snout and posteriorly to a very slender caudal peduncle. **Teeth:** There are 13–28 teeth on each side of the upper jaw, 10–22 on the lower jaw. Vomerine teeth are sometimes present. **Jaw:** The supramaxilla is intermediate in width compared to the other species of bonitos. **Gill rakers:** There are 16–23 gill rakers on the first gill arch. **Fins:** There are 20–23 spines in the first dorsal fin, and the length of the fin base is 29.1–33% of fork length. There are usually 8 dorsal finlets. There are 14–17 rays in the anal fin, usually followed by 7 anal finlets. The pectoral fin has 23–26 rays, usually 24 or 25. **Scales:** The body is covered with very small scales posterior to the corselet. **Vertebrae:** Vertebrae number 26–28 precaudal plus 23–27 caudal, for a total of 50–55, more than in any other species of *Sarda*. The soft anatomy and osteology were described and illustrated by Collette and Chao (1975). **Color:** Body turquoise green to steely blue above, with a silvery hue on the lower parts. Upper part of sides marked with 7–20 narrow dark bluish bands. These dorsal stripes are oblique, with a greater angle than in other species of *Sarda*, and do not extend onto the belly. Juveniles and feeding adults may be transversely barred with 10–12 dark blue bars, but these bars usually disappear before maturity.

GEOGRAPHIC RANGE: Atlantic Bonito live along the tropical and temperate coasts of both sides of the Atlantic Ocean, including the Gulf of Mexico and the Mediterranean and Black seas. In the western Atlantic, they have been found as far north as the outer coast of Nova Scotia, but their usual northern limit is Cape Ann, Massachusetts. Atlantic Bonito are uncommon around southern Florida, present in the northern Gulf of Mexico, and apparently absent from most of the Caribbean Sea. They occur in Colombia and Venezuela and are much more common south of the Amazon River to northern Argentina. In the eastern Atlantic, they have been caught near Oslo, Norway, and in Irish waters south through the Gulf of Guinea to Port Elizabeth, South Africa, as well as in the Mediterranean and Black seas.

SIZE: The maximum reported size of Atlantic Bonito in the Black Sea is 85 cm FL and 5 kg. In the western Atlantic, the largest fish was reported to measure 91.4 cm FL and weigh 5.4 kg. They are common to 50 cm FL and about 2 kg. The IGFA all-tackle game fish record is of an 18 lb, 4 oz (8.3 kg) fish caught off Faial Island in the Azores in July 1953. Atlantic Bonito reach at least 4 years of age in the Mediterranean. Several studies report estimated sizes at age for Atlantic Bonito in the Mediterranean and Gulf of Guinea: age 0: 35–37 cm; 1: 38–52 cm; 2: 45–57 cm; 3: 58–65 cm; and 4: 64–71 cm. Growth curves indicate Atlantic Bonito in Argentina may reach 8 years, with mean lengths at age 1: 42 cm; 2: 50 cm, 3: 54 cm; 4: 58 cm; 5: 62 cm; 6: 69 cm; and 8: 72 cm.

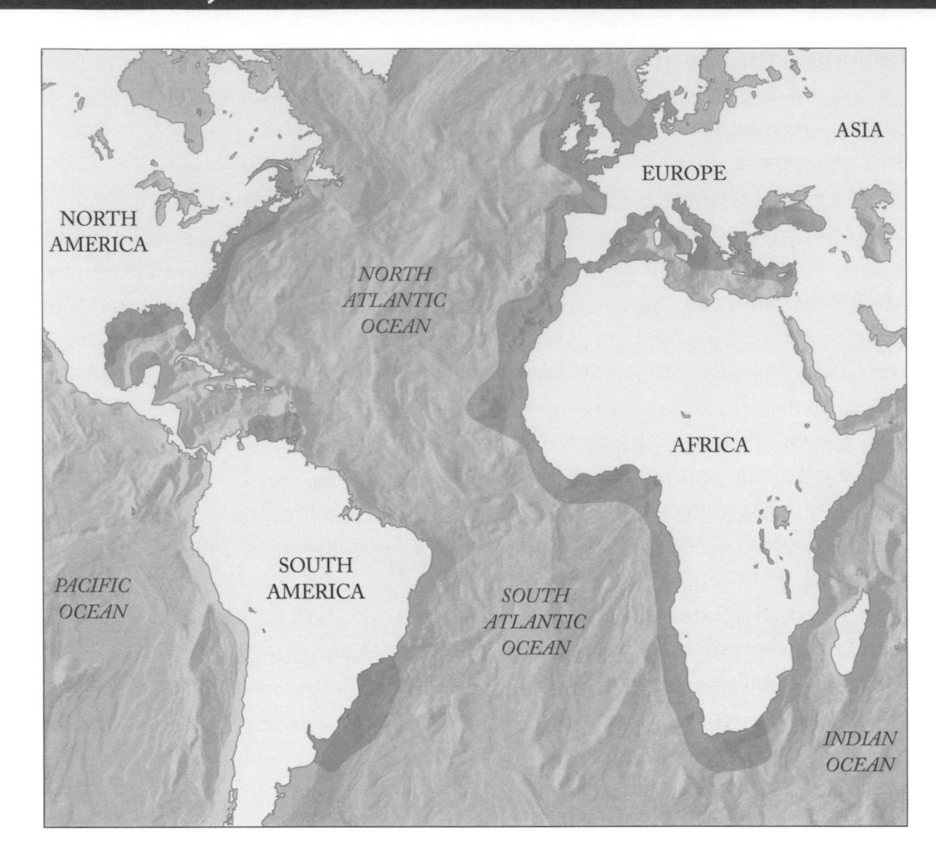

Right: Atlantic Bonito range map.

HABITAT AND ECOLOGY: Atlantic Bonito are a small epipelagic species that occurs in schools. They sometimes enter estuaries and have a depth range of 0–200 m. They are migratory in the eastern Mediterranean Sea, with schools spending the summers in the Black Sea and winters in the Sea of Marmara and the Aegean Sea. They can adapt to gradual but not sudden changes in the environment and may occur in water temperatures between 12 and 27°C and salinities between 14 and 39 ppt, entering estuaries such as Miramichi and the Gulf of St. Lawrence.

FOOD: Adult Atlantic Bonito prey primarily on small schooling fishes such as sardines, anchovies, and mackerel, and also on squids; the choice of species depends on the locality. In the Gulf of Maine, they prey on menhaden, mackerel, and other small fishes. In the Gulf of Mexico, Atlantic Bonito feed on a variety of fishes, squids, and shrimps. They can swallow relatively large prey, and both juveniles and adults are cannibalistic. In the southern Tyrrhenian Sea, the diet is largely herring-like fishes, especially sardines and anchovies. Smaller individuals prefer anchovies and switch to sardines as they increase in size. The diet is similar in the northeast Aegean Sea, with the addition of picarel porgies (Centracanthidae).

REPRODUCTION: Atlantic Bonito are multiple spawners with asynchronous oocyte development. Batch fecundity estimates range from 450,000 to 3 million eggs. Atlantic Bonito mature in their first or second year. Length at first maturity is 35.8–39.5 cm in males and 40.5–42.5 cm in females. They are migratory, and both spawning season and size of maturity vary between populations. There are two main spawning areas, defined both geographically and temporally, within the Mediterranean. The first area is in the western Mediterranean and includes the Alboran Sea, waters south of the Balearic Islands, and waters off the Algerian coast where spawning occurs during early summer. The second area is at the eastern end of the Mediterranean and includes the coastal waters of the Sea of Marmara and the Black Sea, with spawning occurring between May and August with a maximum in June and July. In most parts of the Mediterranean, they spawn between May and July but off Algeria, spawning extends from March to May. In the eastern Atlantic, spawning occurs from December to June, including peaks in January and April off Dakar, and from June to July in Moroccan waters. In the western North Atlantic, Atlantic Bonito spawn in June and July. Gonad development uses lipids derived mostly from the liver, but the lipid contents of the flesh, particularly in females, declines during the spawning season, so the edible part of the fish has a lower nutritive value.

EARLY LIFE HISTORY: The eggs of Atlantic Bonito are pelagic, spherical, and transparent, 1.15–1.57 mm in diameter. They have a variable number of oil globules, 1–9. Incubation takes about 1–1.5 days and the hatch size is about 4 mm. Length at flexion is 6.4 mm. Larvae differ from other scombrids in having a large pigment spot on the caudal peduncle and a supraoccipital crest. There are illustrations of seven larvae 3.7–12.0 mm notochord length in Richards. Juveniles 5.7–8.4 cm were collected in the Zrmanja River estuary, one of the main nursery grounds for pelagic species in the eastern Adriatic Sea.

STOCK STRUCTURE: Significant genetic differences have been demonstrated between Mediterranean and eastern North Atlantic populations of Atlantic Bonito. Genetic differences have also been noted between populations in the Mediterranean Sea, with the Black Sea and Marmara Sea populations comprising one genetic unit, the Aegean Sea and Mediterranean coast of Turkey populations a second unit, and the Adriatic Sea population another genetically distinct unit. Slight morphometric and meristic differences have been reported between northern and southern populations of Atlantic Bonito in the western Atlantic, with those off the coast of Argentina having fewer vertebrae, gill rakers, and fin rays.

FISHERIES INTEREST: Atlantic Bonito are caught in a variety of fisheries throughout most of their range. The species is particularly important in the Mediterranean and Black seas, where they are taken by trap net, ring net, gill net, trammel net, purse seine, beach seine, and hook and line. Reported landings of Atlantic Bonito have generally fluctuated between 15,000 and

50,000 mt since the early 1950s, although catches spiked at 80,000 mt in 2005. The majority of catches come from the Mediterranean Sea, and of the many countries that report landings from this area, Turkey has the largest catches. Fishing in the Black Sea peaks between May and October, while in the Mediterranean it varies from area to area and in some locations may extend throughout the year. Fishing in the eastern tropical Atlantic takes place between October and May, while it extends throughout the year off Morocco. In the Bay of Biscay, the season is much shorter, occurring from mid-April to mid-May; however, Spanish vessels may extend their operation through November. Peak fishing of the Spanish fleet around the peninsula is in late spring and in fall. In the western North Atlantic, Atlantic Bonito are taken from the Gulf of Maine to Venezuela, including the Caribbean Sea. Landings are reported by several nations, with the largest catches reported by Mexico. The fishing season occurs between June and October in the Gulf of Maine and over a longer period in warmer waters. Argentina reports the largest landings of Atlantic Bonito in the western South Atlantic, where the fishing season occurs in the southern summer. Landings begin in the last week of January, increasing up to March and then decreasing sharply. The Argentine bonito fleet caught Atlantic Bonito but no other scombrids the most often, Atlantic Bonito with Skipjack Tuna the next most often, and Atlantic Bonito with Little Tunny the least often.

THREATS: The majority of Atlantic Bonito landed in Turkey are immature. Should the fishery on small fish increase in scope, this stock may be unable to renew or sustain itself long term. This species was listed as Least Concern on the IUCN Red List.

CONSERVATION: The minimum landing size for Atlantic Bonito in Turkey is 25 cm. This is a very widespread species that is fairly fast growing and abundant in many areas. It has been caught in both commercial and recreational fisheries for a long period with no evident overall population declines.

REFERENCES: Ateş et al. 2008; Bloch 1793; Bloch and Schneider 1801; Bonaparte 1831; Boschung 1966; Campo et al. 2006; Cengiz 2013; Collette 2002; Collette and Chao 1975; Collette and Nauen 1983; Collette et al. 2011; Cuvier 1829; Cuvier and Valenciennes 1832; Demir 1963, 1980; Fletcher et al. 2013; Gill 1862; Gündoğdu and Baylan 2017; Hansen 1987, 1988a, 1988b, 1989; IGFA 2018; Jordan and Gilbert 1882; Kahraman et al. 2014; Pallas 1814; A. Postel 1955; Quigley and Flannery 1992, 2008; Rey et al. 1984, 1986; Rafinesque 1810; Rodriguez-Roda 1966, 1981; Rondelet 1554; Sábates and Recasens 2001; Turan 2015; Turan et al. 2015;Türgan 1958; Valeiras et al. 2008; Valenciennes 1844; Viñas et al. 2004, 2010; Yoshida 1980; Zaboukas and Megalofonou 2007; Zaboukas et al. 2006; Zorica and Sinovčić 2008.

Blue Mackerel

Scomber australasicus Cuvier, 1832

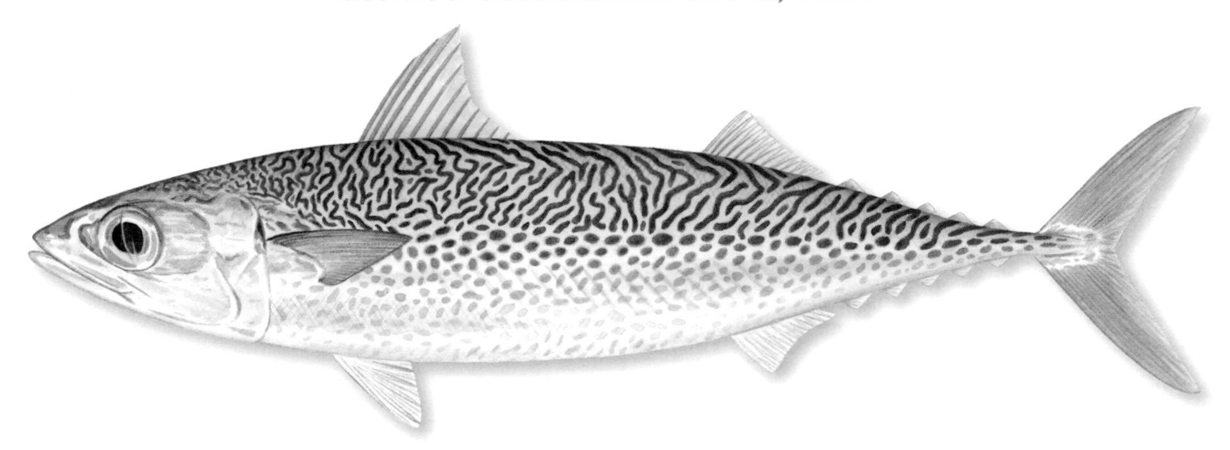

COMMON NAMES: English – Blue Mackerel, Spotted Mackerel, or Spotted Chub Mackerel
French – Maquereau tacheté
Spanish – Caballa pintoja

ETYMOLOGY: Described as *australasicus* by the great French anatomist and ichthyologist Georges Cuvier (1832:49) from specimens from "la Nouvelle-Hollande," as Australia was then known, in the Southern, or Austral, Hemisphere.

SYNONYMS: *Scomber australasicus* Cuvier in Cuvier and Valenciennes, 1832; *Scomber tapeinocephalus* Bleeker, 1854; *Scomber antarcticus* Castelnau, 1872; *Pneumatophorus tapeinocephalus* (Bleeker, 1854); *Pneumatophorus japonicus tapeinocephalus* (Bleeker, 1854); *Scomber indicus* Abdussamad, Sukumaran, and Ratheesh, 2016

TAXONOMIC NOTE: Molecular analyses confirm separation of the Blue Mackerel from the Chub Mackerel. The population of Blue Mackerel in the Red Sea and the Arabian Sea formerly considered to be *Scomber japonicus* was re-identified as *Scomber australasicus*. This population was described as a separate species, *Scomber indicus*, by Abdussamad et al. (2016) but is considered as a subspecies here.

FIELD MARKS:
1 The first dorsal fin has 10–13 spines and is separated from the second dorsal fin by a length approximately equal to length of the first dorsal groove. Other mackerels have fewer dorsal spines (9 or 10), and their interdorsal space is greater than the length of the dorsal groove.
2 Blue to turquoise with black wavy lines dorsally; belly is white and spotted.

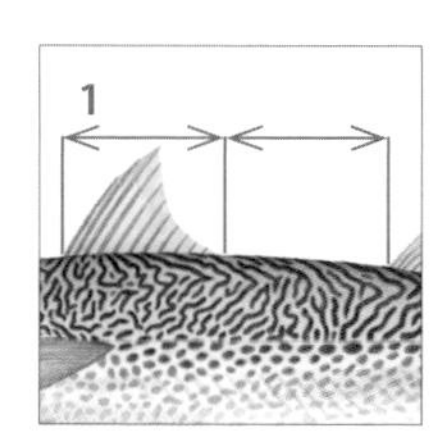

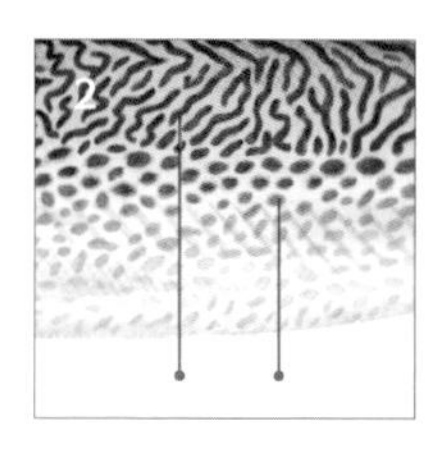

UNIQUE FEATURE: Although originally described in 1832, the validity of Blue Mackerel as a distinct species was not confirmed until 1998.

DIAGNOSTIC FEATURES: Body elongate, rounded, and covered with small cycloid scales. The snout is pointed. Adipose eyelids cover the anterior and posterior margins of the eye. **Teeth:** The teeth in the upper and lower jaws are small, conical, and clearly marked with crenulations. Teeth are also present on the vomer and palatine bones. The palatine teeth are in single or double rows but when double, the rows are close together. **Gill rakers:** The gill rakers are shorter than the gill filaments and are barely visible through the open mouth. There are 25–35 gill rakers on the lower limb of the first arch. **Fins:** The first dorsal fin has 10–13 spines, and the space between the end of the first dorsal-fin groove and the origin of the second dorsal fin is approximately equal to the length of the groove. The origin of the anal fin is clearly more posterior than the origin of the second dorsal fin. The anal-fin spine is separate from the rest of the fin. **Swim bladder:** A swim bladder is present. **Vertebrae:** There are 14 precaudal vertebrae and 17 caudal vertebrae, for a total of 31 vertebrae. The first haemal spine is posterior to the first interneural process. There are 15–20 interneural bones under the first dorsal fin. **Color:** The back is steel blue to turquoise and crossed by oblique lines that zigzag and undulate. The belly is pearly white and marked with thin, wavy, broken lines and spots that look like speckling in places.

GEOGRAPHIC RANGE: Blue Mackerel are present in several geographically isolated populations in the Red Sea and Arabian Sea, the western Pacific from the southern Kuril Islands, China, and Japan, south to Australia and New Zealand, east to the Hawaiian Islands, and into the eastern tropical Pacific in the Revillagigedo Islands.

SIZE: Maximum size is 40 cm FL. The IGFA all-tackle angling record for Blue Mackerel is of a 2.14 kg (4 lb, 11 oz) fish caught off the Hibiscus Coast of New Zealand in November 2011.

HABITAT AND ECOLOGY: Blue Mackerel are pelagic and oceanodromous in coastal waters and also in oceanic waters to depths of 300 m. They are found in large schools near inshore reefs off the coast of New South Wales, Australia. They school by size, and schools may include jack mackerels and Pacific sardines. Around Taiwan, Blue Mackerel migrate from the north to the northeast area for overwintering and to the southwest area for spring spawning. Predators include billfishes and cetaceans such as the pantropical Spotted Dolphin.

FOOD: On the southeastern Australian shelf, Blue Mackerel feed primarily on pelagic invertebrates such as ascidians, pyrosomes, and salps, with fishes comprising less than 40% of the diet.

REPRODUCTION: Blue Mackerel first reach maturity in New Zealand waters at an age of 3 years and a length of 28 cm, while in southern Australia 50% of

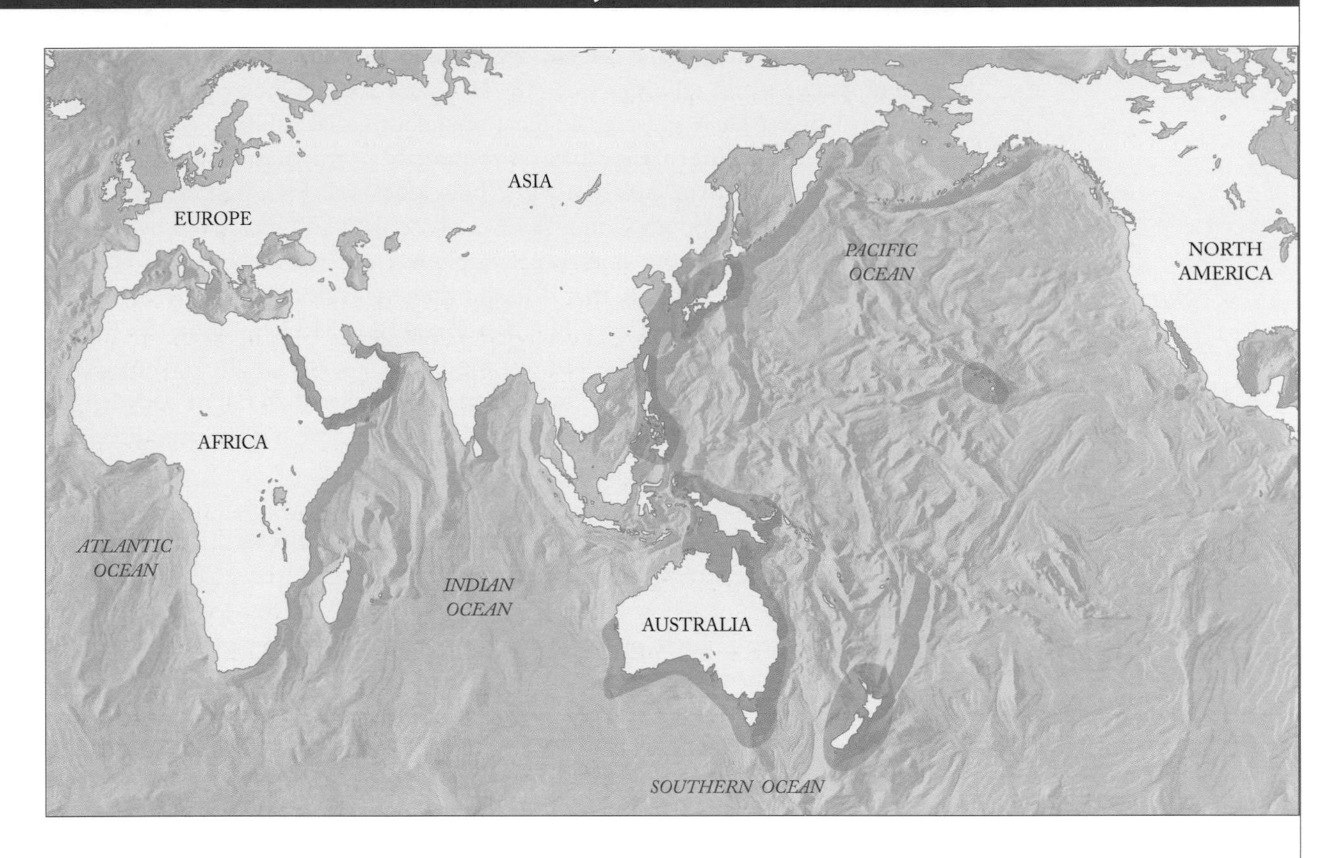

Above: Blue Mackerel range map.

of males and females mature at 24 and 29 cm FL, respectively. They are serial spawners with asynchronous oocyte development and indeterminate fecundity. Blue Mackerel spawn between November and April off southern Australia and between July and October off eastern Australia. In southeastern Australia, the outer shelf is a preferred spawning area in waters 100–125 m deep with mean temperatures of 19–20°C. Mean spawning frequencies ranged from 2 to 11 days off southern Australia. Batch fecundity increased with fish size, and the mean batch size was 69,894 oocytes. Mean batch fecundity in southern Australia was 52,182 eggs, while in eastern Australia it was 22,085 eggs.

Blue Mackerel spawn in the central and southern parts of the East China Sea and the southern coastal area of Kyushu in February, March, and April and the central and southern parts of the sea and the area west of Kyushu in May. Mature Blue Mackerel were observed in the East China Sea in the areas with sea surface temperatures of 17–25°C.

EARLY LIFE HISTORY: Pelagic eggs of Blue Mackerel are spherical and 1.05–1.30 mm in diameter with a smooth chorion, small perivitelline space, and a single oil globule, 0.26–0.31 mm in diameter, that becomes pigmented mid-stage in development. Identification of ethanol-preserved eggs was confirmed by genetic techniques. Photographs of two eggs 1.20 mm in diameter, mid-stage, and 1.25 mm in diameter, prehatching, were published by Neira and

Keane. Most eggs and larvae were collected at stations where mean temperatures and salinities were 18–21°C and 35.35–35.60 ppt, respectively. Larval Blue Mackerel were the most abundant larval fish species in a survey of larval fishes in Yen-Liao Bay on the northeastern coast of Taiwan, much more abundant than larval Pacific Chub Mackerel. Five postlarvae 3.8–9.2 mm SL from the South China Sea were described and illustrated by Ozawa. They possess distinctive dorsal melanophores just below the origin of body muscle on their shoulders that distinguishes them from co-occurring Pacific Chub Mackerel. In the East China Sea, larval Blue Mackerel showed a higher and narrower habitat temperature range than Pacific Chub Mackerel, 20–23°C versus 15–22°C. Larvae of both species smaller than 6 mm body length feed mainly on copepod nauplii and with growth, calanoid copepodites and appendicularians become more important as prey.

STOCK STRUCTURE: Although once considered a subspecies of the Pacific Chub Mackerel, several morphological and molecular studies have confirmed that the Blue Mackerel is a distinct species, genetically separated from Pacific Chub Mackerel. There are at least four geographically separate populations of Blue Mackerel: 1) western North Pacific: Japan, China, and Taiwan; 2) western South Pacific: Australia and New Zealand; 3) eastern tropical Pacific: Revillagigedo Islands, Mexico; and 4) the Red Sea and western Arabian Sea. They have also been recorded from the Hawaiian Islands. Three populations, western North Pacific, western South Pacific, and Red Sea–western Arabian Sea, are genetically distinct, but the population in the Revillagigedo Islands does not differ from that in the western North Pacific. The Red Sea–western Arabian Sea population was described as a separate species, *Scomber indicus*, by Abdussamad et al. but is considered as a subspecies here pending a more complete analysis of the entire species.

In Japan, two stocks of Blue Mackerel are recognized for management purposes: the Pacific stock and the East China Sea stock. There has been a debate regarding stock structure of Blue Mackerel stock around Taiwan. Analyses of genetic characters, morphological characters, and some catch data have supported the presence of two stocks, while other genetic studies and analyses of catch data were consistent with a single stock. Genetic, parasite, and otolith microchemistry data suggest the presence of stock structure within the Australia–New Zealand population.

FISHERIES INTEREST: There are important commercial fisheries for Blue Mackerel in Japan, China, Taiwan, Korea, Australia, and New Zealand. Landings reported to the FAO have typically ranged from 10,000–20,000 mt since the early 1990s, with a reported catch of 21,280 mt in 2015. Both the Blue Mackerel and the Pacific Chub Mackerel are abundant in the East China Sea where they are fished commercially, mainly with purse seines. Purse seining has replaced handlining and become the main fishing method for mackerel in Taiwan since 1977. Fishing grounds are concentrated along the northeast coast of Taiwan in autumn and winter and adjacent to Prata Island, southwest Taiwan, during spring.

Blue Mackerel are exploited by recreational and commercial fishermen off the coast of New South Wales, Australia. Since the late 1980s, they have become targets of an expanding purse seine fishery from small boats targeting schools of fish in inshore waters. Seventy percent of the fishery was based on fish aged 1 year, with most of the rest being fish in their second or third years. It is a key species for recreational anglers throughout Australia, targeted both as live bait for larger pelagic species and as a table fish.

In New Zealand, Blue Mackerel are one of the seven main species targeted by the purse seine fleet and are also taken by midwater trawl vessels. Total annual reported landings increased rapidly from the 1989–90 to the 1992–93 fishing year and have fluctuated between about 6,000 and 10,000 mt subsequently. Reported catches peaked in 1991–92 at about 15,000 mt, of which about 70% was taken by purse seine vessels, mostly from the Bay of Plenty and east Northland.

THREATS: In the western North Pacific, estimated spawning stock biomass for at least one stock is increasing, while the other stock is fluctuating but relatively stable. In some regions of the species' range, landings are not identified to species, and in others there is a lack of reporting from some fisheries. Blue Mackerel are listed as Least Concern on the IUCN Red List. However, more information on the status of this species' population in other parts of its range is recommended.

CONSERVATION: In Japan, Blue Mackerel are managed as a single unit under a total allowable catch system. In Japan and the Tsushima Current, spawning stock biomass for the Pacific Stock of Blue Mackerel has been estimated to be steadily increasing since 1995 from 50,000 to 150,000 mt, with a peak of 300,000 mt in 2006. Estimated spawning stock biomass for the East China Sea fluctuated from 40,000 to 80,000 mt from 1992 to 2007. Blue Mackerel in the New Zealand Exclusive Economic Zone are managed in five separate quota management areas. In Australia, the species is managed with a total allowable catch. In New Zealand and Australia, there are recreational bag limits and catch limits for all mackerel species.

SELECTED REFERENCES: Abdussamad et al. 2016; Baker and Collette 1998; Bleeker 1854; Bulman et al 2001; Castelnau 1872; Catanese et al 2010; Cheng et al. 2011; Collette and Nauen 1983; Collette et al. 2011; Cuvier and Valenciennes 1832; Devine et al. 2009; Infante et al. 2007; Manning et al. 2006, 2007; Matsui 1967; Morrison et al. 2001; Neira and Keane 2008; Ozawa 1984; Rogers et al. 2009; Sassa and Tsukamoto 2010; Sassa et al. 2008; Scoles et al. 1998; Stewart and Ferrell 2001; C. H. Tzeng et al. 2009; T. D. Tzeng 2004, 2007; Wang et al. 2003; Ward et al. 2009; Yukami et al. 2009.

Atlantic Chub Mackerel

Scomber colias Gmelin, 1789

COMMON NAMES: English – Atlantic Chub Mackerel
French – Maquereau Blanc
Spanish – Estornino or Caballa

ETYMOLOGY: Described by the German naturalist Johann Friedrich Gmelin (1789:1329) using the Latin name for mackerel, *colias*, for the species.

SYNONYMS: *Scomber japonicus* (non Houttuyn, 1782); *Scomber colias* Gmelin, 1789; *Scomber scomber lacertus* Walbaum, 1792; *Scomber pneumatophorus* Delaroche, 1809; *Scomber macrophthalmus* Rafinesque, 1810; *Scomber grex* Mitchill, 1814; *Scomber capensis* Cuvier in Cuvier and Valenciennes, 1832; *Scomber maculatus* Couch, 1832; *Scomber undulatus* Swainson, 1839; *Scomber gracilis* Swainson, 1839; *Scomber dekayi* Storer, 1855; *Pneumatophorus colias* (Gmelin, 1789); *Scomber gigas* Fowler, 1935; *Pneumatophorus japonicus marplatensis* López, 1955

TAXONOMIC NOTE: This species is now recognized as distinct from the Pacific Chub Mackerel, *Scomber japonicus* (Collette 1999; Infante et al. 2007). The three species can be distinguished based on morphological and genetic characters.

FIELD MARKS:
1 The belly is pearly white and marked by spotting or wavy broken lines, not unmarked as in the Pacific Chub Mackerel.
2 The back is blue to turquoise with black wavy lines.

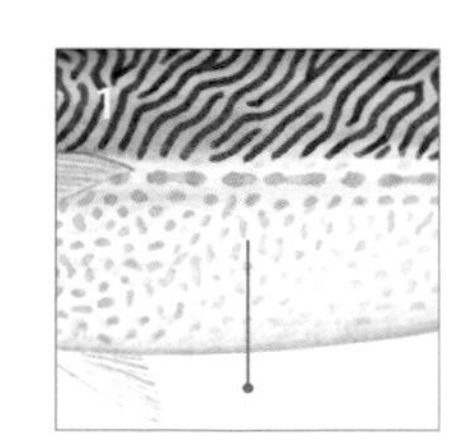

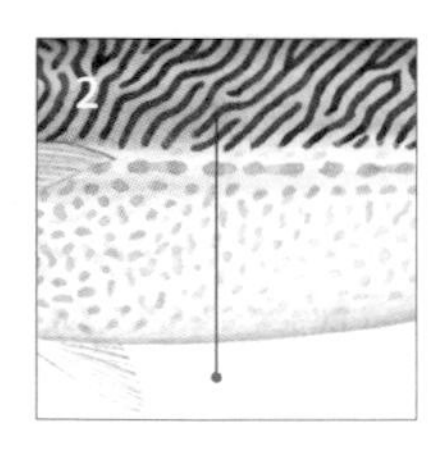

UNIQUE FEATURE: This species of mackerel has been described as a new species 12 times, more than any other mackerel.

DIAGNOSTIC FEATURES: Body elongate, rounded, and covered with small cycloid scales. The snout is pointed. Adipose eyelids cover the anterior and posterior margins of the eye. **Teeth:** The teeth in the upper and lower jaws are small, conical, and clearly marked with crenulations. Teeth are also present on the vomer and palatine bones. The palatine teeth are in single or double rows but when double, the rows are close together. **Gill rakers:** The gill rakers are shorter than the gill filaments and are barely visible through the open mouth. There are 25–35 gill rakers on the lower limb of the first arch. **Fins:** The first dorsal fin has only 9 or 10 spines, and the distance from the last dorsal-fin spine to the origin of the second dorsal fin is less than the distance between the first and last dorsal spines. The origin of the anal fin is opposite to the origin of the second dorsal fin or somewhat more posterior. The anal-fin spine is conspicuous, clearly separate from the rest of the fin. There are 17–21 pectoral fin rays. **Swim bladder:** A swim bladder is present. **Vertebrae:** There are 14 precaudal vertebrae and 17 caudal vertebrae, for a total of 31. The first haemal spine is posterior to the first interneural process. There are 12–15 interneural bones under the first dorsal fin. Several molecular analyses confirm separation of the Atlantic Chub Mackerel from the Pacific Chub Mackerel. **Color:** The back is steel blue to turquoise and crossed by oblique lines that zigzag and undulate. The belly is pearly white and marked by spotting or wavy broken lines.

GEOGRAPHIC RANGE: The Atlantic Chub Mackerel lives in the western North and South Atlantic and the eastern Atlantic, including the Mediterranean and Black seas. In the eastern Atlantic, it occurs commonly from the Bay of Biscay to South Africa, including the Canary, Madeira, Azores, and Saint Helena islands. There are seven scattered records of Atlantic Chub Mackerel from Irish waters, the most northerly off the northwest coast of County Donegal. The recent northward expansion of the Atlantic Chub Mackerel off Portugal may be part of a more general northward expansion of "southern" species, possibly linked to global warming. In the western Atlantic, it occurs from outer Nova Scotia and the Gulf of St. Lawrence south to the Bahamas, Florida, the Gulf of Mexico, and Venezuela. In the western South Atlantic, it is found from Brazil south to Uruguay and Argentina.

SIZE: Atlantic Chub Mackerel can reach a total length of 65 cm and a weight of 2.9 kg, and may reach 20 years of age. Otolith analysis of Atlantic Chub Mackerel from the Adriatic Sea indicated ages ranging from 1 to 13 years old. Length-weight studies and mean lengths at age from several different regions have been compared. Atlantic Chub Mackerel grow rapidly and, off Madeira, reach 40% of their asymptotic length during their first year.

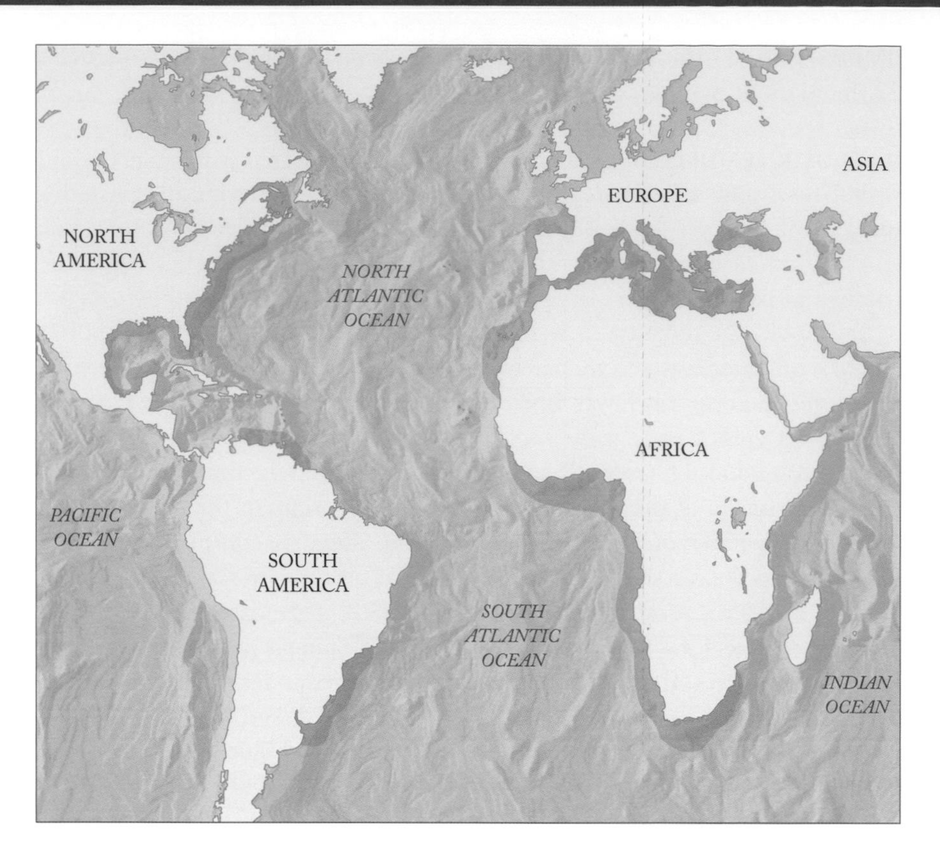

Right: Atlantic Chub Mackerel range map.

HABITAT AND ECOLOGY: The Atlantic Chub Mackerel is a coastal pelagic species, and to a lesser extent it enters epipelagic and mesopelagic environments over the continental slope down to depths of about 400 m. Schooling by size is well developed and initiates at approximately 3 cm. They may also form schools with bonitos, jacks, and clupeids. In the western South Atlantic population, schools migrate from the cooler overwintering shelf waters to the warmer coastal waters to spawn. The migration begins in early October at depths of about 50 m and continues through late January. They are preyed upon by tunas, billfishes, and marine mammals such as Bottlenose Dolphin.

FOOD: Atlantic Chub Mackerel have two strategies for obtaining food—filter feeding and particulate feeding, switching between modes based on the concentration and size of prey. In the North Aegean Sea of Greece, major prey items included fish larvae and chaetognaths. Off the Canary Islands, the diet consisted of 17 taxonomic groups. Small zooplankton, principally copepods, were numerically the most important, followed by appendicularians and mysids, whereas juvenile snipefish were common in the diet of adults. However, by volume, mysids, followed by copepods and small fishes, constituted the bulk of the diet. The diet shifted from small fishes (sardines and atherinids) and copepods in juveniles to mysids and copepods in immatures to mysids and fishes in adults longer than 22 cm. Mysids and euphausiids were of particular importance off the Canary Islands. In the Adriatic Sea, nine taxonomic groups were found in the stomachs of Atlantic Chub Mackerel of which decapods, amphipods, and copepods were the most common prey, and the fishes consumed were mostly eggs, larvae, and small sardines and anchovies. In the Aegean Sea, juveniles (15–20 cm total length) fed mainly on fishes during the summer and autumn, thaliaceans (*Salpa*) during the winter, and planktonic crustaceans (amphipods and copepods) during spring.

In the western South Atlantic off Argentina, the main food item of Atlantic Chub Mackerel is the Argentine Anchovy, but they also consume copepods, cladocerans, euphausiids, and larval decapod crustaceans. A large number of stomachs of adults contained mackerel eggs. On Georges Bank in the western North Atlantic, feeding habits of the Atlantic Chub Mackerel were similar to those for the Atlantic Mackerel in 1896, with both species full of the same species of crustaceans. The few specimens of Atlantic Chub Mackerel from the northwest Atlantic that were examined recently contained copepods, larvaceans, and fishes.

REPRODUCTION: Length at first maturity of Atlantic Chub Mackerel varies with location. In the northern Aegean Sea, length at first maturity was 18 cm TL. Around the Azores, the length at 50% maturity was 27.8 cm TL, much larger than the 18.3 cm FL reported for the Adriatic, where smaller sizes at 50% maturity were observed for males, 16.8 cm TL, than for females, 20.4 cm TL. In Madeiran waters, length and age at first maturity were estimated to be 22.1 cm TL and age 1.05 for males and 21.6 cm TL and age 0.82 for females, with lengths at 50% maturity of 19.8 and 19.9 cm, respectively. Off the Atlantic coast of Morocco, total length at 50% maturity was 22.9 cm for males and 23.0 cm for females.

Spawning takes place from March to August, peaking in June around the Azores, in the northern Aegean Sea, and in the eastern Adriatic. In Madeiran waters, spawning occurs from January to April, reaching a peak in February and March. Off the Canary Islands, the reproductive season extends from November to March, peaking between December and January. Off the Atlantic coast of Morocco, spawning occurs from December to March, with a peak in January. Atlantic Chub Mackerel spawn in winter and early spring off South Africa when northwesterly winds bring in relatively warm water, 14–15°C. Only one spawning area has been detected in the western South Atlantic, off the Mar del Plata coast of Argentina from October to January.

Atlantic Chub Mackerel are batch spawners, and mean batch fecundity off the Atlantic coast of Morocco was 285,704 oocytes, with a range of 77,621 to 465,712. Batch fecundity in the Adriatic ranged from 99,166 to 394,120 oocytes per ovary. Fecundity was estimated at 100,000–400,000 in the western North Atlantic.

EARLY LIFE HISTORY: The egg diameter from western North Atlantic Chub Mackerel was 1.14–1.24 mm, with a single oil globule 0.28–0.32 mm. Length at flexion was about 6 mm. Six larvae, 2.9 mm notochord length to 14.7 mm SL, were illustrated by Richards. Off southern Brazil, they grow 7 cm during their first 2 months of life and 5.5 cm during the following 2 months.

STOCK STRUCTURE: There are at least three main populations of Atlantic Chub Mackerel: western North Atlantic, western South Atlantic, and eastern Atlantic. There are fewer gill rakers, 25–29, on the lower gill arch in the western Atlantic than in the eastern Atlantic, 29–35. Significant genetic differences were noted between Atlantic Chub Mackerel from Florida and Argentina in the western Atlantic, while there was no significant differentiation among samples from the eastern Mediterranean Sea, the Ivory Coast in the Gulf of Guinea, and South Africa. Significant differences in growth were found between Atlantic Chub Mackerel in the western South Atlantic and those in the eastern Atlantic, while no differences in growth rates were observed between samples from the Alboran Sea in the western Mediterranean and the Gulf of Cadiz on the Atlantic coast of Spain. The population in the eastern South Atlantic off Namibia and South Africa might also represent a discrete population, and genetic and morphological data (head size) support the policy of managing northern and southern stocks (north and south of 39° S) of the western South Atlantic population as separate stocks.

FISHERIES INTEREST: Atlantic Chub Mackerel support important commercial fisheries, particularly in the Mediterranean Sea, around Madeira and the Azores, and off the coasts of South Africa and Argentina. The majority of the catch is taken with purse seines, often together with sardines, and they are also caught with midwater trawls, gill nets, traps, beach seines, and trolling lines. Until 1973, catches in the eastern South Atlantic were mainly made by purse seine vessels off the Western Cape of South Africa but subsequently by midwater or demersal trawlers, primarily off Namibia. The fishery off Mar del Plata, Argentina, consists of small purse seiners that provide the raw material for the canning industry. Atlantic Chub Mackerel were tremendously abundant in the waters of the Gulf of Maine toward the end of the eighteenth century and the early nineteenth century through 1830. Then they practically disappeared from the Atlantic coast of the United States between 1840 and 1850. Great schools were found on Georges Bank in 1909, but there have been few recent records of such abundance since then.

Worldwide reported landings for this species show increasing catches from 30,900 mt in 1950 to 57,400 mt in 1960. During the following decade, there was a dramatic increase in the fishery, with reported catches steadily increasing to 391,000 mt in 1970. Over the next 40 years, catches varied from as low as 170,000 mt to a high of 399,000 mt. Since 2010, catches have increased from 315, 241 mt to 472,275 in 2015. The majority of the catch is from the eastern Atlantic. The reported catches may be an underestimate, as many countries do not report their catches, and there may be a high discard rate of small fish by some fleets targeting crustaceans. In some areas, such as off Portugal and the Atlantic coast of Morocco, catches of Atlantic Chub Mackerel may sometimes be combined with Atlantic Mackerel.

THREATS: Since 2003, there has been at least a 50% decline in catches of Atlantic Chub Mackerel in the eastern Atlantic, although there is no information on current effort. While considered fully exploited, there is no evidence of a long-term decline in SSB, and the species is listed as Least Concern on the IUCN Red List. However, there are some indications of regional declines, and the species should be monitored closely. Within the Mediterranean Sea current exploitation levels are intense, and population declines approaching 30% are suspected. Recent decreases in population trends are coincident with increases in Atlantic Mackerel. Because of the steady decline in catches over the past 20 years, this stock is listed as Near Threatened by the IUCN.

CONSERVATION: A 2009 assessment determined that the stock is fully exploited in the eastern Atlantic. A similar conclusion was reached by an FAO-sponsored assessment of Atlantic Chub Mackerel off western Africa in 2015. Within the Mediterranean, Atlantic Chub Mackerel are common and locally abundant and have had fairly high, fluctuating catches. However, there has been a steady decline in landings since the 1980s, when landings peaked at 41,200 mt. Since 2010, reported landings in the Mediterranean Sea have ranged from 10,000–15,000 mt. A targeted management plan for this species is needed.

SELECTED REFERENCES: Baird 1977; Bowman et al. 2000; Carvalho et al. 2002; Castro Hernández and Santana Ortega 2000; Catanese et al. 2010; Cengiz 2012; Cheng et al. 2011; Čikeš Keĉ and Zorica 2012, 2013; Čikeš Keĉ et al. 2012; Collette 1999, 2002; Collette and Nauen 1983; Collette et al. 2011; Couch 1832; Crawford and De Villiers 1984; Cuvier and Valenciennes 1832; Delaroche 1809; Fowler 1935; Gmelin 1789; Infante et al. 2007; Karachle 2017; López 1955; Lorenzo and Pajuelo 1996; Martins et al. 2013; Matsui 1967; Mitchill 1814; Navarro et al. 2012; Perrotta 1993; Perrotta et al. 2001; Quigley and Mullins 2004; Rafinesque Schmaltz 1810; Richards 2006; Roldán et al. 2000; Scoles et al. 1998; Sever et al. 2006; Storer 1853; Swainson 1839; Techetach et al. 2010; Velasco et al. 2011; Walbaum 1792; Zardoya et al. 2004.

Pacific Chub Mackerel

Scomber japonicus Houttuyn, 1782

COMMON NAMES: English – Pacific Chub Mackerel
French – Maquereau Espagnol
Spanish – Macarela Estornino

ETYMOLOGY: The Pacific Chub Mackerel was named *japonicus* by the Dutch naturalist Maarten Houttuyn (1782:331) because his specimens came from Japan.

SYNONYMS: *Scomber japonicus* Houttuyn, 1782; *Scomber scombrus japonicus* Temminck and Schlegel, 1844; *Scomber saba* Bleeker, 1854; *Scomber janesaba* Bleeker, 1854; *Scomber diego* Ayres, 1856; *Pneumatophorus peruanus* Jordan and Hubbs, 1925; *Scomber peruanus* (Jordan and Hubbs, 1925); *Pneumatophorus japonicus* (Houttuyn, 1782); *Scomber japonicus peruanus* (Jordan and Hubbs, 1925)

TAXONOMIC NOTE: The Pacific Chub Mackerel (*Scomber japonicus*) has an Indo-Pacific distribution and is closely related to the Atlantic Chub Mackerel, *S. colias,* and the Blue Mackerel, *S. australasicus.*

FIELD MARKS:

1 The first dorsal fin has only 9 or 10 spines instead of 11–13; the distance between the last dorsal spine and the origin of the second dorsal fin is less than the distance between the first and last dorsal-fin spines.
2 The back is blue to turquoise with black wavy lines; belly is pearly white and unmarked.

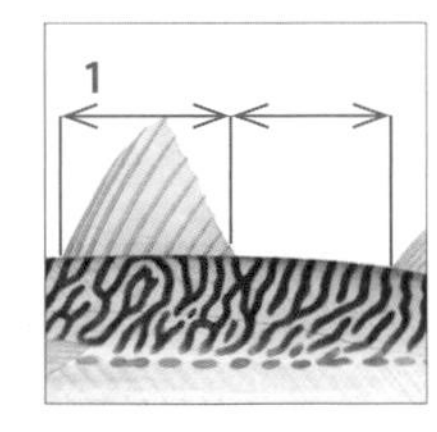

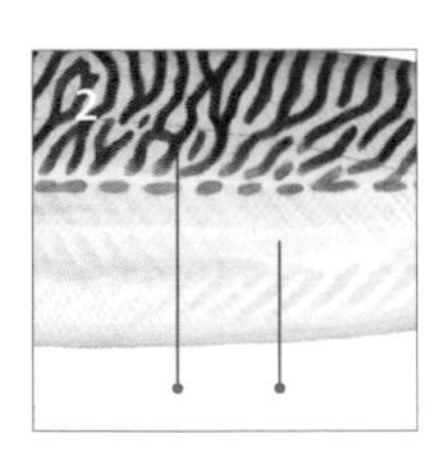

UNIQUE FEATURE: Pacific Chub Mackerel support the largest mackerel fishery in the world, and it is one of the largest single-species fisheries.

DIAGNOSTIC FEATURES: Body elongate, rounded, and covered with small cycloid scales. The snout is pointed. Adipose eyelids cover the anterior and posterior margins of the eye. **Teeth:** The teeth in the upper and lower jaws are small, conical, and clearly marked with crenulations. Teeth are also present on the vomer and palatine bones. The palatine teeth are in single or double rows but when double, the rows are close together. **Gill rakers:** The gill rakers are shorter than the gill filaments and are barely visible through the open mouth. There are 25–35 gill rakers on the lower limb of the first arch. **Fins:** The first dorsal fin has only 9 or 10 spines, and the distance from the last dorsal-fin spine to the origin of the second dorsal fin is less than the distance between the first and last dorsal spines. The origin of the anal fin is opposite the origin of the second dorsal fin or somewhat more posterior. The anal fin spine is conspicuous, clearly separate from the rest of the fin. **Swim bladder:** A swim bladder is present. **Vertebrae:** There are 14 precaudal vertebrae and 17 caudal vertebrae, for a total of 31 vertebrae. The first haemal spine is posterior to the first interneural process. There are 12–15 interneural bones under the first dorsal fin. **Color:** The back is steel blue to turquoise and crossed by oblique black lines that zigzag and undulate. The belly is pearly white and unmarked.

GEOGRAPHIC RANGE: Pacific Chub Mackerel is widespread in the Pacific Ocean. In the eastern Pacific, it ranges from Alaska to the Gulf of California and central Mexico, including the Revillagigedo Islands, where it co-occurs with Blue Mackerel, *S. australasicus*. It also occurs from Panama to southern Chile (45°41′ S), including the Cocos and Malpelo islands and the Galapagos Archipelago. The population from the Red Sea and northern Indian Ocean (the Gulf of Aden and Oman) previously identified as Pacific Chub Mackerel, *S. japonicus*, are Blue Mackerel, *S. australasicus*.

SIZE: The largest Pacific Chub Mackerel recorded was 24.8 in (63 cm) and weighed about 6.3 lb (2.9 kg). The IGFA all-tackle game fish record is a 4 lb, 12 oz (2.17 kg) fish caught off Guadalupe Island, Mexico, in June 1986. Pacific Chub Mackerel have an average longevity of approximately 7 years along the Peruvian coast, although longevity can be as high as 14 years based on size-frequency growth studies in the United States and Mexico. In waters off Taiwan, Pacific Chub Mackerel and Blue Mackerel grow fastest in their first 2 years, after which the growth rate slows down.

HABITAT AND ECOLOGY: Pacific Chub Mackerel are a coastal pelagic species found in warm and temperate waters over the continental shelf and slope from the surface to depths of 300 m. They occur in waters with temperatures of 5–30°C and at relatively high salinity, between 32 and 35.7 ppt. Pacific Chub Mackerel stay near the bottom during the day and go up to surface water at night to feed. They school by size, starting at lengths of approximately 3 cm.

Pacific Chub Mackerel may also form mixed-species schools with bonitos, jacks, and sardines. They are preyed on by tunas, billfishes, dolphinfishes, sharks, sea lions, and sea birds.

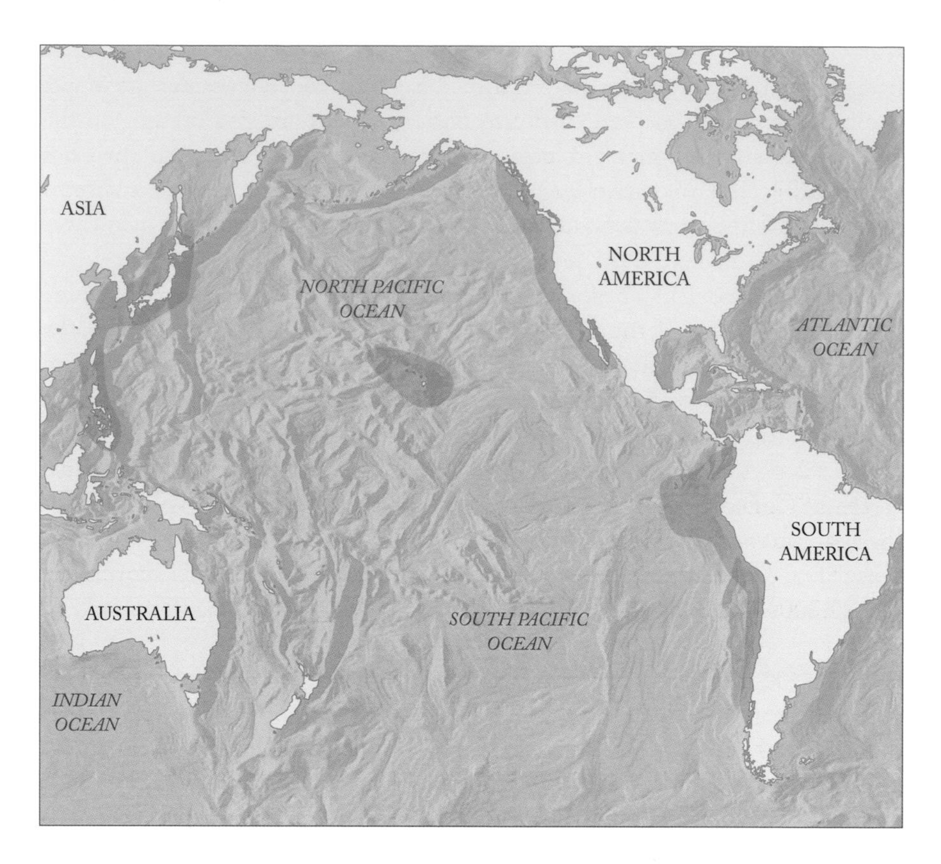

Right: Pacific Chub Mackerel range map.

FOOD: Pacific Chub Mackerel are facultative carnivores that feed on zooplankton, squids, and small fishes, particularly anchovies and sardines. Adults have a small gill-raker gap and will filter feed when they encounter prey of appropriate size at high densities. In the Gulf of California, analysis of stomach contents showed the most meaningful prey categories were fish larvae (anchovies and sardines), crustacean larvae (brachyurans and stomatopods), and calanoid copepods. Similarly, in Chile, they feed on zooplankton and small anchovies. Off the coast of Peru, the diet of Pacific Chub Mackerel consisted of zooplankton, phytoplankton, and some fishes, particularly Peruvian anchoveta. Three groups of invertebrates were most important, heteropod mollusks, copepods, and euphausiids, constituting two-thirds of the total food consumed, and of these, euphausiids were 35.5% of the food by weight. Euphausiids were also an important component of the diet of Pacific Chub Mackerel from the western North Pacific.

REPRODUCTION: Pacific Chub Mackerel normally spawn in the first half of the year in the northern hemisphere and in the second half of the year in the southern hemisphere, but spawning may take place throughout the year in equatorial waters. In two regions of Chile, age of first maturity is approximately 2 to 4 years, and size at first maturity was 25.1 to 29.6 cm FL for males and 25.8 to 28.1 cm FL for females. Generation length is estimated to be 4 to 6 years.

Pacific Chub Mackerel are indeterminate batch spawners. In the eastern North Pacific, spawning takes place from British Columbia south to Punta Abreojos, Mexico. In the Southern California Bight, the average female spawned 8.8 times during a 101-day sampling period from April to July at average intervals of 1.2 days. The average batch fecundity was 68,400 oocytes.

In the western North Pacific, Pacific Chub Mackerel spawn in the central and southern parts of the East China Sea and the area west of Kyushu in February, March, and April. They spawn in the central part of the East China Sea, the area west of Kyushu and Tsushima Strait in May, and in Tsushima Strait and the western part of the Sea of Japan in June. Mature Pacific Chub Mackerel were observed in the East China Sea at sea surface temperatures of 15–22°C. Females larger than 275 mm FL (age 1 and older) were mature. In Japanese waters, females spawned actively from 2200 to 2400 hours about every 5.7 days, although some spawned every day. The average batch fecundity was 89,200 oocytes and was significantly correlated with condition factor.

EARLY LIFE HISTORY: Eggs of Pacific Chub Mackerel are transparent spheres 0.92–1.28 mm in diameter, with a 0.25–0.32 mm oil globule. Incubation times of laboratory-spawned fish ranged from 33 hours at 23°C to 117 hours at 14°C. Larval Pacific Chub Mackerel lack the distinctive dorsal melanophores just below the origin of body muscle on their shoulders that are present in the co-occurring Blue Mackerel in the western North Pacific. The frequency of occurrence of internal melanophores on the nape was significantly lower in larval Pacific Chub Mackerel than larval Blue Mackerel from 2.5–6.5 mm SL, but the difference may not be reliable for species identification.

Larval Pacific Chub Mackerel averaged 3.1 mm SL at hatching. First feeding occurred 46 hours after hatching. The 50% threshold for onset of feeding occurred at about 2 days after hatching, when the larvae were 3.6 mm SL and the eyes were fully pigmented. Growth was slow for the first 10–15 days, followed by a rapid acceleration through metamorphosis. Up to 110 days old, Pacific Chub Mackerel from the Gulf of California grew on average 1.33 mm per day. The caudal fin formed at larval lengths of 5–7 mm SL, earlier than the other fins. Notochord flexion was complete by 9 mm SL. Swimming speeds ranged from 1.3 standard lengths per second for first-feeding larvae to 3.8 standard lengths per second for larvae at metamorphosis. Larger juveniles, 140–240 mm FL, had maximum swimming speeds of 70–120 cm/s (3.8–5.8 FL/s). The chief food items of sea-caught larvae were larval stages of copepods. Gut contents of larvae changed with growth from 4 mm to 9 mm; the proportion of Appendicularia and *Oithona* increased while the proportions of crustacean eggs and *Evadne* decreased. Early juveniles that were plankton-fed did not show any schooling behavior by day 22 (at 25 mm SL) whereas those that were fish-fed completed schooling behavior on day 18, at 19 mm SL.

In the East China Sea, larval Pacific Chub Mackerel showed a lower and broader habitat temperature range than Blue Mackerel, 15–22°C versus 20–23°C. Larvae of both species smaller than 6 mm SL feed mainly on copepod nauplii and with growth, calanoid copepodites and appendicularians become more important as prey.

STOCK STRUCTURE: There are three geographically isolated populations of Pacific Chub Mackerel: western North Pacific, eastern North Pacific, and eastern South Pacific, and there is evidence for stock structure within each of these three populations.

Eastern North Pacific: As with the western North Pacific, a variety of stock structure scenarios have been proposed for the eastern North Pacific. Some studies support the existence of two stocks along the Pacific coast of the United States and Mexico: one from southern California to Vizcaino Bay and the other in the Gulf of California, while others propose three stocks: one in the Gulf of California, one in the vicinity of Cabo San Lucas, and one extending along the Pacific coast north of Punta Abeojos, Baja California.

Eastern South Pacific: Life history parameters support the existence of two stocks off southeastern Chile.

Western North Pacific: Genetic analyses indicate the presence of multiple stocks, but there is disagreement regarding the number and range of the stocks. One analysis suggested separate stocks in the East China and South China seas: one including the East China Sea, extending to the coast of Hainan Island in the South China Sea, and the other off the southern coast of Hainan Island. Other analyses support the existence of three Chinese stocks: one from the Yellow Sea to Hainan Island, the second off the southern coast of Hainan Island, and the third in the middle of the East China Sea.

FISHERIES INTEREST: Pacific Chub Mackerel support important commercial fisheries throughout their range. They are typically taken with purse seines, often together with sardines or other coastal pelagics, and are also caught with midwater trawls, gill nets, beach seines, traps, and trolled lines. They are marketed fresh, frozen, smoked, salted, and occasionally canned. Pacific Chub Mackerel are also taken in sport fisheries.

Pacific Chub Mackerel support one of the largest commercial fisheries in the world, and as with many coastal pelagic species, there is considerable variation in stock biomass in response to natural environmental fluctuations on a decadal level. Reported landings to FAO increased from approximately 250,000 mt in 1950 to 460,000 mt in 1960 and 1,625,000 mt in 1970 and peaked at 3,103,728 mt in 1978. Catches decreased through the 1980s to 829,844 mt in 1992. Since that time, annual catches have ranged from about 1,300,000 to 2,100,000 mt.

Eastern North Pacific: FAO-reported landings increased from less than 15,000 mt in 1950 to more than 34,000 mt in 1980 and peaked at 78,089 mt in 1999. Since that time, landings have fluctuated between 11,000 and 66,000 mt without trend. During the early part of the time series the majority of landings were from the United States, but for the past 20 years, Mexico has had the largest reported landings.

Eastern South Pacific: Chub Mackerel landings rose dramatically in the 1970s, increasing from 51,800 mt in 1970 to 731,958 mt in 1980 and peaking at 735,807 mt in 1979. Since that time, catches have averaged around 200,000 mt with considerable year-to-year variation. Ecuador, Peru, and Chile have significant harvests that have varied over time in response to changes in stock biomass and distribution.

Western North Pacific: The western North Pacific stock has supported the highest landings of Pacific Chub Mackerel, with current (2015) catches close to 1,500,000 mt. FAO-reported landings increased from just over 250,000 mt in 1950, to 470,000 mt in 1960, to 1,625,000 mt in 1970, peaking at 3,102,728 mt in 1978. Catches decreased somewhat in the 1990s and have ranged between 1,100,000 and 1,800,000 mt since 2000. China, Japan, South Korea, and Taiwan all have substantial fisheries for Pacific Chub Mackerel, with China and Japan having the greatest landings.

THREATS: Natural fluctuations in stock biomass in conjunction with high fishing pressure can result in local depletion. For example, the catch of Pacific Chub Mackerel in the East China Sea decreased in 1997 and stabilized at a low level after 2000. In some areas, the age distribution has been severely truncated, with age-0 and age-1 fish comprising the vast majority of catches. The lack of fisheries data limits stock assessments and management advice in some regions. The Pacific Chub Mackerel is listed as Least Concern on the IUCN Red List.

CONSERVATION: Stocks of Pacific Chub Mackerel are assessed on a regional basis and some countries have implemented management measures and local regulations to protect the stocks.

Eastern North Pacific: A 2015 assessment of the eastern North Pacific stock that extends north of Punta Abreojos, Baja California, north to southeastern Alaska, showed a period of low abundance from 1940–77, increasing in the late 1970s, and peaking at more than 500,000 mt in early 1980s. Since 1982, SSB has declined, reaching an estimated 13,000 mt in 2004 and 2005.

Recently SSB has increased and was estimated to be over 47,000 mt in 2015. The United States has implemented harvest control rules for the management of Pacific Chub Mackerel, with limits set every two years. In recent years, catch levels have been well below harvest guidelines.

Eastern South Pacific: Based on acoustic sampling, biomass since 1999 has been reduced, likely due to a shift in abundance and changes in water temperature. There are few management measures in place, but Peru prohibits the capture of Pacific Chub Mackerel for reduction fisheries.

Western North Pacific: Spawning stock biomass of Pacific Chub Mackerel in the Japan Current peaked in 1979 at 1,400,000 mt and then declined to less than 38,000 mt in 2002. Biomass remained low but stable until 2004, when it increased, reaching 3000,000 in 2006, with a subsequent decline. In the Tsushima Current, SSB has averaged around 350,000 mt since 1973. It peaked at 550,000 mt in 1989, fell to 100,000 mt in 2004, and subsequently increased. Both stocks have been managed by Japan as a single unit under a total allowable catch system since 1997. A recent assessment based on length data indicated that the East China Sea stock is at least fully fished, with most of the catch being less than 2 years of age.

REFERENCES: Ayres 1856; Baker and Collette 1998; Bleeker 1854; Bolaños and Tzeng 1994; Canales 2006; Caramantin-Soriano et al. 2008; Castro Hernández and Santana Ortega 2000; Catanese et al. 2010; Cerna and Plaza 2014; J. Cheng et al. 2011; Q. Cheng et al. 2014; Collette and Nauen 1983; Crone and Hill 2015; Dickerson et al. 1992; Dorval et al. 2007; Fry 1936; Gluyas-Millán and Quiñonez-Velázquez 1996; Gluyas-Millán et al. 1998; Houttuyn 1782; Hunter and Kimbrell 1980; Infante et al. 2007; Jordan and Hubbs 1925; Kohno et al. 1984; Kramer 1960; Matsui 1967; Mendo 1984; Molina et al. 1996; Nakayama et al. 2003; Nishimura 1959; Ojeda and Jaksić 1979; Ozawa 1984; Ozawa et al. 1991; Parrish and MacCall 1978; Sassa and Tsukamoto 2010; Sassa et al. 2008; Sassa et al. 2014; Schaefer 1980; Scoles et al. 1998; Sepulveda and Dickson 2000; Sezaki et al. 2001; Takahashi 1967; Tzeng et al. 2009; Wang et al. 2013; Watanabe 1970; Yamada et al. 1998; Yan et al. 2015; Yukami et al. 2009; Zeng et al. 2012; Zhu et al. 2016.

Atlantic Mackerel

Scomber scombrus Linnaeus, 1758

COMMON NAMES: English – Atlantic Mackerel
French – Maquereau Commun
Spanish – Caballa del Atlántico

ETYMOLOGY: The Swedish naturalist Carl Linnaeus (1758:297), who devised the binomial nomenclature system we use, selected the name *scombrus* from the Greek *scombros*, meaning "mackerel," for his original description, although he changed the spelling to *scomber*, making it the same as the generic name in a later edition (1766).

SYNONYMS: *Scomber scombrus* Linnaeus, 1758; *Scomber scomber* Linnaeus, 1766; *Scomber glauciscus* Pallas, 1814; *Scomber vernalis* Mitchill, 1815; *Scomber vulgaris* Fleming, 1828; *Scomber punctatus* Couch, 1849; *Scomber scriptus* Couch, 1863

TAXONOMIC NOTE: Some authors have placed the other three species of the genus *Scomber* in a separate genus, *Pneumatophorus*, based on their possession of a swim bladder that is absent in the Atlantic Mackerel, but all four species are currently included in *Scomber*.

FIELD MARKS:

1 The space between the end of the first dorsal-fin groove and the origin of the second dorsal fin is clearly longer than the length of the groove.
2 Atlantic Mackerel differ from Atlantic Chub Mackerel in lacking spots on their belly.

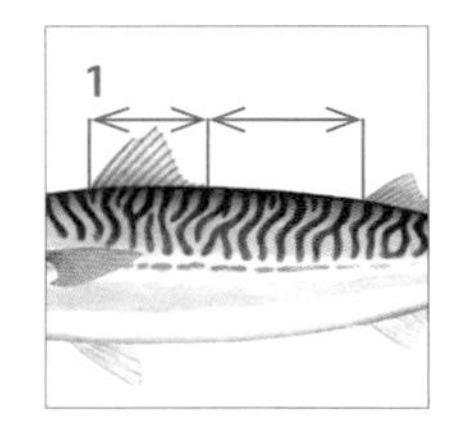

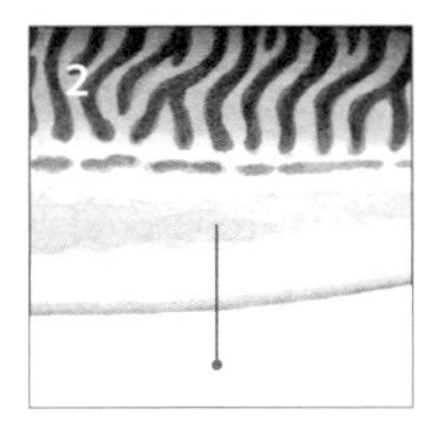

UNIQUE FEATURE: The Atlantic Mackerel is the only member of its genus without a swim bladder.

DIAGNOSTIC FEATURES: Body elongate, rounded, and covered with small cycloid scales. The snout is pointed. Adipose eyelids cover the anterior and posterior margins of the eye. **Teeth:** The teeth in the upper and lower jaws are small, conical, and faintly marked with crenulations. Teeth are also present on the vomer and palatine bones. The palatine teeth are in two widely spaced rows. **Gill rakers:** There are 25–35 gill rakers on the lower limb of the first arch. The gill rakers are shorter than the gill filaments and are barely visible through the open mouth. **Fins:** The first dorsal fin has 11–13 spines. The space between the end of the first dorsal-fin groove and the origin of the second dorsal fin is greater than the length of the groove, about 1.5 times as long. The second dorsal fin has 9–15 rays, usually 12, and is followed by 5 dorsal finlets. The anal fin has 11–12 rays, followed by 5 anal finlets. The origin of the anal fin is opposite to that of the origin of the second dorsal fin or nearly so. The anal fin spine is conspicuous, joined to the fin but clearly independent of it. There are 17–21 pectoral-fin rays. **Swim bladder :** The swim bladder is absent. **Vertebrae:** There are 13 precaudal vertebrae and 18 caudal vertebrae, for a total of 31. The first haemal spine is anterior to the first interneural process. There are 21–28 interneural bones under the first dorsal fin. **Color:** The back is steel blue and crossed by oblique to nearly vertical lines that zigzag and undulate. The belly is unmarked.

GEOGRAPHIC RANGE: Atlantic Mackerel are restricted to the North Atlantic Ocean. In the western Atlantic, they are present from Labrador to Cape Lookout, United States. In the eastern Atlantic, they are present from Iceland to Mauritania, including the southwestern Baltic Sea and the Mediterranean and Black seas. They are widespread in the Mediterranean basin.

SIZE: Atlantic Mackerel can reach a total length of 70 cm and a weight of 3.2 kg, although fish greater than 50 cm are uncommon. The IGFA all-tackle game fish record is of a 3 lb, 8 oz (1.6 kg) fish caught off L'Ampolla, Spain, in August 2015. Age determination based on otolith annual growth rings of 16–49 cm fish from the west coast of Portugal found ages between 0 and 15 years. Maximum longevity is estimated to be 17 years.

HABITAT AND ECOLOGY: Atlantic Mackerel are pelagic and oceanodromous, mainly diurnal, and most abundant in cold and temperate shelf areas. They are sensitive to changes in temperature and prefer water warmer than 5°C. Atlantic Mackerel overwinter in deeper waters and move closer to shore in spring, when water temperatures range between 11° and 14°C. In the Mediterranean, juveniles were most likely to be present on the continental shelf at a depth of less than 130 m in upwelling areas and in deeper waters in areas with downwelling. Stocks migrate great distances on a seasonal basis, and conventional tag returns indicate displacements as great as 1,200 km in

13 days. Atlantic Mackerel form large schools near the surface and fall prey to a host of larger sea animals. Whales and porpoises, sharks and bony fishes such as tunas, and bonito all take a heavy toll. Spiny Dogfish, Atlantic Cod, and Silver Hake are among the most important fish predators in the western North Atlantic.

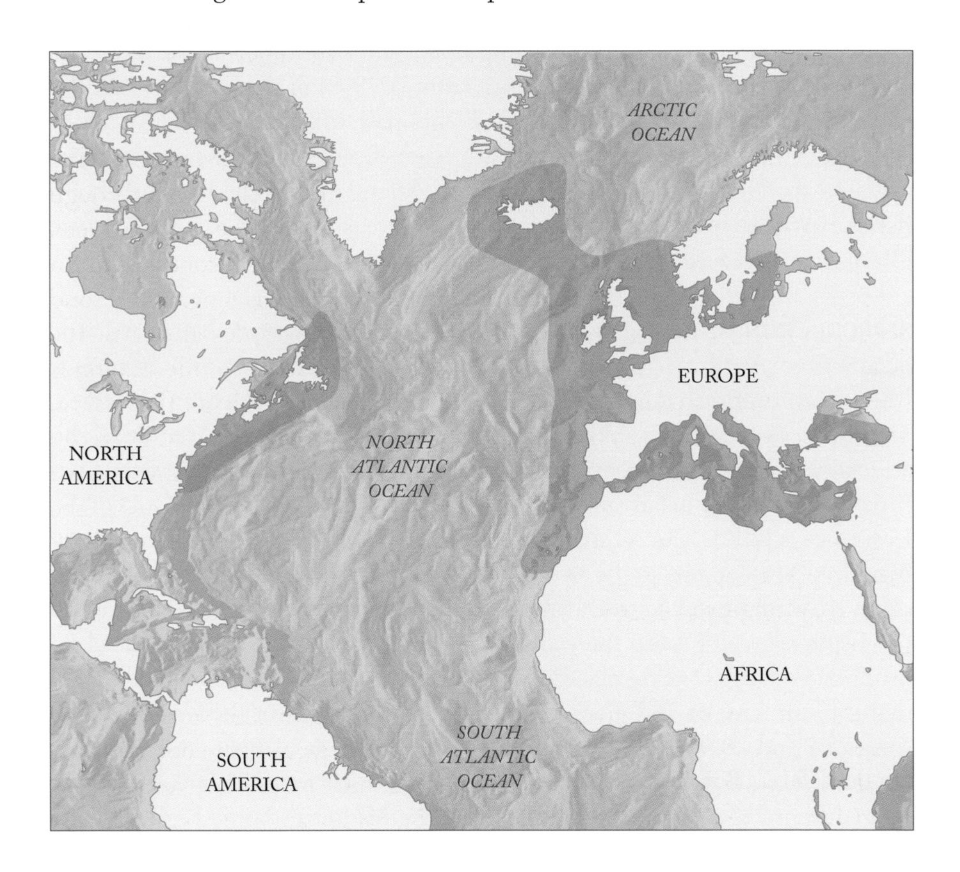

Left: Atlantic Mackerel range map.

FOOD: Atlantic Mackerel have two methods of feeding, either by filtering smaller pelagic organisms from the water using their gill rakers or by selecting individual animals by sight and then swallowing them whole. Practically all floating animals that are neither too large nor too small regularly serve to nourish them. Their diet changes markedly during growth, switching from copepods, amphipods, and larval fishes to larger prey such as squids and larger fishes like hakes. First-feeding larvae (3.5 mm) feed on phytoplankton. The diet of 4.5 mm larvae consists of larval crustaceans. Larvae larger than 6.5 mm are cannibalistic. In addition to conspecifics, larvae prey on other larval fishes, such as Yellowtail Flounder, Silver Hake, and Redfish, on the Scotian Shelf. Adults feed on the same prey as juveniles but their diet includes a wider array of organisms and larger prey items. Euphausiid, pandalid, and crangonid shrimps are common prey, and chaetognaths, larvaceans, pelagic polychaetes and the larvae of many different marine species have been identified from stomachs. Analysis of stomach contents of 3,617 Atlantic Mackerel collected from off the northeastern United States showed that the type of prey eaten varies enormously depending upon the year, the season, and the area.

During the latter part of the twentieth century, the feeding grounds of the northeast population of Atlantic Mackerel were mostly limited to the North Sea and the Norwegian Sea, but they have been extending their summer feeding distribution north and west, including around Iceland, since around 2006. In Icelandic waters, copepods constitute 70–80% of their diet by weight. Euphausiids and hyperiid amphipods are also important, making the combined crustacean diet weight range from 70–99% of the total diet, similar to the diet of the co-occurring herring. Fishes such as sand lances (*Ammodytes*) and Capelin (*Mallotus villosus*) were also a significant part of the diet in some areas. In the North Aegean Sea of Greece, Atlantic Mackerel feed largely on fishes, particularly anchovies and sardines, whereas the co-occurring Atlantic Chub Mackerel feed largely on fish larvae and crustaceans.

REPRODUCTION: Spawning of Atlantic Mackerel in the western North Atlantic occurs from April to August and progresses from south to north as surface waters warm and fish migrate. Two spawning contingents have been identified. The southern contingent spawns from April to July in the Mid-Atlantic Bight, and the northern contingent spawns in the southern Gulf of St. Lawrence in June and July. The mean peak of spawning in the southwestern Gulf of St. Lawrence is July 1, which coincides with the maximum abundance of summer plankton. Most spawn in the shoreward half of the continental shelf, although some spawning extends to the shelf edge and beyond. Atlantic Mackerel do not begin spawning until the water has warmed to about 8°C, with the chief production of eggs taking place at temperatures of 9–14°C. Median lengths at maturity for females and males from the northeast coast of the United States were 25.7 and 26.0 cm FL, respectively. Median age at maturity was 1.9 years for both sexes. Batch fecundity estimates for fish between 31 and 44 cm FL ranged from 285,000 to 1,980,000 eggs in the Middle Atlantic Bight.

In the eastern North Atlantic, more than half of Atlantic Mackerel are mature at age 2, with 100% maturity at age 7, although 90% of fish in the North Sea reach maturity at age 2. The generation length is estimated at 6.9 years. The spawning distribution has shifted northward in the last three decades, probably in response to global sea warming.

In the Mediterranean, knowledge of spawning mainly comes from the Adriatic Sea. Atlantic Mackerel migrate in the autumn/winter from the northern portion to the central portion to reach a single central spawning ground in winter, mainly from January to March.

EARLY LIFE HISTORY: Atlantic Mackerel eggs are 1.09–1.39 mm in diameter, with one oil globule 0.19–0.53 mm in diameter. The eggs generally float in the surface water layer above the thermocline or in the upper 10–15 m. Incubation depends primarily on water temperature, requiring 7.5 days at 11°C, 5.5 days at 13°C, and 4 days at 16°C. Newly hatched larvae are 3.1–3.3 mm long and have a large yolk sac. The yolk is absorbed, the mouth is formed, teeth are visible, and the first traces of the caudal fin form by the time the larva is about 6 mm long. Fin rays and finlets are formed by 22 mm. Illustrations of six early stages of 2.9 mm notochord length to 15.1 mm SL are provided by Fahay.

STOCK STRUCTURE: There are two main populations of Atlantic Mackerel, one in the western North Atlantic and the other in the eastern North Atlantic. Genetic and tagging data are consistent with a lack of connectivity between the populations. The range of the western North Atlantic population of Atlantic Mackerel extends from North Carolina north to Canada, and two spawning groups have been identified within the population. During the 40-year period 1968–2008, the distribution of this population shifted about 250 km to the north and east.

Three stocks have been proposed for the eastern North Atlantic population of Atlantic Mackerel (Southern, Western and North Sea) based on spawning location, but genetic, tagging, and parasite data are consistent with mixing between these units and do not support separate stocks. The southern component spawns in Spanish and Portuguese waters, the western component spawns in the Bay of Biscay and northward around Ireland and west/northwest of the UK, and the North Sea component spawns in the North Sea and Skagerrak. The eastern Mediterranean populations (Greece, Italy) are genetically distinct from the western Mediterranean populations (Barcelona) that are genetically similar to eastern Atlantic populations.

FISHERIES INTEREST: Atlantic Mackerel have high commercial importance throughout their range. Pelagic trawls and purse seines account for the majority of the landings, but significant catches are also taken in artisanal fisheries using gill nets, beach seines, and handlines. Atlantic Mackerel are also an important recreational species, fished with hook and line. Overall, FAO-reported landings for the species increased from about 116,000 mt in 1950 to 163,000 mt in 1960. Catches soared in the 1960s, reaching 1,023,000 mt in 1967 and 1,072,000 mt in 1972. Annual landings generally fluctuated between 600,000 and 700,000 mt through the mid-2000s and have recently increased, reaching 1,421,000 mt in 2014.

Eastern North Atlantic: Prior to the early 1960s, Atlantic Mackerel were caught mainly by trawl, gill net, and hook and line in the North Sea, with a minor portion of the catch taken by beach seine and small Norwegian purse seiners operating in coastal waters. The fishery grew rapidly in the 1960s owing to the development of the Norwegian purse seine fishery, increasing from 146,000 mt in 1960 to 934,000 mt in 1967. The catch decreased, falling to 430,000 mt by 1970, and fluctuated between roughly 500,000 and 700,000 mt through 2010. Landings have increased since 2010, reaching 1,382,000 mt in 2014, with increased catches by Norway and Iceland. Climate change appears to be affecting the distribution of this species, with movement toward the north, but it is not yet known if contraction will occur in the southern part.

Western North Atlantic: Atlantic Mackerel were not heavily fished in the western North Atlantic prior to the 1970s, with annual landings of less than 25,000 mt. Catches increased rapidly in the 1970s, due in large part to the participation of foreign vessels in the fishery, and reached more than 420,000 mt in 1973. Landings dropped dramatically with the establishment of the US and Canadian exclusive economic zones (EEZs) and have fluctuated between 10,000 and 70,000 mt since the late 1970s.

Mediterranean Sea: Catches of Atlantic Mackerel in the Mediterranean Sea were greater prior to 1965, peaking at just over 23,000 mt in 1954. Since the mid-1960s, catches have fluctuated between 5,000 and 15,000 mt, with higher catches of that time period occurring in the mid-2000s.

THREATS: This is a common and locally abundant species that exhibits large fluctuations in population size and fishery landings. The world catch declined from about 1.1 million mt in 1975 to about 610,000 mt in 1981, and increased to more than 1.4 million mt in 2014. The Atlantic Mackerel is listed as Least Concern on the IUCN Red List.

CONSERVATION:

Eastern North Atlantic: Atlantic Mackerel are assessed as a single stock in the eastern North Atlantic although different spawning groups are recognized. In addition to fishery-dependent information, assessments incorporate data from egg surveys and trawl surveys, as well as estimates of mortality from tagging data. The 2014 benchmark assessment indicated that spawning stock biomass increased from roughly 2 million mt in the late 1990s/early 2000s to 5 million mt in 2009. Strong recruitment in 2002 and 2006 contributed to the increase in spawning stock biomass. Management measures for the stock include areas restricted to fishing, seasonal closures, minimum landing size (30 cm in the North Sea and 20 cm in Skagerrak), as well as prohibitions on high grading and discards. Several eastern Atlantic countries have minimum landing sizes for Atlantic Mackerel: EU (18 cm), Ukraine (15 cm), Turkey (20 cm), Bulgaria (22 cm), and Romania (23 cm).

Western North Atlantic: Atlantic Mackerel are managed as a single stock in the western North Atlantic by the United States and Canada, with annual quotas established for each country. The last benchmark assessment was conducted in 2005, and a new assessment should be available in 2018. Spawning stock biomass recovered from an estimated 663,000 mt in 1976 to 2,300,000 mt in 2004, and at the time of the 2005 assessment the stock was not considered to be overfished, and overfishing was not occurring. Current management measures for the US commercial fishery include trip limits for various permit types and a few restricted areas for bottom gear. There are no size or possession limits for the recreational fishery in federal waters.

REFERENCES: Astthorsson et al. 2012; Berrien 1975, 1978, 1982; Bowman et al. 2000; Bruge et al. 2016; Collette 2002; Collette and Nauen 1983; Collette et al. 2011; Costa et al. 2017; Couch 1849, 1863; Fahay 2007; Fleming 1828; Fortier and Villeneuve 1995; Fritzsche 1978; Giannoulaki et al. 2017; Hamre 1978; IGFA, 2018; Jansen et al. 2013; Karachle 2017; Linnaeus 1758, 1761; Matsui 1967; Meneghesso et al. 2013; Mitchill 1815; Morse 1980; Navarro et al. 2012; Nesbø et al. 2017; O'Brien et al. 1993; Óskarsson et al. 2016; Overholtz et al. 2011; Pallas 1814; Papetti et al. 2013; Peterson and Ausubel 1984; Scoles et al. 1998; Sette 1943, 1950; Sinovčič 2001; Sinovčič et al. 2004; Skagen 1989; Ware 1977; Ware and Lambert 1985; Worley 1933; Zardoya 2004.

Serra Spanish Mackerel

Scomberomorus brasiliensis Collette, Russo & Zavala-Camin, 1978

COMMON NAMES: English – Serra Spanish Mackerel
French – Thazard tacheté du sud
Spanish – Serra

ETYMOLOGY: The Serra Spanish Mackerel was named *brasiliensis* by the US ichthyologists Bruce Collette and Joseph Russo and their Brazilian colleague Alberto Zavala-Camin (Collette et al. 1978) because of the importance of this fish in Brazil.

SYNONYMS: There are no synonyms for the Serra Spanish Mackerel although the name *Scomberomorus maculatus* was used for the species before it was recognized as distinct.

TAXONOMIC NOTE: Literature records previous to 1978 for *Scomberomorus maculatus* from the Caribbean and the Atlantic coasts of Central and South America apply to *Scomberomorus brasiliensis*, which was considered *Scomberomorus maculatus* by previous authors (Collette and Russo 1985, Banford et al. 1999).

FIELD MARKS:

1 The Serra Spanish Mackerel is very similar to the Atlantic Spanish Mackerel in having rows of yellow-bronze spots along its sides.
2 It differs mainly in having fewer vertebrae and second dorsal-fin rays that are difficult to count.
3 Perhaps they are best differentiated by locality—any spotted Spanish mackerel from the coast of the United States and throughout the Gulf of Mexico are Atlantic Spanish Mackerel; those from Yucatan south to Brazil are Serra Spanish Mackerel.

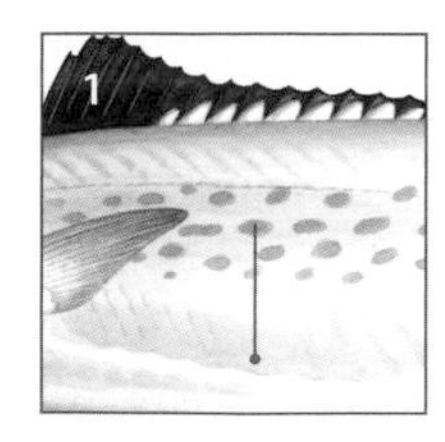

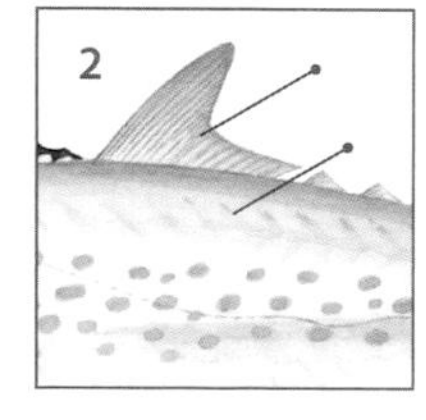

UNIQUE FEATURE: Formerly confused with the Atlantic Spanish Mackerel.

DIAGNOSTIC FEATURES: Gill rakers: There are a moderate number of gill rakers on the first gill arch: 1–3 on the upper limb; 9–13 on the lower limb; 11–16 total (usually 13–15). **Fins:** The first dorsal fin has 17 or 18 spines, rarely 19; the second dorsal has 15–19 rays, usually 17 or 18, followed by 8–10 finlets. The anal fin has 16–20 rays, usually 17–19, followed by 7–10 finlets, usually 9. There are 21–24 pectoral-fin rays, usually 22 or 23. The pelvic fins are relatively short, 3.6 to 5.9% of fork length. **Vertebrae and intestine:** Vertebrae number 19–21 precaudal plus 27–29 caudal, for a total 47–49, usually 48. The intestine has 2 folds and 3 limbs. Descriptions and illustrations of the osteology and soft anatomy of the Serra Spanish Mackerel are included in the revision of the Spanish mackerels (Collette and Russo 1984). **Color:** The sides of the body are silvery, with several rows of round yellowish bronze (in life) spots, the number of spots increasing with size of fish from about 30 at 20 cm FL to between 45 and 60 at 50 to 60 cm FL. The first dorsal fin is black anteriorly (membranes between the first 7 spines) and along the upper edge of the posterior portion. The basal portion of posterior membrane of the first dorsal fin is white. The pectoral fin is dusky; the pelvic and anal fins are light colored.

GEOGRAPHIC RANGE: Serra Spanish Mackerel are restricted to the western Atlantic along the Caribbean and Atlantic coasts of Central and South America from Belize to Rio Grande do Sul, Brazil. A seasonal model of distributional changes due to global warming predicts that the southern border will shift from Rio Grande do Sul to the northern coast of the Santa Catarina region in northern summer, and in northern winter the boundary will shift from Belize to Nicaragua. This prediction is supported by the abundance of Serra Spanish Mackerel increasing along the Brazilian coast in March but decreasing continuously throughout July, August, and September. Seasonal movements have also been noted off Trinidad, where they are most abundant from May to September.

SIZE: Maximum size is 125 cm FL. The IGFA all-tackle angling record is a 14 lb, 13 oz (6.71 kg) fish caught off Mangaratiba, Brazil, in June 1999. Off Brazil 60% of the fish in large samples taken in the period from 1962 to 1968 ranged between 40 and 65 cm. Aging based on otoliths showed 9 age classes. Serra Spanish Mackerel may live up to 13 years, reaching more than 1 m TL.

HABITAT AND ECOLOGY: Serra Spanish Mackerel are an epipelagic, neritic species, occurring from the surface to 130 m, most commonly in a depth range of 20–60 m. They are concentrated in coastal areas and are common off rocky coasts, open beaches, and islands. Serra Spanish Mackerel do not migrate extensively, although seasonal movements occur in some areas. They tend to form schools and enter tidal estuaries.

REPRODUCTION: Off Brazil, Serra Spanish Mackerel first mature at a average length of 41.1 cm TL for females and 44.3 cm TL for males, corresponding to ages of 3 years for females and 4 years for males. Length at first maturity of males has decreased from 41 cm in the 1970s to 28 cm TL in 2007. Serra Spanish Mackerel spawn over the continental shelf, probably between 15 and 36 m of depth, although some spawning takes place offshore. In the Gulf of Paria, spawning occurs throughout the year, peaking from October to April. On the Guyana shelf, ripe fish are encountered in September. The peak breeding season in northeastern Brazil takes place in a sequential manner: off Natal the spawning season is from March to June; in Maranhão coastal waters, spawning occurs in October and from June to November; and in Rio Grande do Norte, the rainy season influences maturation, and individuals mature later during the rainy season, with a peak of reproduction from March to June. The spawning period on the western coast of Maranhão is March to June.

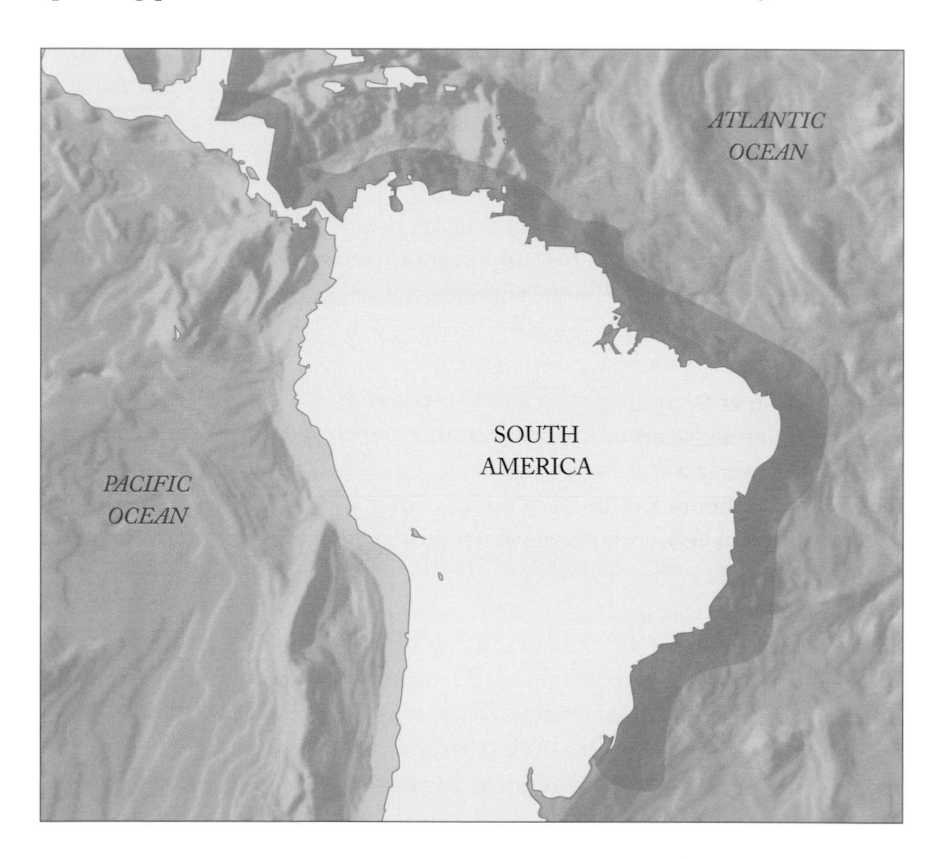

Right: Serra Spanish Mackerel range map.

FISHERIES INTEREST: Serra Spanish Mackerel is fished with gill nets, handlines, and beach seines. It is one of the most important commercial marine fishes from northeastern Brazil available throughout the year. In Trinidad, it is an important commercial and recreational species targeted by artisanal fisheries. It was considered "the most important fin fish in Trinidad since data collection started in 1964" and is one of the most important targets of recreational fishermen. Most of the catch is consumed fresh, but in Brazil some is salted, canned, or prepared as fish burgers.

Reported landings for Serra Spanish Mackerel increased from 300 mt in the early 1950s to over 10,000 mt in 1989. Catches remained strong through the 1990s and slowly decreased to about 4,000 mt by 2010. There is considerable variability in annual catches, but they have averaged around 4,000 mt for the past 5 years. Most of the catch that was previously reported as *S. maculatus* from the western central Atlantic by Colombia, Trinidad and Tobago, and Venezuela (4,120 mt in 1981) was in fact *S. brasiliensis.*

THREATS: It is unlikely that there has been a decline of 30% or more in the global population of Serra Spanish Mackerel over the past three generation lengths (approximately 15–20 years), so the species was listed as Least Concern on the IUCN Red List. However, there is an urgent need for a more recent stock assessment for this species in Brazil and for more comprehensive assessments in the Caribbean portion of its range.

CONSERVATION: A stock assessment in 1991 categorized this species in the waters of Trinidad as fully exploited, and a more recent assessment indicated that the stock biomass was below that necessary to support MSY and that F (fishing mortality) was greater than FMSY. However, this most recent stock assessment was based on two different models with some conflicting results. In general, there is uncertainty in these results, and the recommendation is continued fishing at current levels.

Stock assessments from north and northeastern Brazil indicate that the species is fully exploited, and the decrease in average body length and catches in both areas suggest that overfishing is occurring. Fishing effort has increased (boat size, number of fishermen per trip, and average length of gill nets) and fishing effort has moved farther away from landing ports, suggesting further depletion of stocks. This species is also caught as bycatch. Juveniles up to 2 years old are often caught on the northeastern coast of Brazil in small meshed nets for sardine (*Opisthonema oglinum*).

Since it is estimated that one-third of individuals caught in northeastern Brazil are immature, there are recommendations to regulate mesh size to 3.5 cm to protect juveniles. There is one no-take protected area, the Marine Biological Reserve of Arvoredo, which protects a variety of fishes including Serra Spanish Mackerel, but it is located in southern Brazil, south of the main fisheries for the species.

Fishing effort is not controlled in Trinidad but there are regulations that specify maximum length and depth and minimum mesh size for gill nets (11 cm). Similar regulations are imposed for seines, with maximum dimensions for the nets and minimum mesh size requirements. Individuals less than 305 mm may not be taken or sold.

REFERENCES: Banford et al. 1999; Batista and Fabré 2001; Chellappa et al. 2010; Collette and Nauen 1983; Collette and Russo 1985; Collette et al. 1978; Collette et al. 2011; Fonteles-Filho 1988; Fonteles-Filho and Alcantara-Filho 1977; Gesteira 1973; Gold et al. 2010; Gonçalves et al. 2003; Henry and Martin 1992; IGFA 2018; Isaac et al. 2013; Julien et al. 1984; Lam et al. 2008; Lessa 2006; Lima 2004; Lucena et al. 2004; Manooch et al. 1978; Martin and Nowlis 2004; Menezes 1970; Mike and Cowx 1996; Morales-Nin 1989; Nóbrega and Lessa 2009a, 2009b; Nóbrega et al. 2009; Quadros and Bolini 2015; Quadros et al. 2015; Silva et al. 2005; Sturm 1978; Ximenes de Lima 2007.

King Mackerel

Scomberomorus cavalla (Cuvier, 1829)

COMMON NAMES: English – King Mackerel
French – Thazard Serra
Spanish – Carite lucio
Portuguese – Cavala

ETYMOLOGY: Named *cavalla* by the distinguished French anatomist and ichthyologist Georges Cuvier (1829:200), apparently based on its Portuguese common name, *cavala*.

SYNONYMS: *Cybium cavalla* Cuvier, 1829; *Cybium caballa* Cuvier in Cuvier and Valenciennes, 1832; *Cybium clupeoideum* Cuvier in Cuvier and Valenciennes, 1832; *Cybium acervum* Cuvier in Cuvier and Valenciennes, 1832; *Cybium immaculatum* Cuvier in Cuvier & Valenciennes, 1832.

TAXONOMIC NOTE: Redescribed three times in the same publication by Cuvier (1832).

FIELD MARKS:

1. King Mackerel have the fewest gill rakers of the four western Atlantic species of Spanish mackerels, usually 11 or fewer compared to 12 or more in the other species.
2. Unlike the other three western Atlantic Spanish mackerels, the King Mackerel has no black area in the anterior part of the first dorsal fin.
3. This is the only species of Spanish Mackerel in the western Atlantic that has a deep dip in the lateral line under the second dorsal fin.

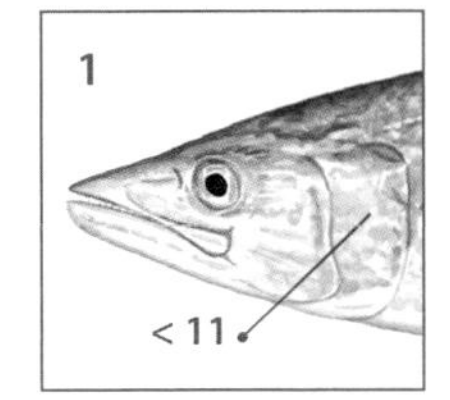

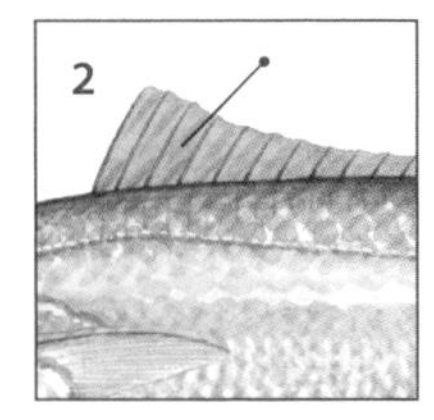

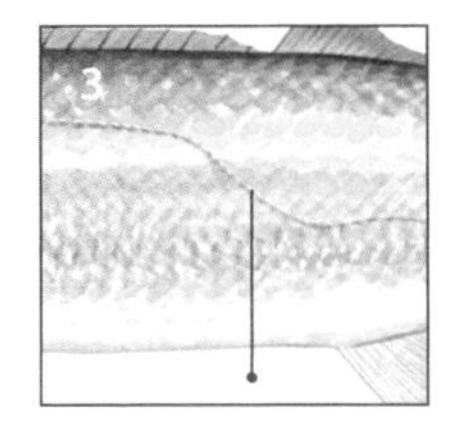

UNIQUE FEATURE: The King Mackerel is the largest Spanish mackerel in the western Atlantic and the only one with a deep dip in the lateral line.

DIAGNOSTIC FEATURES: Gill rakers: There are relatively few gill rakers on the first arch, 1–3 on the upper limb and 6–10 on the lower limb, for a total of 7–13, usually 9 or 10. **Fins:** The first dorsal fin has 12–18 spines, usually 15. The second dorsal has 15–18 rays, followed by 7–10 finlets, usually 9. The anal fin has 16–20 rays, usually 18 or 19, followed by 7–10 finlets, usually 8. There are 21–23 pectoral-fin rays. **Lateral line:** The lateral line curves abruptly downward below the second dorsal fin. **Vertebrae and intestine:** Vertebrae number 16 or 17 precaudal plus 24–26 caudal, for a total of 41 to 43, usually 42. The intestine has 2 folds and 3 limbs. The osteology of King Mackerel has been described and illustrated by Mago Leccia (1958) and Collette and Russo (1985). **Otoliths:** Sagittal otoliths were illustrated by Baremore and Bethea (2010). **Color:** The sides of the body are plain silver without bars or spots. Juveniles have five or six irregular rows of bronze spots that are smaller than the pupil of the eye. Adults lack any black on the anterior part of the first dorsal fin, unlike many species of Spanish mackerels.

Below: Juvenile King Mackerel, approx. 18 in FL.

GEOGRAPHIC RANGE: King Mackerel are found in the western Atlantic from Massachusetts, United States, to Santa Catarina State, Brazil. A record exists for Saint Paul's Rocks (approximately 940 km off the coast of Brazil), however, this may represent a vagrant, as there have been no records from there in the last 15 years despite close monitoring.

SIZE: The reported maximum size for King Mackerel is 173 cm FL and 45 kg weight, and the IGFA all-tackle angling record is a 93 lb (42.18 kg) fish caught off San Juan, Puerto Rico, in 1999. Off northeastern Brazil, length in the commercial catches ranges mostly between 50 and 90 cm FL. Samples from the sport fishery in southeast Florida ranged in size from 58.5–150 cm FL and in weight from 1.47–32.09 kg. Females live longer and grow larger than males: a maximum of at least 26 years and 158 cm FL for females and 24 years and 127 cm FL for males from the Atlantic and Gulf of Mexico coasts of the United States, and 32 years for females and 26 years for males in northeastern Brazil.

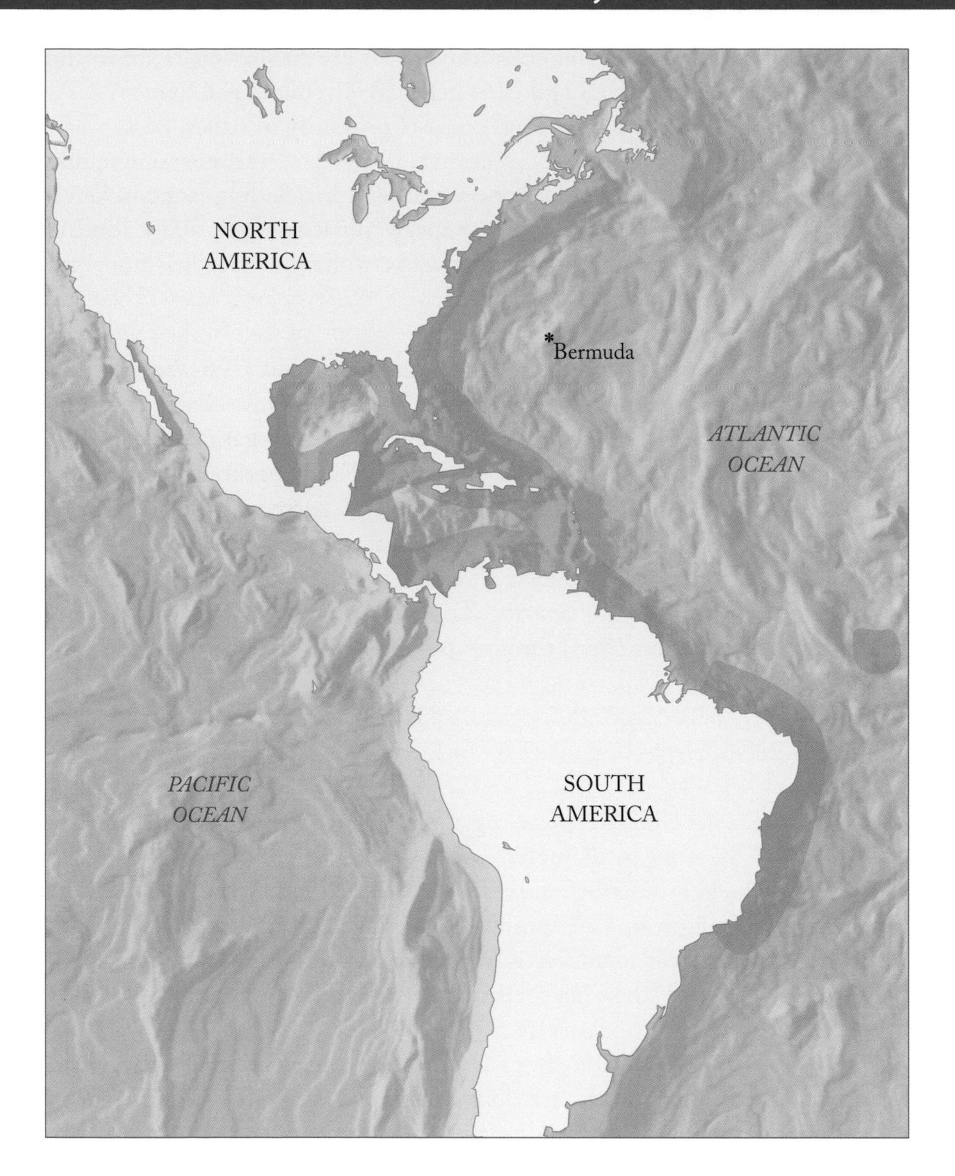

Left: King Mackerel range map.

HABITAT AND ECOLOGY: King Mackerel inhabit coastal areas out to the edge of the continental shelf but mostly occur in waters shallower than 80 m. They appear to avoid turbid regions, possibly because they are visual predators. They occur singly or in small groups, often in outer reef areas. Large schools have been found to migrate over considerable distances along the US Atlantic coast within water temperatures of 20–26°C. King mackerel appear to be present throughout the year off Louisiana and off the state of Ceará in northeastern Brazil. There also seem to be some resident populations in south Florida waters, as fish are available to the recreational fishery throughout the year. Bottlenose Dolphins prey on King Mackerel in the Florida King Mackerel troll fishery.

FOOD: King Mackerel are piscivorous throughout life, beginning as larvae and post-larvae (2.8–22 mm SL) and continuing as juveniles (9–42 cm FL). As in other members of the genus, food consists primarily of fishes, particularly herring-like species, with smaller quantities of penaeid shrimps and squids. Thirty-one families and at least 62 species of fishes, including jack mackerels (Carangidae), snappers (Lutjanidae), grunts (Pomadasyidae), and halfbeaks (Hemiramphidae), were found in stomachs of King Mackerel from the southeastern United States.

REPRODUCTION: In northwest Florida, first maturity occurs at about 45–50 cm FL, 50% were mature at 55–60 cm FL, and all females were mature at 95–90 cm FL. In the southeastern United States, males mature at 2–3 years and females at 3–4 years, while in Brazil, females mature at about 77 cm at 6–7 years. In the southeastern United States, King Mackerel spawn from May to November and in the western Gulf of Mexico, from May through September. Spawning occurs at depths between 35 and 180 m over the middle and outer continental shelf. In Puerto Rico and Trinidad, spawning occurs throughout the year, with peaks from April through August and October through March. In Ceará State, Brazil, spawning occurs from October to March. Fecundity ranged from 69,000–12,207,000 eggs in the southeastern United States and 345,000–2,280,000 eggs in females 63 to 123 cm long in Brazil.

EARLY LIFE HISTORY: King Mackerel eggs have one oil globule and range in size from 0.9–1.3 mm in diameter. Larvae are 2.56 mm notochord length at hatching. Wollam described and illustrated seven larvae ranging from 3.3 mm notochord length to 17.1 mm standard length, and the illustrations are reproduced in Richards' identification guide to larval western central Atlantic fishes. King Mackerel larvae have fewer vertebrae and myomeres (41–43) than other western Atlantic Spanish mackerels and lack the pigmentation in the gular region that is characteristic of larval Cero. King Mackerel larvae have been found from April to October along the Atlantic coast of the southeastern United States, primarily at depths greater than 40 m, and in the northwestern Gulf of Mexico in surface waters of 26.3–31.0°C and salinities of 26.9–35 ppt. Daily otolith growth rings imply a moderately high average growth rate for early larvae followed by very rapid growth (2.9 mm/day) for the first 1–3 months of age.

STOCK STRUCTURE: Extensive tagging efforts in the 1970s and 1980s and growth studies identified three migratory groups of King Mackerel in United States waters: western Gulf of Mexico, eastern Gulf of Mexico, and Atlantic. No genetic differences have been found between the two Gulf of Mexico populations, and very slight genetic differences between the Gulf of Mexico and the western North Atlantic are consistent with some gene flow across this region. The species is currently managed in the United States as two migratory stocks: Gulf of Mexico and southeastern US coast. Winter migrations occur from both stocks to south Florida, where they occur in a mixed-stock fishery.

FISHERIES INTEREST: King Mackerel are an important species for commercial, artisanal, and recreational fisheries throughout their range. They are caught primarily with purse seine, gill net, and hook and line. Landings for this species are likely underreported, as King Mackerel may be included as "unclassified *Scomberomorus* species." Catches of King Mackerel ranged from 6,000–8,000 mt in the 1960s, 8,000–10,000 mt in the 1970s, 9,000–15,000 mt in the 1990s, and 11,000–15,000 mt since 2000. The United States, Mexico, and Brazil have been the major harvesters, although several other nations in the western North Atlantic report landings.

In the United States, sport fishing with hook and line is carried out from April to December (but mostly in spring and fall) in North Carolina, and all year-round (with local seasonal peaks) in Florida. Commercial fisheries operate in the same areas, as well as off Louisiana and Mississippi.

King Mackerel is the main Spanish mackerel species of interest to the commercial fishery that extends throughout the year off northeastern Brazil. The major Brazilian fishing grounds are located some 6 to 16 mi off the coastline. Most of the catch is generally processed into steaks or sold fresh, but some has been salted and canned in coconut milk and cottonseed and babassu oils in northeastern Brazil. In northeastern Brazil an assessment of the exploitation status of the stock estimated a mean annual biomass of 12,742 mt for a mean yield of 3,307 mt/year, indicating that, despite being underexploited, the stock is near its maximal exploitation limit.

THREATS: In the 1980s, this species was overfished throughout its range in the United States but is now considered to have recovered to a healthy level. Lack of reporting from artisanal fisheries and reporting of mixed species assemblages remain issues for this species. The Gulf of Mexico population, which has experienced an estimated 2.5-fold increase in spawning stock biomass since the early 1990s, displayed a decline in size-at-age for ages 2–7, while the Atlantic population, which has experienced an approximately 45% decline in estimated spawning stock biomass of approximately 45% over the same time period, displayed an increase in size-at-age for ages 4–10. The Fishery Management Plan for Coastal Migratory Pelagic Resources established a number of conservation measures that have helped to rebuild King Mackerel fisheries, including quotas, minimum sizes, bag limits, and trip limits. Drift gill nets were banned in 1989. King Mackerel are under a conservative management regime in the western North Atlantic, and recent landings appear stable throughout its range. There may be some local depletions; however, the biomass of the species is estimated to be stable. This species is listed as Least Concern on the IUCN Red List.

CONSERVATION: Management measures have been effective in rebuilding the stocks to currently healthy levels. The estimated SSB is currently higher than the SSB necessary for MSY, and F (fishing mortality) is lower than FMSY for both the US south Atlantic and Gulf of Mexico stocks. In the United States, estimates of the SSB for the South Atlantic stock have declined, ranging from a peak of 12.8 million fish in 1981 to 5.9 million in 2001–2. For the Gulf of Mexico stock, the SSB has generally increased from 4 million fish in 1984–85 to 17.2 million in 2006–7. The results of a 2014 assessment indicate that the US south Atlantic and Gulf of Mexico King Mackerel stocks are not overfished, and it is uncertain whether overfishing is occurring on either stock. In the United States the management bodies are the South Atlantic Fishery Management Council (SAFMC) and the Gulf of Mexico Fishery Management Council (GMFMC).

In northeastern Brazil Lessa et al. assessed the exploitation status of the stock and estimated a mean annual biomass of 12,742 mt for a mean yield of 3,307 mt per year, indicating that, despite being underexploited, the stock is near its maximal exploitation limit. There are no specific conservation measures in place in Brazil; however, there is a restriction on the length of gill nets, which may not exceed 2.5 km, but this is poorly enforced. The distribution of this species in Brazilian waters may coincide with some marine protected areas where further fishing regulations may apply.

REFERENCES: Alcantara Filho 1972; Baremore and Bethea 2010; Bastos et al. 1973; Beardsley and Richards 1970; Beaumariage 1973; Broughton et al. 2002; Brusher and Palko 1987; Chávez and Arreguín-Sánchez 1995; Clardy et al. 2008; Collette and Nauen 1983; Collette and Russo 1985; Collette et al. 2011; Collins and Stender 1987; Collins and Wenner 1988; Collins et al. 1989; Cuvier 1829; DeVane 1978; DeVries and Grimes 1997; Figuerola-Fernandez et al. 2007; Finucane et al. 1986, 1990; Gold et al. 2002; Hogarth and Martin 2006; IGFA 2018; Ivo 1972, 1974; Johnson et al. 1994; Lubbock and Edwards 1981; Mago Leccia 1958; Manooch 1979; Manooch et al. 1978; Martin and Nowlis 2004; Mayo 1973; McEachran et al. 1980; Menezes 1969; Naughton and Saloman 1981; Nóbrega and Lessa 2009; Nomura and Rodrigues 1967; Ortiz 2004; Paiva and Costa 1966; Richards 2006; Saloman and Naughton 1983; Shepard et al. 2010; Sturm and Salter 1990; Wall et al. 2009; Wollam 1970; Zollett and Read 2006.

Narrow-barred Spanish Mackerel

Scomberomorus commerson (Lacepède, 1800)

COMMON NAMES: English – Narrow-barred Spanish Mackerel
French – Thazard rayé
Spanish – Carite estriado

ETYMOLOGY: The Narrow-barred Spanish Mackerel was named *commerson* for the French explorer and naturalist Philibert Commerson by the French naturalist Bernard Germaine Etienne de la Ville, Comte de Lacepède (1800:598).

SYNONYMS: *Scomber commerson* Lacepède, 1800; *Scomber maculosus* Shaw, 1803; *Cybium commersoni* (Lacepède, 1800); *Cybium Konam* Bleeker, 1851; *Scomberomorus commersonii* (Lacepède, 1800); *Cybium multifasciatum* Kishinouye, 1915

TAXONOMIC NOTE: The species name should simply be *commerson*, as it was originally described, without adding a terminal *i* or *ii*.

FIELD MARKS:
1 There are only one to eight gill rakers on the first arch.
2 The lateral line curves abruptly downward below the end of the second dorsal fin.
3 In life, the body is covered with alternating, wavy silvery gray bars.

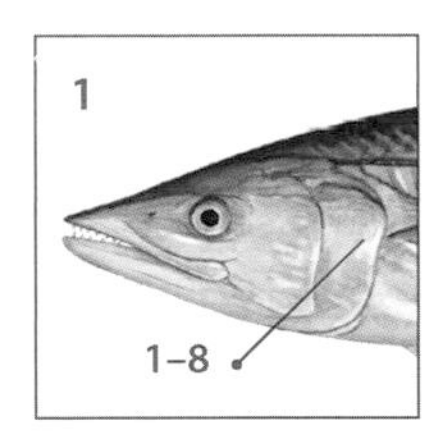

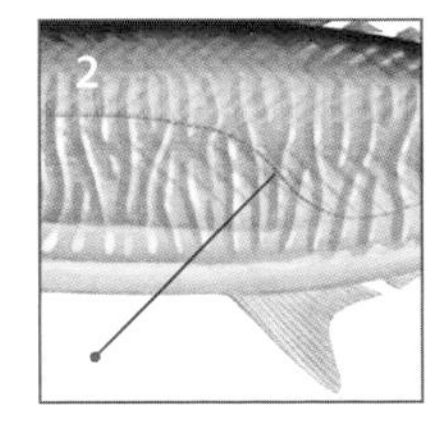

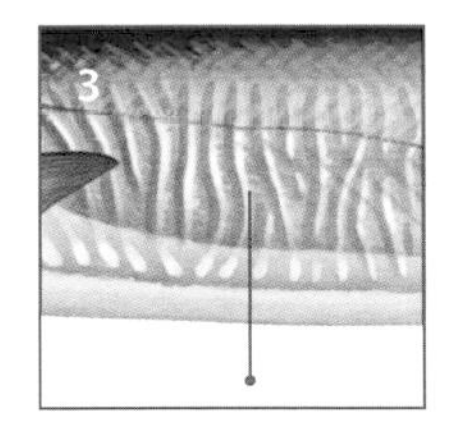

UNIQUE FEATURE: The Narrow-barred Spanish Mackerel has the greatest natural geographic distribution of any of the Spanish mackerels and, in addition, has successfully invaded the Mediterranean Sea by way of the Suez Canal.

DIAGNOSTIC FEATURES: Teeth: The teeth in the upper and lower jaws are large, triangular, and laterally compressed, 5–38 (mean 14) in the upper jaw and 4–29 (mean 11) in the lower jaw. **Gill rakers:** There are very few gill rakers on the first arch, 0–2 on the upper limb and 1–8 on the lower limb, for a total of 1–8, usually 3–5. **Fins:** The first dorsal fin has 15–18 spines, usually 16 or 17. The second dorsal fin has 15–20 rays, usually 17 or 18, followed by 8–11 finlets, usually 9 or 10. The anal fin has 16–21 rays, usually 18 or 19, followed by 7–12 finlets, usual1y 9 or 10. There are 21–24 pectoral fin rays, usually 22 or 23. **Lateral line:** The lateral line curves abruptly downward below the end of the second dorsal fin. **Vertebrae:** Vertebrae number 19 or 20 precaudal plus 23–27 caudal, for a total of 42 to 46. The intestine has 2 folds and 3 limbs. The osteology of Narrow-barred Spanish Mackerel was described and illustrated by Collette and Russo (1985). **Color:** The sides of the body are silvery gray and marked with transverse vertical bars of a darker gray, sometimes breaking up into spots ventrally. There are 40–50 bars in adults but usually fewer than 20 in juveniles up to 45 cm FL. Cheeks and lower jaw are silvery white. The first dorsal fin is bright blue but rapidly turns to a blackish blue when the fish is removed from water. The lobes of the caudal fin, the second dorsal fin, the anal fin, and the dorsal and anal finlets are pale grayish white. In juveniles, the anterior membranes of the first dorsal fin are jet black, contrasting with pure white posteriorly.

GEOGRAPHIC RANGE: The Narrow-barred Spanish Mackerel is widespread throughout the Indo-West Pacific from the Red Sea and South Africa to Southeast Asia, north to China and Japan and south to southeast Australia, and east to New Caledonia, Fiji, Tonga, and Samoa. Its current distribution in Australia extends south to Geographe Bay in southwestern Australia and to Sydney on the east coast. It is an immigrant to the eastern Mediterranean Sea by way of the Suez Canal and it has been found westward to at least Tunisia on the southern coast of the Mediterranean Sea and to Rhodes in the Aegean Sea along the northern coast. In the southeast Atlantic, it has been reported from St. Helena as a vagrant.

SIZE: The maximum length of the Narrow-barred Spanish Mackerel is about 230 cm FL, commonly 60–120 cm. The IGFA all-tackle game fish record is of a 99 lb (44.91 kg) fish taken off Scottburgh, Natal, South Africa, in March 1982. They live for up to 16 years, and maybe as long as 22 years. In north Queensland, Australia, the oldest male was 10 years (127 cm, 19.0 kg) and the oldest female 14 years (155 cm, 35 kg). A generation length would be at least 6 to 9 years, but possibly longer. Females grow larger than males. In India, they attain an average size of 40.2 cm TL at age 1, 72.6 cm at 2, 99.5 cm at 3, and 118.6 cm at 4. Length-weight data was reported from the Persian Gulf along with summaries of similar data from other areas in the Indian Ocean.

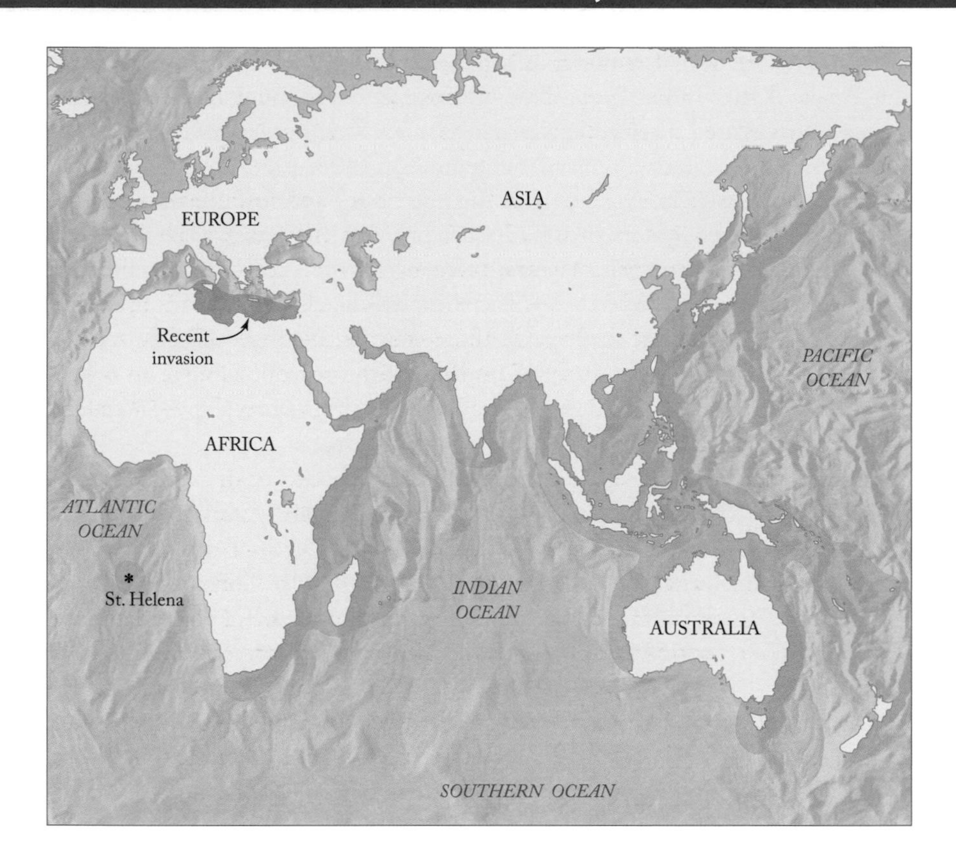

Left: Narrow-barred Spanish Mackerel range map.

HABITAT AND ECOLOGY: The Narrow-barred Spanish Mackerel is pelagic and oceanodromous. It is distributed from near the edge of the continental shelf to shallow coastal waters, often of low salinity and high turbidity. It is also found in drop-offs and shallow or gently sloping reef and lagoon waters. They inhabit coastal waters to depths of 200 m but are more frequently caught in areas of less than 100 m depth. They usually hunt solitarily as adults and often swim in shallow water along coastal slopes. They are known to undertake lengthy long-shore migrations, but permanent resident populations also seem to exist. Sometimes, they are found in small schools when actively feeding.

FOOD: Like other Spanish mackerels, the Narrow-barred Spanish Mackerel feeds primarily on small fishes like anchovies, clupeids, and jacks, but also on squids and penaeoid shrimps, as shown in studies from many regions: Madagascar, South Africa, India, Persian Gulf, Cambodia, and Australia. In India, two species of sardines (*Sardinella*) were the dominant food items, comprising 97% of the diet. Narrow-barred Spanish Mackerel change from feeding on larval fishes to feeding on sardines when they reach a length of about 16 cm. Feeding is intense between 0830 and 1130 hours and between 1900 and 2200 hours. In the Mediterranean region, the main food items of this invasive species are an anchovy (*Engraulis encrasicolus*), sardines, and shrimps, overlapping with the diet of the Largehead Hairtail Ribbonfish.

REPRODUCTION: Sexual maturity is attained at 70–80 cm FL in Madagascar, the Persian Gulf, India, Papua New Guinea, and Fiji, but not until 90–100 cm in South Africa. They generally spawn in early summer: from December–February in Madagascar, January–September in India, October–December in Queensland, Australia, April–July in the northern Persian Gulf, and in May–June in the coastal waters of Oman, just prior to the onset of the summer oceanographic regime. The spawning season may be shortened or extended depending on water temperatures. The trimodal distribution of ova in Indian specimens indicates that they spawn more than once a season. In Australian waters, each female spawns several times over the season, about 2 to 6 days apart, depending on the locality. Spawning occurs off the reef slopes and edges, and spawning aggregations may form in specific areas.

EARLY LIFE HISTORY: Average hatching time of artificially fertilized eggs of Narrow-barred Spanish Mackerel takes 25 or 26 hours at 26°C, similar to other species of Spanish mackerels. These eggs averaged 1.23 mm in diameter, with a 0.31 mm oil globule. Larval development was described and illustrated from six larvae 3.7–9.1 mm. Larvae were captured in the Great Barrier Reef lagoon region off Townsville, Australia, from November to January. Small larvae feed on larvaceans and then feed on larval fishes with growth.

STOCK STRUCTURE: Regional genetic studies of the widespread Narrow-barred Spanish Mackerel indicate that there may be five or more populations: Arabian Sea and Gulf of Oman; Arabian Sea and Bay of Bengal; Indo-Malay-Papua Archipelago; Australia (with several subregions); and New Caledonia. Mitochondrial DNA analyses of Narrow-barred Spanish Mackerel suggest there is a single genetic stock in the Arabian Gulf, Gulf of Oman, and Arabian Sea, and a study of Indian samples from the Arabian Sea and the Bay of Bengal concluded that fish from these areas comprised a single population. Mitochondrial DNA haplotypes from Narrow-barred Spanish Mackerel sampled in the Indo-Malay-Papua Archipelago and the western South Pacific coalesced into a clade that was deeply separated from a second clade consisting of all haplotypes from the Persian Gulf and Sea of Oman. Further phylogeographic partitioning was evident between the Indo-Malay-Papua Archipelago and regions sampled to the east and south of it, northern Australia, West Papua, and the Coral Sea. Genetic analyses revealed an east-west division along Wallace's Line, separating Australian populations from those in the East Indies. Cluster analysis of stable isotopes in the sagittal otolith separated Australian samples into four subregions: central Western Australia, northwestern Australia, northern Australia and the Gulf of Carpentaria, and eastern Australia. Later genetic studies confirmed the presence of three stocks of Narrow-barred Spanish Mackerel across northern Australia: northwest, Torres Strait, and east coast. The population around New Caledonia seems to be separate from other populations and comprises two distinct subpopulations, one from the east coast and northern Belep Islands and another from the west and south coast of the main island.

FISHERIES INTEREST: Narrow-barred Spanish Mackerel are taken throughout their range by commercial, artisanal, and recreational fisheries. They are caught primarily with gill nets in southern Asia and India, but are also caught with purse seines, bamboo stake traps, midwater trawls, rod and reel and by trolling. They are also taken as bycatch in longline, purse seine and gill net gear targeting larger scombrids. They are one of the most sought-after game fishes in southern African coastal waters and in Australia. They are marketed fresh, dried-salted, frozen, smoked, and canned; they are commonly made into fish balls.

Catch estimates for Narrow-barred Spanish Mackerel are highly uncertain due to a lack of identification to species level in some fisheries or a lack of reporting in others. Reported global catches increased steadily over the past 65 years, with landings of approximately 11,000 mt in 1950, 80,000 mt in 1980, 170,000 mt in 2000, and 305,000 mt in 2015.

THREATS: Landings data are lacking for some fisheries, and in some instances, landings of Narrow-barred Spanish mackerels are included with other species of Spanish Mackerel. Effort data are also lacking for many fisheries. However, regional overfishing poses the most serious threat to this species. Several sub-regional stock assessments in the western Indian Ocean report this species to be heavily overexploited, with some estimates of fishing mortality in the Arabian Gulf between two and four times FMSY. In Oman, the fishery is overfished, with a high chance of recruitment failure in the near future. High fishing mortality of juveniles has been reported from the Arabian Sea and Persian Gulf, and non-selective gears such as the drift and set net fisheries that catch small fish should be avoided in those regions to prevent growth overfishing.

In Indian waters the minimum size at maturity in the region is 70 cm, so nearly all the fish caught by trawl and purse seine are juveniles. Measures are needed to increase the size of the mesh in purse seines and nets used by the trawlers, to allow escape of juveniles. Effort data is needed to interpret national and regional catch landings. Narrow-barred Spanish Mackerel are listed as Near Threatened on the IUCN Red List.

CONSERVATION:

Western Indian Ocean: In South Africa, recent reports note that there is no indication of overfishing of Narrow-barred Spanish Mackerel. However, several sub-regional stock assessments in the western Indian Ocean report this species to be overexploited, with some estimates of fishing mortality in the Arabian Gulf between two and four times FMSY. However, one assessment in Saudi waters of the Gulf concluded that fishing was occurring at sustainable levels. In Oman, it is estimated that the current fishing mortality rate is 16% for females and 27% for males, indicating that overfishing is occurring. Similarly, in the Arabian Sea, the fishing mortality rate is four times greater than the FMSY, and the resource is heavily exploited. Landings in the region increased to 35,000 mt in 1988, followed by a precipitous decline associated with a collapse of the fishery and a drop in the annual catch to 10,662 mt by 1991. High fishing mortality rates of juveniles have been reported from the Arabian Sea and Persian Gulf.

India: In 1996, a stock assessment of Narrow-barred Spanish Mackerel along the western coast of India indicated that the exploitation rate should be reduced by 60%, based on an estimated MSY. During fishing in Karnataka, India, in 2004–9, the mean length of Narrow-barred Spanish Mackerel caught by drift gill net was 69 cm, whereas it was only 39 cm and 44 cm in trawl and purse seine catches, respectively. The minimum size at maturity in the region is 70 cm, so nearly all the fish caught by trawl and purse seine were juveniles. Measures are needed to increase the size of the mesh in purse seines and nets used by the trawlers, to allow escape of juveniles.

Australia: Stock assessments of Narrow-barred Spanish Mackerel in eastern Australia indicate the stock is presently at 40–50% of the unfished biomass. In 2007, this species was classified as fully exploited, with declines in landings but with stable CPUE from 1997 to 2007. Assessments have also been carried out in northern Australia, the Torres Strait, and in Western Australia. A number of models were explored and concluded that the Western Australia stock may be approaching full exploitation, fully exploited, or somewhat overexploited. Narrow-barred Spanish Mackerel in Northern Territory waters were heavily exploited by Taiwanese drift netters from around 1974, but these vessels were excluded in the mid-1980s, and the annual catch since that time (including illegal fishing) has varied between 600 and 1,100 mt. Stock assessments in the Northern Territory of Australia concluded that overfishing is not occurring and that the stock is not overfished. In the Torres Strait, there has been an estimated 40–50% decline in biomass since 1980. Based on assessments in 2000 and 2002, it was acknowledged that there was a significant degree of uncertainty in fisheries models and that the fisheries for this species in Queensland were overfished and in danger of collapsing.

Narrow-barred Spanish Mackerel comprises a number of stocks throughout its range, and conservation measures may be better implemented by national agencies rather than regional management organizations. It has been suggested that banning fishing for Narrow-barred Spanish Mackerel in the northern Persian Gulf in June and July, the peak of the spawning season there, would help protect brood stocks. There are no known species-specific conservation measures for this species, except in Australia, where there are minimum size and bag limits. Regulations include a minimal size limit of 75 cm that applies to both commercial and recreational fishers on the east coast of Queensland. This size is below the minimum size of first maturity of 89 cm TL. Recreational fishers are limited to 10 Narrow-barred Spanish Mackerel per angler. These licenses also regulate fishing practices and gear. There is a recreational possession limit of two fish in Northern Territory waters. A TAC of 410 mt was set for all zones in Western Australia in 2003 as the fishery moved from an interim-managed fishery to a fully managed fishery in 2011.

REFERENCES: Bakhoum 2007; Begg et al. 2006; Ben Meriem et al. 2006; Ben Souissi et al. 2006; Bleeker 1851; Buckworth and Clarke 2001; Chacko et al. 1968; Claereboudt et al. 2005; Collette and Nauen 1983; Collette and Russo 1985; Collette et al. 2011; Corsini-Foka and Kalogirou 2008; Darvishi et al. 2012; Devaraj 1981, 1983, 1999; Devaraj and Mohamad Kasim 1998; Devaraj et al. 2000; Dineshbabu et al. 2012; Fauvelot and Borsa 2011; Govender et al. 2006; Grandcourt 2013; Grandcourt et al. 2005; Grant 1982; Grubert et al. 2013; Jayabalan et al. 2011; Jenkins et al. 1984a, 1984b, 1985; Kaymaram et al. 2013, 2014; Kishinouye 1915; Kumaran 1964; Lacepède 1800; Lewis et al. 1974, 1983; Mackie et al. 2003; Mansourkiaei et al. 2016; McPherson 1992; Merceron 1970; Molony 2015; Motlagh and Shojaei 2009; Munro 1942; Newman et al. 2009; Niamaimandi et al. 2015, 2017; Prado 1970; Roa-Ureta 2015; Shaw 1803; Streftaris and Zenetos 2006; Sulaiman and Ovenden 2010; Tobin and Mapleston 2004; Van der Elst 1981; Vineesh et al. 2017; von der Heyden et al. 2014; Welch et al. 2002.

Monterey Spanish Mackerel

Scomberomorus concolor (Lockington, 1879)

COMMON NAMES: English – Monterey Spanish Mackerel, Gulf Sierra
French – Thazard de Monterey
Spanish – Carite de Monterey

ETYMOLOGY: The Monterey Spanish Mackerel was named *concolor*, the Latin for "uniformly colored," by the California zoologist William N. Lockington (1879a) because, as he stated later (Lockington 1879b), it lacks the dark oblique streaks on the flanks of the Eastern Pacific Bonito (*Sarda chiliensis*), from which fish Lockington was differentiating it.

SYNONYMS: *Chriomitra concolor* Lockington, 1879

TAXONOMIC NOTE: This species is frequently confused with another species of Spanish mackerel, the Pacific Sierra, to which it is closely related (Banford et al. 1999) and with which it co-occurs.

FIELD MARKS:

1 The Monterey Spanish Mackerel has more gill rakers on the first arch, 21–27, which differ entiates it from the co-occurring Pacific Sierra, which has only 12–17 gill rakers.
2 The uniformly colored males are more easily distinguished from the Pacific Sierra.
3 The females of Monterey Spanish Mackerel and Pacific Sierra have golden spots on their sides: two to four rows (up to eight in large specimens) in Pacific Sierra, but only two rows in female Monterey Spanish Mackerel.

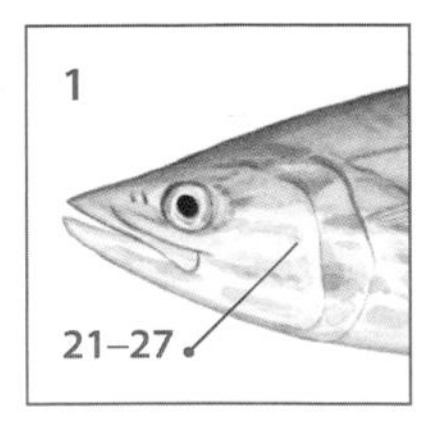

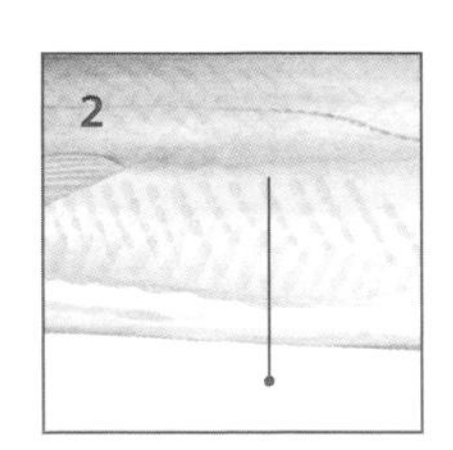

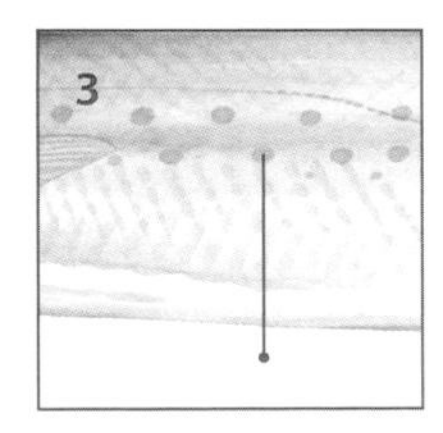

UNIQUE FEATURE: This is the most endangered species of Spanish Mackerel.

DIAGNOSTIC FEATURES: Gill rakers: Gill rakers on the first arch are numerous: 4–8 on upper limb, 15–21 on lower limb, total 21–27. **Fins:** The first dorsal fin has 15–18 spines, usually 17; the second dorsal fin has 16–20 rays, usually 18 or 19, followed by 6–9 finlets, usually 8. The anal fin has 19–23 rays, usually 20, followed by 6–8 finlets. The pectoral fin has 19–22 rays, usually 21. **Lateral line:** The lateral line curves down toward the caudal peduncle. **Vertebrae and intestine:** Vertebrae number 18–20 precaudal plus 27–29 caudal, total 46–48, usually 19 + 28 = 47. The intestine has 2 folds and 3 limbs. Descriptions and illustrations of the osteology and soft anatomy of the Monterey Spanish Mackerel are included in the revision of the Spanish mackerels (Collette and Russo 1984). **Otoliths:** Sagittal otoliths were illustrated and distinguished from other eastern Pacific tunas and mackerels (Fitch and Craig 1964). **Color:** Males are steel blue dorsally, silvery on sides and below, without streaks or spots; females are darker, with two alternate rows of brown spots (gold in life) on sides.

GEOGRAPHIC RANGE: Monterey Spanish Mackerel is endemic to the eastern Pacific. Historically, it probably had a continuous distribution from Monterey Bay, California, around Baja California and into the Gulf of California. It was caught commercially in the 1870s and 1880s in Monterey Bay, California. The initial diminution of range began between 1880 and 1920, perhaps associated with gradual reduction in water temperature along the California coast. There are only 10 recent records from the California coast, and its occurrence on the west shore of the Gulf of California is extremely rare, with only three records since 1968. They were abundant in the Gulf of California in the 1970s and were observed in Guaymas in the early 1980s. The present distribution is limited to the central and northern part of the Gulf of California. Over the last 20 years very few specimens have been found outside of this area, although there are records from Loreto in the lower Gulf of California as late as 1993.

SIZE: Maximum size 77 cm FL. Otolith aging indicates a maximum age of 8 years. Based on their length-weight curve, length at first maturity is 36.5 cm at 3 years. Generation length is estimated to be 3.8 years. In a study in the Gulf of California, 442 specimens represented 7 age groups, 0–5 for males, 0–6 for females. Maturity for males averaged 36.6 cm FL and 1.8 years and for females, 38.3 cm FL and 2 years.

HABITAT AND ECOLOGY: Monterey Spanish Mackerel are residents of the coastal pelagic environment of the upper Gulf the year-round. There is a single homogeneous population in the Gulf that moves seasonally to spawn and feed. Their biology is poorly known. They occur along the upper east coast of the Gulf of California in the fall months in shallow estuaries; spawning occurs in late spring and early summer. They may be colder water fish than *S. sierra*, retreating to deep waters in summer.

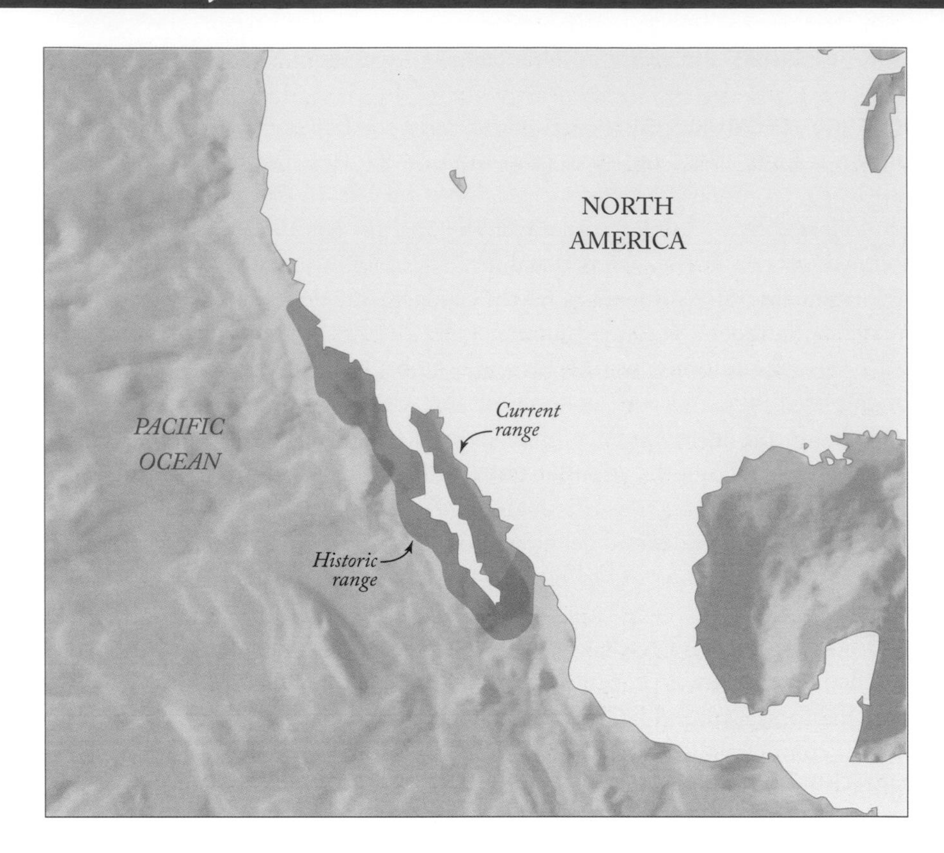

Right: Monterey Spanish Mackerel range map.

FOOD: Monterey Spanish Mackerel feed on euphausiids and clupeids. In the winter when cooler conditions predominate in the upper Gulf, they move to the central region to feed on large concentrations of sardines and anchovies.

REPRODUCTION: The reproductive season for Monterey Spanish Mackerel begins in April, peaks from April through July, and ceases by September.

FISHERIES INTEREST: The Monterey Spanish Mackerel is among the top four species taken by small-scale fishing operations inside the upper Gulf of California Biosphere Reserve. In the upper Gulf of California, it is caught with another species of Spanish mackerel, *Scomberomorus sierra*, mainly using gill nets close to the coast. The nets most commonly used are 400 m long and 2 m wide and set at an average of about 15 m deep. The commercial fishery operates from November to April in shallow coastal waters, bays, and estuaries, and the combined catch of both species is about 4,500 mt per year. Sport fishers also catch both species of Spanish mackerels.

THREATS: Because catches are reported as a combined group of both species of Spanish mackerels ("sierras"), it is not possible to determine a population trend for this species. The fishery is apparently having severe effects on the remnant of the *S. concolor* population. Intense fishing pressure on spawning aggregations that are predictable in time and space appear to have led to rapid declines, and the species is at risk of collapse. It is important to note that the northern Gulf of California has been heavily altered by the reduction of flow of the Colorado River. Two other endangered species occur in the area, the Totoaba (*Totoaba macdonaldi*, family Sciaenidae) and the Vaquita (*Phocoena sinus*), a porpoise.

CONSERVATION: There are no specific conservation measures for this species. Restrictions on the gill net fishery for Totoaba in 1975 may have helped all three endangered species survive. Establishment in 1993 of the Gulf of California Biosphere Reserve as a marine protected area to protect the Totoaba and the Vaquita may also help protect the Monterey Spanish Mackerel, but the gill net fishery is continuing. In order to specify reference points for sustainable exploitation, population parameters for both co-occurring species of Spanish mackerels are needed.

The extent of occupancy of this species is inferred to have been reduced by more than 80%, based on both historical and new information. Prior to 1961 there were records of this species in California, and there is strong evidence that this species no longer occurs outside of the Gulf of California. Furthermore, within the Gulf of California, at least since the 1980s, the range has retracted further to its present limits in the northeastern part of the Gulf of California. The greater than 80% reduction in the population inferred over the past 40 years based on the reduction in range plus the present high levels of exploitation led to an evaluation of Vulnerable on the IUCN Red List.

REFERENCES: Banford et al. 1999; Collette 1995; Collette and Nauen 1983; Collette and Russo 1985; Collette et al. 2011; D'Agrosa et al. 1994; Dominguez-López et al. 2015; Erisman et al. 2015; Espinoza-Tenorio et al. 2010; Fitch and Craig 1964; Fitch and Flechsig 1949; Goode 1884; Instituto Nacional de la Pesca 2002; Jaramillo-Legorreta and Taylor 2010; Jefferson 2010; Lockington 1879a, 1879b; Quiñónez-Velázquez and Montemayor-López 2002; Philips 1932; Ramírez-Pérez et al. 2015; Roedel 1939; Romero et al. 1994; Valdovinos-Jacobo 2006.

Indo-Pacific King Mackerel

Scomberomorus guttatus (Bloch & Schneider, 1801)

COMMON NAMES: English – Indo-Pacific King Mackerel
French – Thazard ponctué
Spanish – Carite del Indo-Pacific

ETYMOLOGY: Named *guttatus*, the Latin for "spotted," by the German physician and naturalist Marcus Élieser Bloch and the German scholar Johann Gottlob Theanus Schneider, who corrected and expanded *Sistema Ichthyologiae* (1801:23-24).

SYNONYMS: *Scomber guttatus* Bloch and Schneider, 1801; *Scomber leopardus* Shaw, 1803; *Cybium interruptum* Cuvier, 1832; *Cybium kuhlii* Cuvier, 1832; *Cybium crookewitii* Bleeker, 1851; *Cybium guttatum* (Bloch and Schneider, 1801); *Scomberomorus crookewiti* (Bleeker, 1851); *Scomberomorus guttatus* (Bloch and Schneider, 1801)

TAXONOMIC NOTE: The Indo-Pacific King Mackerel was misidentified as Korean Seerfish in India until the Indian ichthyologist Muthiah Devaraj (1976) recognized the distinctness of the two species.

FIELD MARKS: 1 Both the Indo-Pacific King Mackerel and the Korean Seerfish are spotted and have numerous fine auxiliary branches from the anterior part of the lateral line. However, the body of the Indo-Pacific King Mackerel is not as deep as that of the Korean Seerfish, 22.8–25.2% FL versus 24.4–26.7% FL.

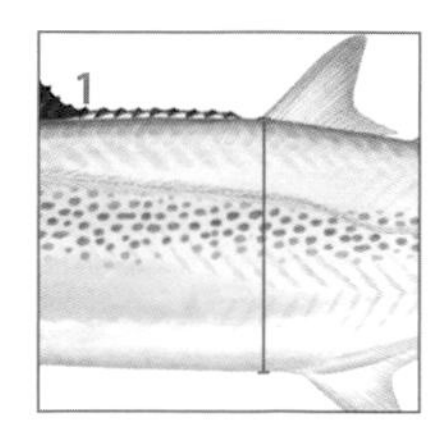

UNIQUE FEATURE: Two species of Spanish mackerels have many small auxiliary branches to the anterior part of the lateral line, but the Indo-Pacific Spanish Mackerel has the usual two loops and three limbs to the intestine.

DIAGNOSTIC FEATURES: Gill rakers: There are relatively few gill rakers on the first gill arch: 1 or 2 on the upper limb, 7–12 on the lower limb, for a total of 8–14, usually 10–12. **Teeth:** There are 11–36 teeth on each side of the upper jaw and 12–34 teeth on each side of the lower jaw. **Fins:** The first dorsal fin has 15–18 spines; the second dorsal fin has 18–24 rays, followed by 7–10 finlets. The anal fin has 19–23 rays, also followed by 7–10 finlets. There are 20–23 pectoral-fin rays, usually 21. **Lateral line:** The lateral line has auxiliary branches anteriorly and gradually curves down toward the caudal peduncle. **Vertebrae and intestine:** Vertebrae number 19–22 precaudal plus 28–31 caudal, for a total 47–52, usually 50 or 51. The intestine has 2 folds and 3 limbs. Descriptions and illustrations of the osteology of the Indo-Pacific King Mackerel are included in two papers, Devaraj (1977) and Collette and Russo (1984). **Color:** Fresh specimens from Wakasa Bay in the Sea of Japan were grayish blue dorsally and silvery white laterally and ventrally. There were several longitudinal rows of small brownish dots scattered rather densely along the lateral median line. The first dorsal-fin membrane is black. The pectoral, second dorsal, and caudal fins were dark brown. The pelvic and anal fins were silvery white. Recent fresh specimens appeared iridescent greenish dorsally, white laterally and ventrally, with dense brownish spots along the median line.

GEOGRAPHIC RANGE: Indo-Pacific King Mackerel are restricted to the Indo-West Pacific from the Persian Gulf, Arabian Sea, and Sri Lanka to southeast Asia, Borneo, and the East Indies north to Hong Kong, Taiwan, and Amoy and Swatow, China. The range extends farther out in the East Indies than the ranges of the Streaked Seerfish or Korean Seerfish, at least to Bali. The northernmost record is from Wakasa Bay, Sea of Japan.

SIZE: Adult Indo-Pacific King Mackerel can reach a maximum length of 76 cm FL. Length-frequency analysis of fish from Waltair on the east coast of India indicates that they reach lengths of 28.0, 42.5, 53.0, 61.0, 67.0, 72.0 and 77.0 cm at 1 to 7 years of age, respectively. They may live for 16 years. Length-weight relationships have been published for populations from West Bengal and the Karachi coast of Pakistan.

HABITAT AND ECOLOGY: This is a pelagic migratory fish inhabiting coastal waters of depths less than 200 m. It sometimes enters turbid estuarine waters and is usually found in small schools.

FOOD: In India, adult and juvenile Indo-Pacific King Mackerel feed mainly on teleost fishes, particularly on about five small species of coastal pelagic fishes, mostly sardines. The dietary shift from small juvenile fishes, which constitute the exclusive food of the young, to sardines takes place at about a length of 31.5 cm. Indo-Pacific King Mackerel are crepuscular feeders, most active between 1900 and 2200 hours and between 0600 and 1000 hours. Juveniles 16–161 mm TL feed mostly on juveniles of small fishes, particularly anchovies, and also crustaceans such as copepods.

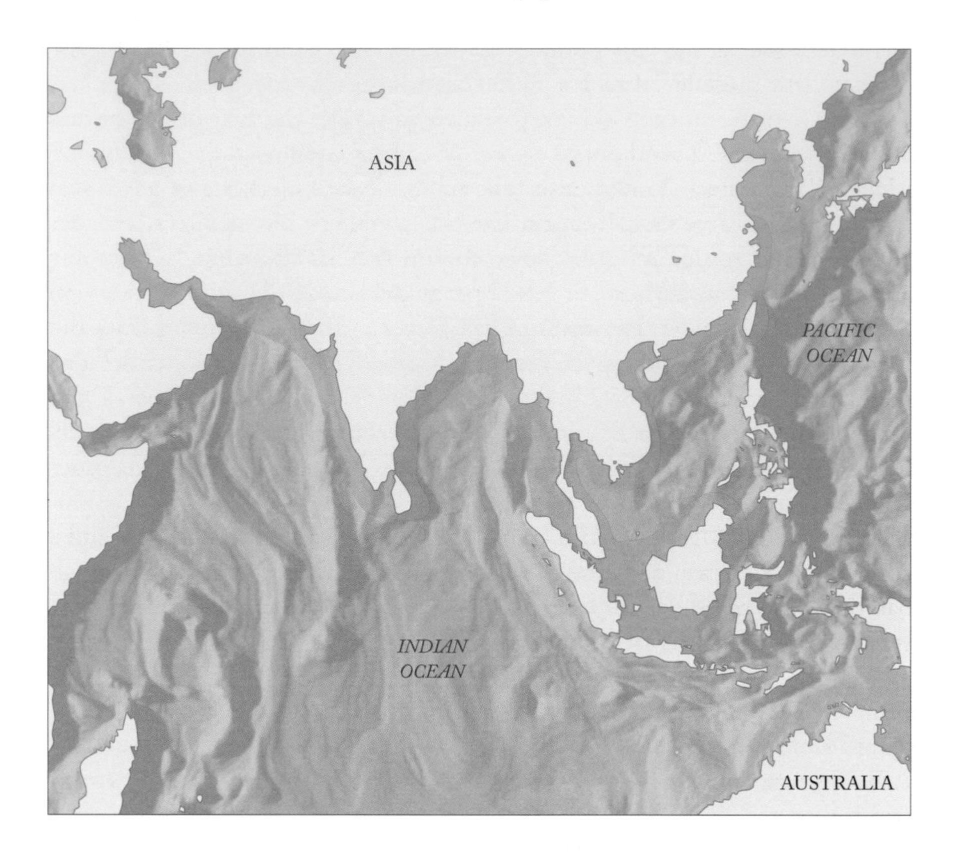

Right: Indo-Pacific Spanish Mackerel range map.

REPRODUCTION: Sexual maturity is reached at around 48–52 cm TL or 1–2 years old in southern India and at about 40 cm TL in Thailand and the Gulf of Mannar. Based on the occurrence of ripe females and the size of maturing eggs, spawning probably occurs from April to July in southern India and in May in Thailand waters. Spawning takes place at depths of 20–60 m in the northern Gulf of Mannar. Fecundity increases with age in the Indian waters, ranging from around 400,000 eggs at age 2 years to over 1 million eggs at age 4 years. Fully ripe eggs measure 1.2 mm in diameter, with a single oil-globule of 0.16–0.34 mm.

EARLY LIFE HISTORY: Indo-Pacific King Mackerel larvae hatch at an average length of 2.8 mm, and first feeding and transformation of prolarvae into postlarvae occur about 96 hours after hatching at a length of about 3.1 mm. Postlarvae 3.8 to 14.5 mm long caught in plankton tows fed on copepod nauplii, and immature fishes of 114–62 mm fed largely on juvenile crustaceans. Eight larvae and juveniles, 2.8–28.5 mm body length, that were caught in plankton tows, were described and illustrated by Vijayaraghavan. Nine larvae

4.15–8.75 mm body length collected in plankton tows from the Gulf of Tonkin and the northern part of Nan Hai in 1963 and 1965 were described and illustrated. In early stages, Indo-Pacific King Mackerel can be distinguished from Narrow-barred King Mackerel by the greater number of gill rakers, the length of the preopercular spines, and the position of the anal fin with relation to the second dorsal fin. They differ from Streaked Seerfish by the characteristic presence of bluish green coloration.

STOCK STRUCTURE: Morphometric comparisons of juvenile Indo-Pacific King Mackerel from five Indian localities, four on the east coast and one on the west coast, revealed highly significant differences among all locations. A broader comparison of morphometric and meristic data from five major geographic areas found no significant differences between samples from the Arabian Sea and Bay of Bengal but significant differences between samples from the Bay of Bengal, East Indies, Gulf of Thailand, and China. The Indian Ocean population had a mode of 50 vertebrae while the other populations all had modes of 51. Genetic analysis revealed that the Indo-Pacific King Mackerel in the South China Sea was genetically diverse.

FISHERIES INTEREST: The Indo-Pacific King Mackerel is a highly valued commerical species and is caught with gill nets, midwater trawls, purse seines, bamboo stake traps, set nets, and hand lines and by trolling. Mesh selectivity studies concluded that a 52 mm mesh bar size was the most efficient at catching commercial-sized fish. In India, Indo-Pacific King Mackerel comprise 23.5–39.4% of the seerfish landings, less than those of the Narrow-barred King Mackerel but much greater than those of the Streaked Seerfish. Landing estimates for Indo-Pacific King Mackerel are projected from very small amounts of information and are therefore highly uncertain, and landings data for the species are often combined with landings of Indo-Pacific Seerfish and Korean Seerfish. Estimated catches have increased steadily since the mid 1960s, reaching around 10,000 mt in the early 1970s and over 30,000 mt by 1989. Landings have remained between 40,000 and 50,000 mt since 2000. In recent years, the countries with the highest catches are Indonesia, India, and Iran. They are marketed mainly fresh but also dried-salted, smoked, and frozen. Small quantities of frozen product are exported to Europe and North America. A proximate analysis of the meat was conducted on fish caught in the Persian Gulf. A traditional fermented fish product called "budu" is made from Indo-Pacific King Mackerel and produced and distributed in an area in the province of West Sumatra.

THREATS: Trawls and small meshed gill nets in the Indian fishery exploit Spanish mackerels at small sizes, below minimum size at first maturity. This results in growth overfishing and may ultimately lead to depletion of spawning stocks and recruitment overfishing. This species is listed as Data Deficient on the IUCN Red List.

CONSERVATION: Indo-Pacific King Mackerel were considered overexploited by the trawl fishery in three Indian states and marginally overexploited by the gill net fishery at Veraval, Gujarat state, India. A more recent analysis concluded it was not overfished in the upper Bay of Bengal because the fishing mortality was lower than the natural mortality. A recent assessment of Indo-Pacific King Mackerel in the Indian Ocean used data-poor catch-based methods, and the stock was found to be fully fished but not overfished, and overfishing was not occurring. There is a lack of reporting for this species, and there is considerable uncertainty associated with the most recent assessment, which indicates the Indian Ocean stock is fully fished. Catch statistics for the Indo-Pacific King Mackerel are often mixed with that for Streaked Seerfish and Korean Seerfish in at least half of this species' range. Effort information is needed to better interpret the increase in landings. Reducing fishing intensity by trawls and small meshed gill nets was considered impractical, but a good alternative would be to increase the mesh size of those nets to allow juveniles to escape.

REFERENCES: Ahmed et al. 2014; Bloch and Schneider 1801; Collette and Russo 1985; Collette et al. 2011; Devaraj 1976, 1977, 1981, 1987, 1998; Devaraj and Mohamad Kasim 1998; Devaraj et al. 1999; Dutta et al. 2012; Ghosh et al. 2009; Krishnamoorthi 1958; Kumaran 1964; Mohamad Kasim et al. 2002; Moini et al. 2012; Muthiah and Pillai 2003; Muthiah et al. 2002; Nakamura and Nakamura 1982; Rashid et al. 2010; Somvanshi et al. 1998; Sreekrishna et al. 1972; Srinivasa Rao 1964, 1975; Srinivasa Rao and Ganapati 1977; Tongyai 1966, 1970; Venkataraman 1961; Vijayaraghavan 1955; Ye 2012; Yusra et al. 2013; Zhang 1985.

Korean Seerfish

Scomberomorus koreanus (Kishinouye, 1915)

COMMON NAMES: English – Korean Seerfish
French – Thazard Coréen
Spanish – Carite Coreano
Japanese – Hira-sawara

ETYMOLOGY: *Scomberomorus koreanus* was named by the distinguished Japanese biologist Kamakichi Kishinouye (1915) for the locality of his original specimens, Korea.

SYNONYMS: *Cybium koreanum* Kishinouye, 1915; *Sawara koreanum* (Kishinouye, 1915); *Scomberomorus guttatus koreanus* (Kishinouye, 1915)

TAXONOMIC NOTE: In India this species was confused with *Scomberomorus lineolatus* until Devaraj (1976) clearly separated the two species.

FIELD MARKS:

1 The lateral line has fine auxiliary branches extending out from the anterior portion of the lateral line.
2 There are several rows of small round to elliptical brownish dark spots scattered along the sides of the body.
3 The body of the Korean Seerfish is deeper than that of the Indo-Pacific King Mackerel, 24.4–26.7% FL versus 22.8–25.2% FL.

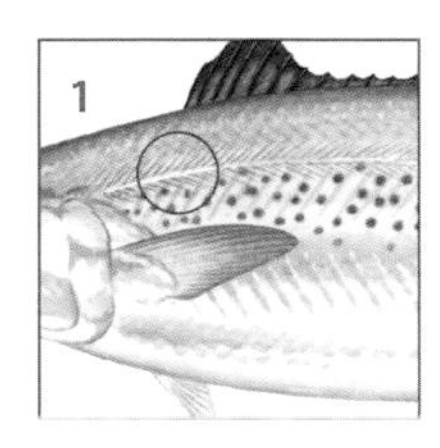

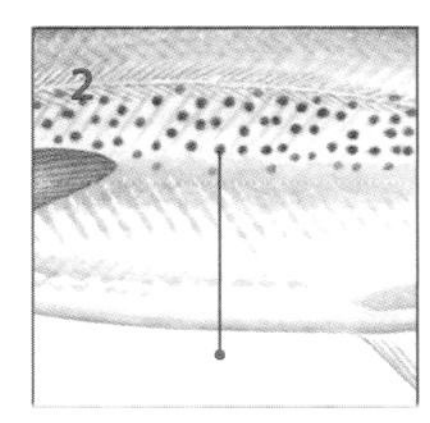

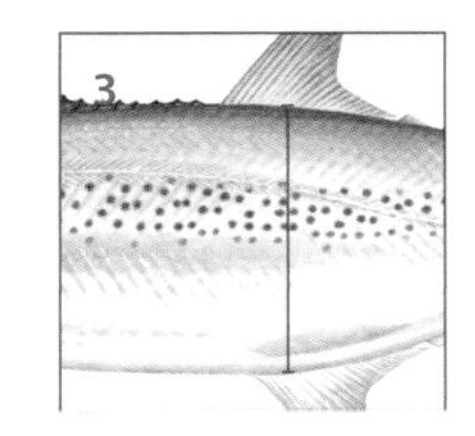

UNIQUE FEATURE: Two species of Spanish mackerels have many small auxiliary branches to the anterior part of the lateral line, but the Korean Seerfish is the only Spanish mackerel that has four loops and five limbs in the intestine; other Spanish mackerels have only two loops or have a straight intestine.

DIAGNOSTIC FEATURES: Gill rakers: There are a total of 11–15 gill rakers on the first gill arch, usually 1 or 2 on the upper limb, 11 or 12 on the lower limb. **Fins:** The first dorsal fin has 14–17 spines; the second dorsal has 20–24 rays, followed by 7–9 finlets. The anal fin has 20–24 rays, followed by 7–9 finlets. The pectoral fin has 20–24 rays. **Lateral line:** The lateral line gradually curves down toward the caudal peduncle. As in *S. guttatus*, there are numerous fine auxiliary branches from the anterior part of the lateral line. **Vertebrae and intestine:** Vertebrae number 21 or 22 precaudal plus 30 or 31 caudal, for a total of 51 to 53. This is the only species of Spanish mackerel with four loops and five limbs to the intestine. The osteology of the Indian species of Spanish mackerels was described and illustrated by Devaraj (1977). Descriptions and illustrations of the soft anatomy and osteology are included in the revision of Spanish mackerels by Collette and Russo (1984). **Color:** Fresh specimens are grayish blue dorsally, silvery white laterally and ventrally. There are several rows of small round to elliptical brownish dark spots scattered along the sides of the body. The first dorsal fin membrane is black. The pectoral, second dorsal, and caudal fins are dark brown. The pelvic and anal fins are silvery white.

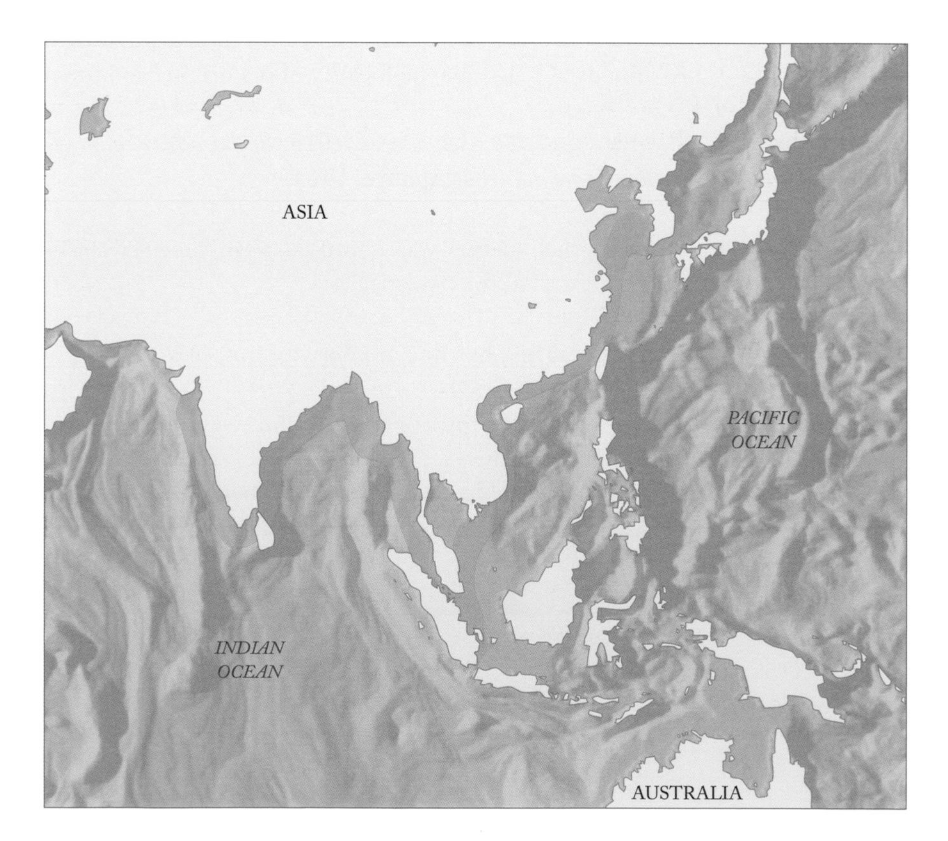

Right: Korean Seerfish range map.

GEOGRAPHIC RANGE: The Korean Seerfish is widespread in continental south and southeast Asia. It is found in the Indo-West Pacific from Pakistan, Sri Lanka, and the west coast of India along the continental shelf eastward to Singapore and out only as far as Sumatra, Indonesia. The range extends north to China, Korea, and Wakasa Bay, Sea of Japan. The second record from Japan, from Kagoshima Province, was published only recently.

SIZE: Maximum size is 150 cm FL and 15 kg.

HABITAT AND ECOLOGY: The Korean Seerfish is pelagic and oceanodromous and occurs to at least 100 m depth.

FOOD: The Korean Seerfish feeds on small schooling fishes, such as sardines and anchovies, and also on shrimps.

REPRODUCTION: Korean Seerfish mature at 72 cm FL and 2.25 kg. They have been reported to spawn in July in the inlet of the Taedong River near Nampo, North Korea.

FISHERIES INTEREST: The fishery for Korean Seerfish began in Daidoko, Korea, by Japanese fishermen in 1917 using drift nets and pound nets. It is not common and is a minor commercial species in parts of its range. It forms a significant portion of the catch of the set net fishery off the Dadohae Marine National Park in Korea and is an important part of the drift net fishery in Palk Bay and the Gulf of Mannar between southeastern India and Sri Lanka but only accounts for 13% of the four species of Spanish mackerels taken in the Indian fishery. Available landing data are often mixed with other species of Spanish mackerels.

THREATS: Species-specific landing and catch effort data are needed from throughout its range, especially as it is considered rare and the impact of fisheries on its population is not known. It is listed as Least Concern on the IUCN Red List. Recommended research includes more information on catch landings and effort for this species. More research is needed on this species biology because this is one of the least-known species of Spanish mackerels.

CONSERVATION: There are no species-specific assessments or conservation measures for Korean Seerfish.

REFERENCES: Cheong and Park 2009; Collette 2001; Collette and Nauen 1983; Collette and Russo 1985; Collette et al. 2003; Collette et al. 2011; Devaraj 1976, 1977; Hata et al. 2015; Kishinouye 1915, 1923; Naik et al. 1998; Nakamura and Nakamura 1982; Siddiqui and Amir 2012.

Streaked Seerfish

Scomberomorus lineolatus (Cuvier, 1829)

COMMON NAMES: English – Streaked Seerfish
French – Thazard Cirrus
Spanish – Carite Rayado

ETYMOLOGY: Named *lineolatum* by the distinguished French anatomist Georges Cuvier (1829) in reference to the distinctive lines along the sides of this species.

SYNONYMS: *Cybium lineolatum* Cuvier, 1829; *Indocybium lineolatum* (Cuvier, 1829); *Scomberomorus lineolatus* (Cuvier, 1829)

TAXONOMIC NOTE: In Madagascar and east Africa, this species name has been used for the Queen Mackerel, *Scomberomorus plurilineatus*. *Scomberomorus lineolatus* was considered a hybrid between *S. commerson* and *S. guttatus* by Srinivasa Rao and Lakshmi (1993, 1996) but Collette (1994) refuted this.

FIELD MARKS:

1 The Streaked Seerfish is the only species of Spanish mackerel that has a pattern of short horizontal lines on its sides. All other Spanish mackerels have some spots, blotches, or bars or are plain and without any markings.

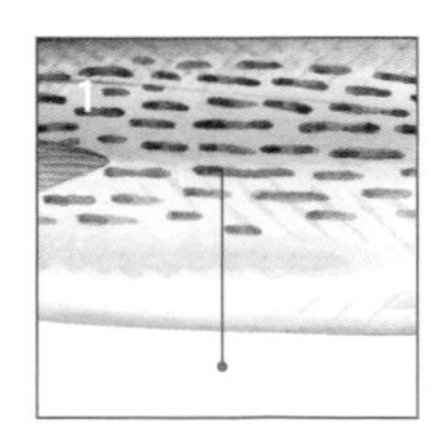

UNIQUE FEATURE: The only Spanish mackerel with a pattern of short, narrow lines on its sides.

DIAGNOSTIC FEATURES: Gill rakers: There are relatively few gill rakers on the first gill arch: 1 or 2 on the upper limb and 6–11 on the lower limb, for a total of 7–13, usually 10 or 11. **Fins:** The first dorsal fin has 15–18 spines; the second dorsal fin has 15–22 rays, followed by 7–10 finlets. The anal fin has 17–22 rays, followed by 7–10 finlets. There are 20–24 pectoral-fin rays, usually 23. **Lateral line:** The lateral line lacks auxiliary branches anteriorly and gradually curves down toward the caudal peduncle. **Vertebrae and intestine:** Vertebrae number 18–20 precaudal plus 25–28 caudal, for a total of 44–46, usually 19 + 27 = 46. The intestine has 2 folds and 3 limbs. Descriptions and illustrations of the osteology of the Streaked Seerfish are included in two papers: Devaraj (1977) and Collette and Russo (1984). **Color:** The upper sides of the body are dark blue dorsally, silvery white ventrally, marked with several rows of elongate lines. Juveniles have spots but develop the adult pattern of interrupted lines by the time they reach a length of 40 cm. The upper areas of the caudal peduncle and the median keel are black, the lower areas dusky. The first dorsal fin is black anteriorly, white posteriorly.

GEOGRAPHIC RANGE: The Streaked Seerfish is an Indo-West Pacific species that is found along the west coast of India and Sri Lanka eastward to Thailand, Malaysia, and Java. It does not extend farther out in the East Indies beyond Wallace's Line.

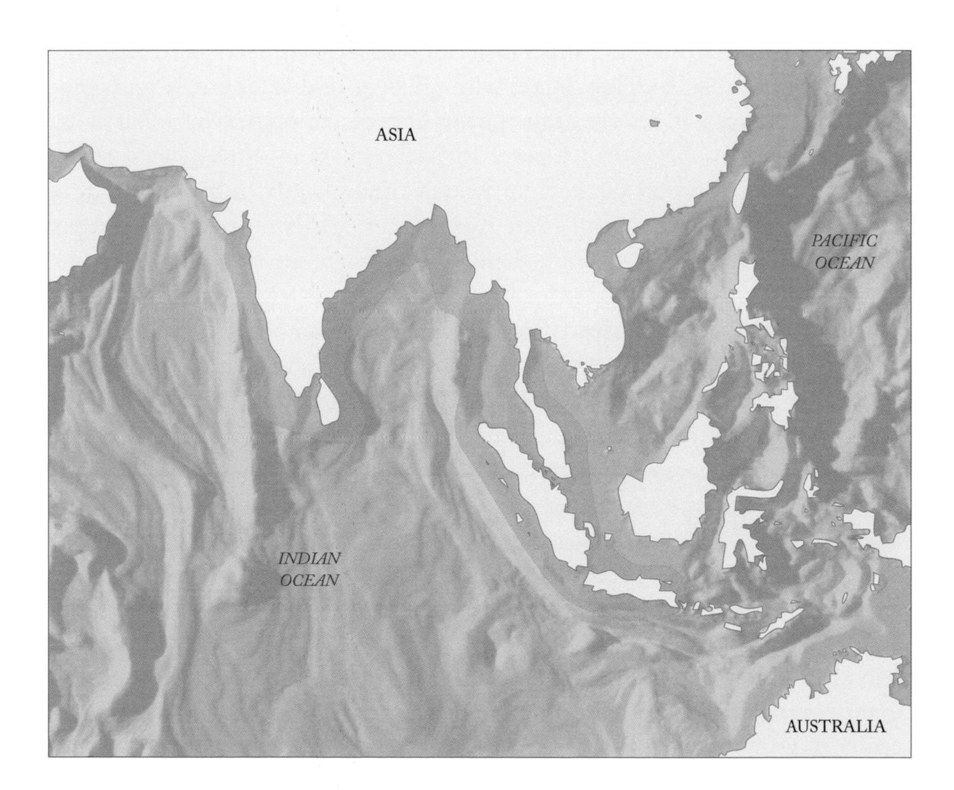

Left: Streaked Seerfish range map.

SIZE: The maximum length of male Streaked Seerfish in Indian catches is 94 cm TL, with a weight of 4.26 kg, and of females, 98.0 cm and 4.55 kg. Length-frequency analysis indicates that Streaked Seerfish reach lengths of 35 cm, 71.3 cm, 83.5 cm, and 96.5 cm TL at 1, 2, 3, and 4 years of age, respectively.

HABITAT AND ECOLOGY: Streaked Seerfish are pelagic and oceanodromous. Unlike the Narrow-barred Spanish Mackerel and the Indo-Pacific King Mackerel, Streaked Seerfish are not found in very turbid waters or waters with greatly reduced salinity. They are more abundant than Korean Seerfish but less abundant than the Indo-Pacific King Mackerel.

FOOD: In India, adult Streaked Seerfish feed primarily on small pelagic fishes, particularly sardines, while juveniles shorter than 20 cm feed on "whitebait" (small pelagic fishes), croakers, and lizardfishes. Feeding is crepuscular, between 0500 and 0900 hours and around 1900 hours.

REPRODUCTION: Length at first maturity is 70 cm TL at age 2. Spawning takes place in inshore waters out to a distance where the water is about 25 m deep.

EARLY LIFE HISTORY: Five postlarvae and juveniles (18.4–99.5 mm) were described from Waltair on the east coast of India and differentiated from Narrow-barred Spanish Mackerel and Indo-Pacific King Mackerel. Early stages were taken in shore seines in February–April, more advanced stages from boat seine catches in July–September.

FISHERIES INTEREST: There are small fisheries for Streaked Seerfish in Thailand, Malaysia, and India. They are caught with gill nets, midwater trawls, and purse seines and by trolling. They are taken from October through November along the Thai coast in the Indian Ocean; in Malaysia from November through February along the west coast, from March through July in the south, from February through March and from August through November in the east; and in India from May through September with other seerfishes. Landing data are often mixed with those for Indo-Pacific Seerfish and Korean Seerfish. Estimated landings of Striped Seerfish in the Indian seerfish fishery from 1991–97 ranged from 45–1,027 mt. The Striped Seerfish is the least important of the four species in the Indian fishery, comprising 0.1–2% of the seerfish catch. During 1991–93, the Streaked Seerfish constituted only 1.3% of the reported Indian Ocean landings of seerfishes. FAO worldwide reported landings are generally low and quite variable. Reported annual landings increased to 200–300 mt during the 1970s and 1980s and varied greatly in the late 1980s and 1990s, exceeding 1,000 mt in 1997. Reported landings have been less than 50 mt since 2002.

THREATS: Trawls and small meshed gill nets in the Indian fishery exploit Spanish mackerels at small sizes, below minimum size at first maturity. This results in growth overfishing and may lead to depletion of spawning stocks and recruitment overfishing. It is listed as Least Concern on the IUCN Red List. However, more data are needed on this species' biology, species-specific landings, and catch effort throughout its range, especially as landing data are often mixed with that for Indo-Pacific King Mackerel and Korean Seerfish.

CONSERVATION: The Streaked Seerfish was considered overexploited in Indian waters. There are no species-specific conservation measures for Streaked Seerfish. Reducing fishing intensity by trawls and small-meshed gill nets was considered impractical, but a good alternative would be to increase the mesh size of those nets.

REFERENCES: Collette 1994; Collette 2001; Collette and Nauen 1983; Collette and Russo 1985; Collette et al. 2011; Devaraj 1977, 1981, 1986, 1998; Devaraj et al. 1999; Mohamad Kasim et al. 2002; Muthiah and Pillai 2003; Muthiah et al. 2002; Somvanshi et al. 1998; Srinivasa Rao 1964; Srinivasa Rao and Ganapati 1977; Srinivasa Rao and Lakshmi 1993, 1996; Tongyai 1970.

Atlantic Spanish Mackerel

Scomberomorus maculatus (Mitchill, 1915)

COMMON NAMES: English – Atlantic Spanish Mackerel
French – Thazard Atlantique, Thazard Tacheté du Sud
Spanish – Carite Atlántico, Sierra Pintada

ETYMOLOGY: The Atlantic Spanish Mackerel was named *maculatus*, meaning black, by the American physician and naturalist Samuel Latham Mitchill (1815) but with no specific indication of why he chose this name. Perhaps it was in reference to the anterior part of the dorsal fin, which is black.

SYNONYMS: *Scomber maculatus* Mitchill, 1815; *Cybium maculatum* (Mitchill, 1815)

TAXONOMIC NOTE: Three species have often been confused with *S. maculatus*, namely: *Scomberomorus tritor* in the eastern Atlantic; *S. sierra* in the eastern Pacific; and *S. brasiliensis* in the Caribbean and along the Atlantic coast of Central and South America (Banford et al. 1999).

FIELD MARKS:

1 The Atlantic Spanish Mackerel is very similar to the Serra Spanish Mackerel in having rows of yellow-bronze spots along its sides.
2 It differs mainly in having more vertebrae and second dorsal-fin rays that are easier to count.
3 Perhaps they are best differentiated by locality—any spotted Spanish mackerel from the coast of the United States and throughout the Gulf of Mexico are Atlantic Spanish Mackerel; those from Yucatan south to Brazil are Serra Spanish Mackerel.

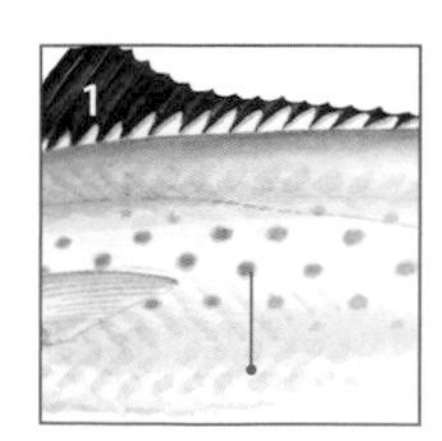

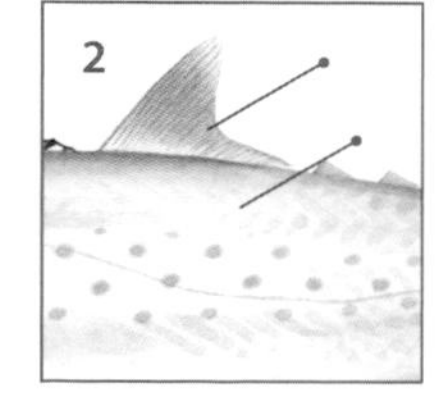

UNIQUE FEATURE: The western Atlantic Spanish mackerel with the most vertebrae, 51–53.

DIAGNOSTIC FEATURES: Gill rakers: Gill rakers on first arch moderate, 1–4 on the upper limb, 8–13 on the lower limb, usually 10 or 11 for a total of 10 to 16. **Fins:** The first dorsal fin has 17–19 spines; the second dorsal fin has 17–20 rays, usually 18 or more, followed by 7–9 finlets. The anal fin has 17–20 rays, followed by 7–10 finlets. The pectoral fin has 20–23 rays, modally 21. The pelvic fins average slightly longer, 4.6–5.8% FL, compared to *S. brasiliensis*, 3.6–5.9% FL. **Lateral line:** The lateral line gradually curves down toward the caudal peduncle. **Vertebrae and intestine:** Vertebrae number 21 or 22 precaudal plus 30 or 31 caudal, for a total of 51 to 53. The intestine has 2 folds and 3 limbs. The osteology of the Atlantic Spanish Mackerel has been described and illustrated by Mago Leccia (1958) and Collette and Russo (1985). Sagittal otoliths were illustrated by Powell (1975) and Baremore and Bethea (2010). **Color:** The dorsal surface is metallic greenish and the sides of the body are silvery, marked with about three rows of round to elliptical dark spots (orange in life). The first dorsal fin is black anteriorly and at distal margin posteriorly. The basal parts of the posterior first dorsal-fin membranes are white.

GEOGRAPHIC RANGE: Atlantic Spanish Mackerel occur in the western Atlantic from Cape Cod to Miami (United States) and along the Gulf of Mexico coast from Florida, United States, to Yucatan, Mexico.

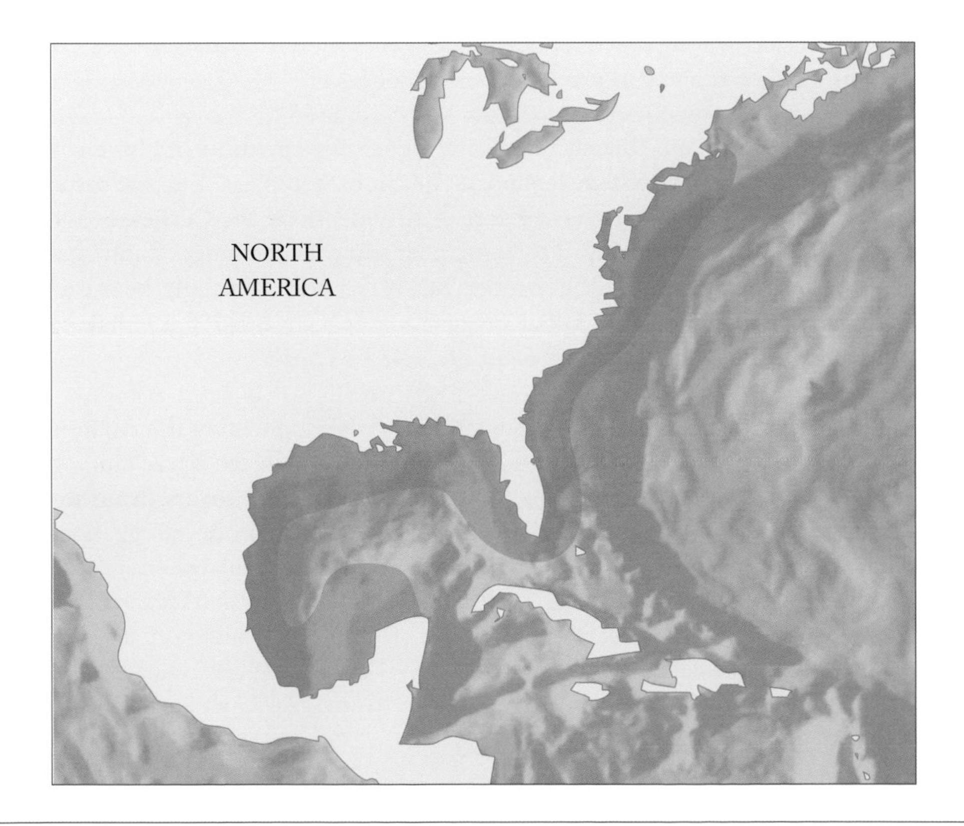

Left: Atlantic Spanish Mackerel range map.

SIZE: Maximum size is 91 cm FL. The IGFA all-tackle angling record is of a 13 lb (5.89 kg) fish taken in Ocracoke Inlet, North Carolina, in November 1987. Samples from the sport fishery in southeast Florida ranged in size from 33–77 cm FL and in weight from 0.45–4.76 kg. This species lives to 9 years in the Gulf of Mexico and to 11 years in the Atlantic. Females grow faster and attain greater size than males. Generation length is estimated to be 4 years.

HABITAT AND ECOLOGY: An epipelagic, neritic species known to migrate in large schools over great distances along the shore. With increasing water temperatures, Atlantic Spanish mackerel move northward, from Florida along the US Atlantic coast to Narraganset Bay, Rhode Island, between late February and July. They move back in the fall and overwinter off Florida. In the northern Gulf of Mexico, schools migrate westward in early spring, reaching Texas in late March. North-south movements along the Mexican coast occur between August and November and in March and April.

FOOD: Like other species of Spanish mackerels, Atlantic Spanish Mackerel feed mainly on small fishes, with lesser quantities of penaeoid shrimps and cephalopods. Eleven families and 24 species of fishes were recorded from stomachs of Atlantic Spanish Mackerel caught off North Carolina, Florida, Louisiana, and Texas. Herring-like fishes such as menhaden (*Brevoortia*), alewives, Thread Herring (*Opisthonema*), and anchovies (*Anchoa*) are particularly important forage, even for juveniles. The percentage of anchovies consumed is higher for juveniles than for adults. Atlantic Spanish Mackerel are piscivorous throughout life, beginning as larvae and postlarvae (2.8–22 mm SL) and continuing as juveniles (9–42 cm FL).

REPRODUCTION: Atlantic Spanish Mackerel attain sexual maturity in Florida by age 2, at 25–32 cm FL for females and 28–34 cm FL for males. They are batch spawners. They spawn in waters of less than 40 m depth off the southeast coast of the United States from April to September and over the inner continental shelf of Texas from May to September. Batch fecundity generally increased with fish size, from as low as 100,000 eggs for a 32.8 cm FL, 295 g female to a high of 2,113,000 eggs for a 62.6 cm FL, 2,415 g female.

EARLY LIFE HISTORY: The earliest accounts of larval development of the Atlantic Spanish Mackerel were in 1882 and 1883. More recently, seven larvae ranging from 3.1 mm notochord length to 17.4 mm standard length were described and illustrated by Wollam. These illustrations, plus a figure of an egg from Mayo, are reproduced in Richards' identification guide to larval western central Atlantic fishes. Small larvae (1.8–3.0 mm) of Atlantic Spanish Mackerel lack the 1–5 melanophores over the posterior portion of the midbrain that are characteristic of small larvae of King Mackerel. Larvae have been encountered in surface waters with temperature ranging between 19.6°C and 29.8°C and salinities of 28.3 to 37.4 ppt.

STOCK STRUCTURE: Genetic analysis of samples from widely spaced geographic regions (Chesapeake Bay, North Carolina, South Carolina, Georgia, and Gulf of Mexico) was consistent with the hypothesis that Spanish Mackerel comprise a single intermingling genetic stock. However, it is believed that there is limited connectivity between fish from the US south Atlantic and Gulf of Mexico regions, and Atlantic Spanish Mackerel from these regions are managed as separate stocks.

FISHERIES INTEREST: The Atlantic Spanish mackerel is a valued fish for recreational and commercial fisheries throughout its range. The fisheries along the Atlantic US coast north of southern Florida and in the Gulf of Mexico are seasonal between spring and late summer or fall, depending on the species migrations, while in southern Florida operations are concentrated in the winter months. The vast majority of the total US catch has been landed in Florida. Early commercial fisheries in the United States utilized trolling lines, gill nets, and pound nets. In the 1990s landings from the cast net fisheries increased as those from gill net fisheries decreased. Artisanal fishermen in Veracruz, Mexico, employ beach seines, gill nets, trolling lures, and trap nets. Nearly all the catch is consumed fresh, frozen, or smoked. A few attempts have been made at canning Atlantic Spanish Mackerel but the product has not been widely accepted.

Recreational anglers target Spanish mackerel from boats while trolling lures or drifting baits and from piers, jetties, and beaches using live baits or lures. Spanish Mackerel were among the ten most abundant species caught by trolling by charter boats in south Florida and the US Caribbean in 1984 and 1985.

The total catch of Atlantic Spanish Mackerel is underestimated due to reporting of unclassified *Scomberomorus* species captures as well as inadequate reporting of artisanal and recreational catches. The International Commission for the Conservation of Atlantic Tunas (ICCAT) reported that annual catches reached 16,725 mt in 1996, dropping off to landings of 8,000–10,000 mt through 2003, and have been averaging about 6,000 mt for the past 10 years. In the United States, catches of Atlantic Spanish Mackerel averaged about 4,000 mt until 2005 and have been about 2,000 mt per year since that time. The species was second in volume among Mexico's Gulf of Mexico fisheries between 1968 and 1976, with an average annual production of 4,900 mt. Since that time catches have averaged about 4,000 mt per year, with most of the landings occurring near Veracruz.

THREATS: Atlantic Spanish Mackerel experience relatively high levels of fishing mortality from directed commercial and recreational fisheries, and juveniles are taken as bycatch in shrimp trawl fisheries. There is underreporting of catches, especially from artisanal fisheries, and there is also misidentification of the related species. This species is listed as Least Concern on the IUCN Red List.

CONSERVATION: In the 1980s, Atlantic Spanish Mackerel was considered overfished throughout its US range, and the stock continued to decline into the 1990s. Subsequent management measures have been effective in rebuilding the stocks to healthy levels. The SSB is currently higher than the SSB necessary for MSY, and fishing mortality is lower than FMSY for both the US south Atlantic and Gulf of Mexico stocks.

In Mexico, a 1994 assessment found that the stock on the Mexican side was slightly underexploited. More recent data from the Institute Nacional de Pesca (2004) show this species to be fully exploited. Catches have been in decline since 1994 in Mexico; however, there is uncertainty surrounding the causes of the decline.

In the United States Atlantic Spanish Mackerel fisheries are managed with a combination of gear limitations, seasonal closures, and trip harvest limits. Drift gill nets were banned in 1989, and the state of Florida banned the use of gill nets in inshore waters in 1995. The recreational fishery is managed with a combination of minimum size limits and bag limits, measures that vary from state to state. Management of the species is as two stocks. The Atlantic stock is managed by the South Atlantic Fishery Management Council and the Gulf stock by the Gulf of Mexico Fishery Management Council. Uncertainties in the stock assessments of the Gulf of Mexico population were addressed by Ehrhardt and Legault (1997). There are no known species-specific conservation actions in place in Mexico. Recent data from the southern Gulf of Mexico and Yucatan indicate that the species is fully exploited, not overfished, and not undergoing overfishing.

REFERENCES: Arreguín-Sánchez et al. 1992; Banford et al. 1999; Baremore and Bethea 2010; Beardsley and Richards 1970; Brusher and Palko 1988; Buonaccorsi et al. 2001; Chávez 1994; Collette and Nauen 1983; Collette and Russo 1985; Collette et al. 2011; Collins and Stender 1987; Doi and Mendizabal 1979; Earll 1883; Ehrhardt and Legault 1997; Fable et al. 1987; Finucane and Collins 1986; Finucane et al. 1990; IGFA 2018; Johnson 1996; Klima 1959; Mago Leccia 1958; Manooch et al. 1978; McBride 2014; McEachran et al. 1980; Mitchill 1815; Naughton and Saloman 1981; Palko et al. 1988; Powell 1975; Richards 2006; Richardson and McEachran 1981; Ryder 1882; Saloman and Naughton 1983; Schmidt et al. 1993; Trent and Anthony 1979; Wollam 1970.

Papuan Seerfish

Scomberomorus multiradiatus Munro, 1964

COMMON NAMES: English – Papuan Seerfish
French – Thazard papou
Spanish – Carite papuense

ETYMOLOGY: The Papuan Seerfish was named *multiradiatus*, meaning "many rayed," by the Australian ichthyologist Ian S.R. Munro (1964) because of its many anal-fin rays.

SYNONYMS: There are no synonyms for this species.

TAXONOMIC NOTE: The Papuan Seerfish is the least known of all the Spanish mackerels and has the fewest references since its original description was created.

FIELD MARKS:

1 The sides of the body are silvery without any spots, bars, or blotches.
2 It has very few gill rakers, none on the upper limb of the gill arch, and only one to four on the lower limb.
3 It has the most anal-fin rays, 25–29, of any species of Spanish mackerel.
4 This is a small species, with a maximum size of 35 cm fork length.

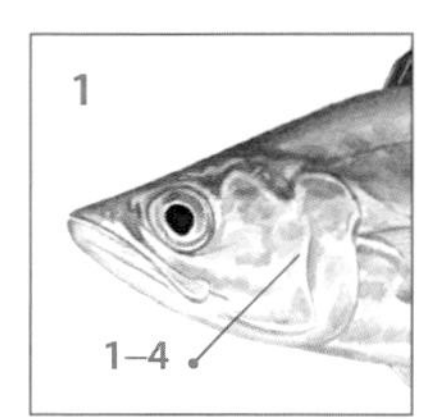

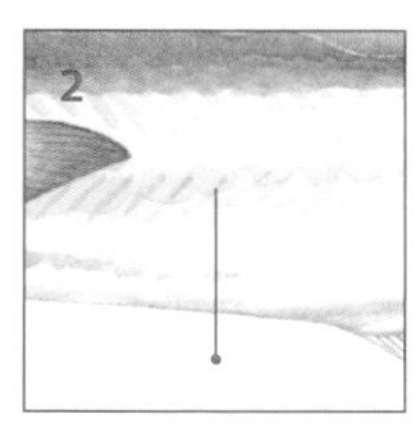

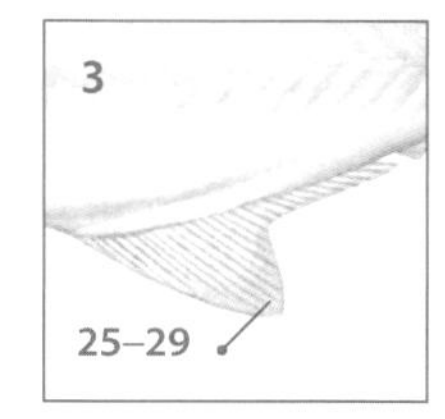

UNIQUE FEATURE: This scombrid is positively known only from off the mouth of the Fly River in Papua New Guinea.

DIAGNOSTIC FEATURES: Gill rakers: There are very few gill rakers on the first arch, none on the upper limb and only 1–4 on the lower limb. **Fins:** First dorsal fin with 16–19 spines, usually 18; second dorsal fin with 21–25 rays, followed by 7–9 finlets; anal fin with 25–29 rays, the highest number in the genus, followed by 6–9 finlets, usually 6; pectoral fin rays 20–23. **Lateral line:** The lateral line gradually curves down toward the caudal peduncle. **Vertebrae and intestine:** This species has the most vertebrae in the genus, 20 or 21 precaudal plus 34–36 caudal, total 54–56. Intestine with 2 folds and 3 limbs. **Color:** The sides are silvery without spots, blotches, or bars. The first dorsal fin is black anteriorly and along the distal edge posteriorly, with some white at the posterior base of the fin.

GEOGRAPHIC RANGE: The Papuan Seerfish is endemic to the Gulf of Papua off the mouth of the Fly River, but it may be more widespread, because there are some records from the Timor Sea.

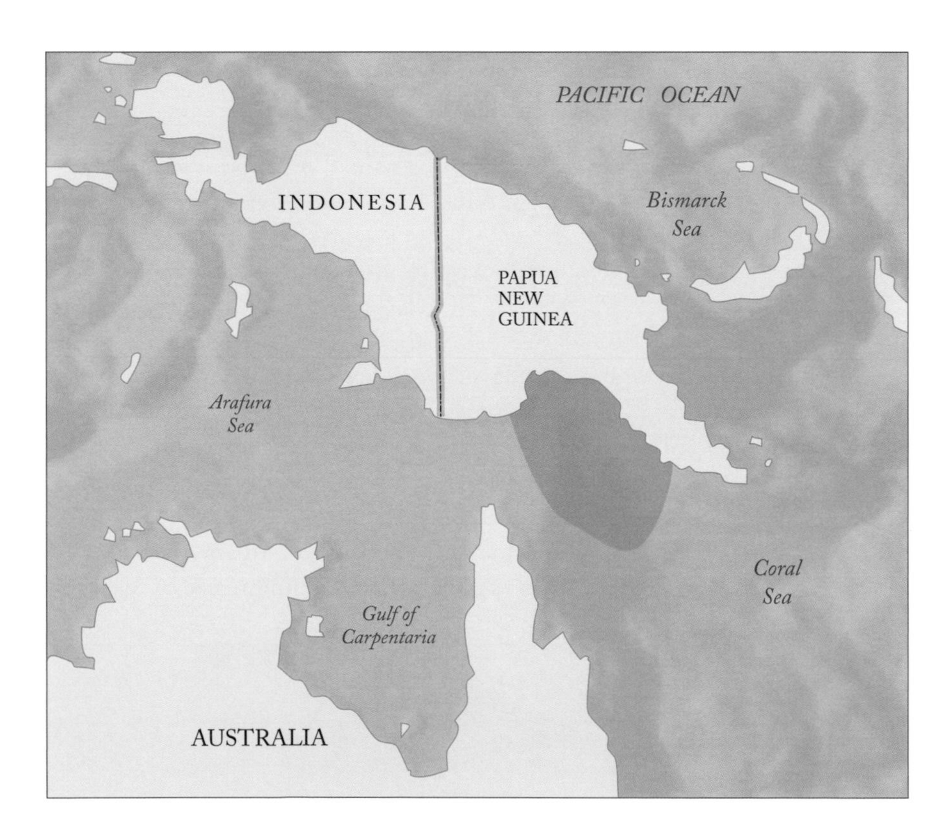

Right: Papuan Seerfish range map.

SIZE: The Papuan Seerfish is the smallest species in the genus, maximum size 35 cm FL. Sexual maturity is attained at less than 30 cm fork length.

HABITAT AND ECOLOGY: This species is pelagic and oceanodromous. It is a neritic species found in turbid waters, likely to a depth of 50 m.

FISHERIES INTEREST: Papuan Seerfish are trawled in the Gulf of Papua but lack commercial significance. There is little population information for this species.

CONSERVATION: There are no species-specific conservation measures. This species is not known to occur in any marine protected areas. Given its small body size, it is not targeted but is caught as bycatch from inshore nets, line fisheries, and shrimp trawlers within its range. It is listed as Least Concern on the IUCN Red List. However, more research is needed on this species' distribution and biology.

REFERENCES: Buckworth and Clarke 2001; Collette 2001; Collette and Nauen 1983; Collette and Russo 1985; Collette et al. 2011; Munro 1964.

Spotted Mackerel

Scomberomorus munroi Collette & Russo, 1980

COMMON NAMES: English – Spotted Mackerel
French – Thazard Australien
Spanish – Carite Australiano

ETYMOLOGY: Collette and Russo (1980) named this species after the distinguished Australian ichthyologist Ian S. R. Munro, who first recognized the distinctiveness of the Spotted Mackerel but mistakenly thought it was conspecific with the Japanese Spanish Mackerel, *Scomberomorus niphonius*.

SYNONYMS: There are no synonyms for this species.

TAXONOMIC NOTE: Spotted Mackerel were confused with the Japanese Spanish Mackerel (*S. niphonius*) from Munro (1943) until it was described by Collette and Russo (1980).

FIELD MARKS:

1 Spiny dorsal fin is bright steely blue in life, with white blotches at posterior base in some specimens.

2 As its common name implies, the Spotted Mackerel is the only Australian Spanish mackerel with many rows of small spots on its sides.

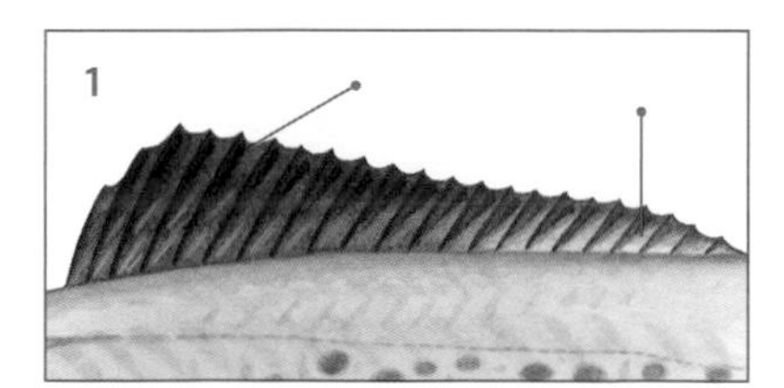

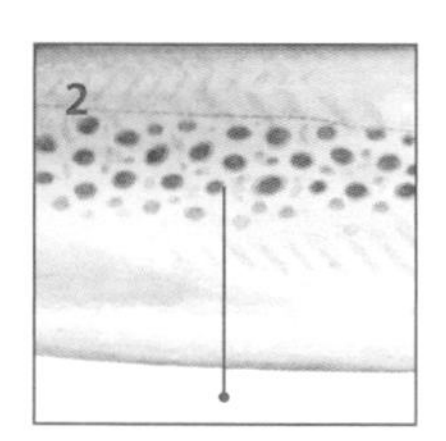

UNIQUE FEATURE: This is the most recently described Spanish Mackerel.

DIAGNOSTIC FEATURES: Gill rakers: Gill rakers on the first arch moderate: 2 on the upper limb, 8–10 on the lower limb, 10–12 total. **Fins:** First dorsal fin with 20–22 spines; second dorsal fin with 17–20 rays, followed by 8–10 finlets; anal fin with 17–19 rays, followed by 8–10 finlets; pectoral fin rays 21–23. **Lateral line:** The lateral line gradually curves down toward the caudal peduncle. **Vertebrae and intestine:** Vertebrae 21 or 22 precaudal plus 28–30 caudal, total 50–52. Intestine with 2 folds and 3 limbs. Descriptions and illustrations of the osteology and soft anatomy of the Spotted Mackerel are included in the revision of the Spanish mackerels, Collette and Russo (1984). **Color:** The sides have several poorly defined rows of round spots, larger than the pupil of the eye, but smaller than the diameter of the whole eye. The inner surface of the pectoral fin is dark blue, the cheeks and belly silvery white, the anal fin and the anal finlets are light silvery gray. The first dorsal fin is black (bright steely blue in fresh specimens) with blotches of white toward the bases of the most posterior membranes in some individuals. Most other species of Spanish mackerels have more extensive white areas on the posterior half of the middle third of the first dorsal fin.

GEOGRAPHIC RANGE: Spotted Mackerel are restricted to the northern coast of Australia, from the Abrolhos Islands region of Western Australia to at least Jervis Bay in southern New South Wales. They also occur in southern Papua New Guinea from Kerema to Port Moresby.

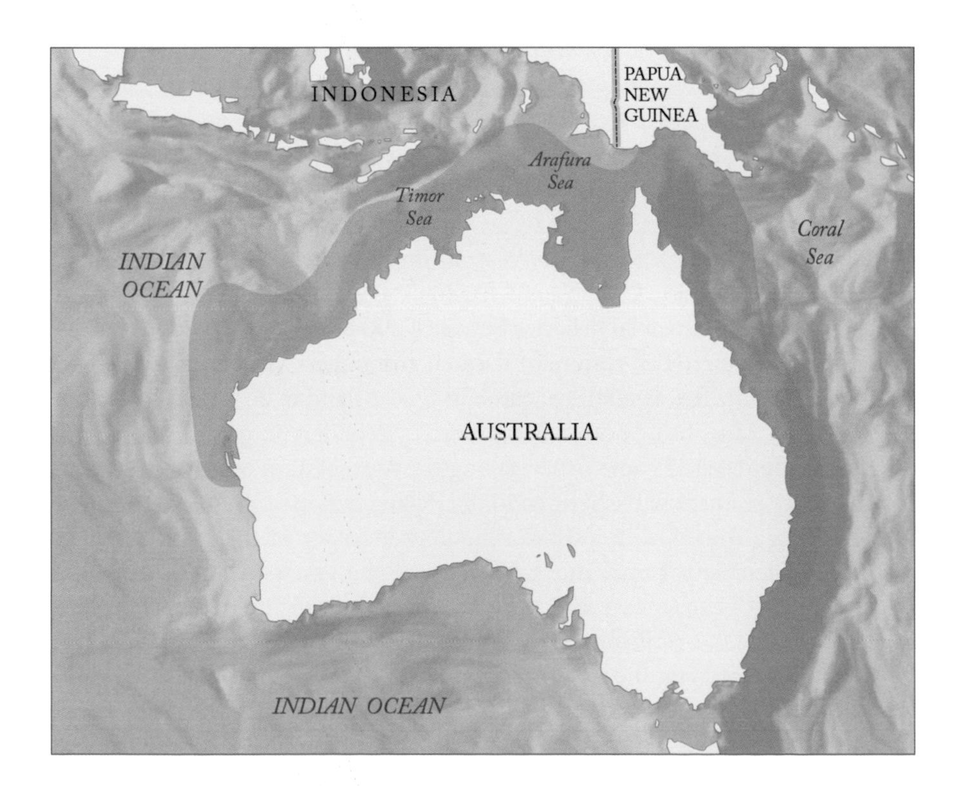

Left: Spotted Mackerel range map.

SIZE: The maximum total length of Spotted Mackerel is at least 105 cm. The IGFA all tackle sport fish record is 20 lb, 6 oz. (9.25 kg) from a fish caught at the South West Rocks, New South Wales, Australia, in July 1987. The largest recorded Spotted Mackerel was 10.21 kg and was taken in the Moorebank area of New South Wales. Spotted Mackerel grow quickly for the first 3 years of life and demonstrate sex-specific growth rates, with females tending to grow faster and to larger sizes.

HABITAT AND ECOLOGY: Spotted Mackerel are neritic and are rarely caught more than a few kilometers offshore. They are most often taken over reefs. For example, Palm Beach Reef and Brunswick River Reef in Queensland/New South Wales are among the best places to find them, especially large fish. They form large schools that move close inshore along the coast of Queensland, commonly between December and April or May. Tagged Spotted Mackerel move long distances, with about 39% recaptured over 100 km from the release site. The longest movement was 1,100 km over 228 days. These movements, together with spatial and temporal patterns of tagging effort and commercial fishing harvest, indicate a single eastern stock that undertakes a seasonal migration associated with spawning and feeding along the Queensland and New South Wales coasts.

FOOD: Spotted Mackerel feed largely on fishes, particularly anchovies and sardines, with smaller quantities of shrimps and squids.

REPRODUCTION: Spotted Mackerel spawn between August and October in northern Queensland waters, with peak spawning occurring in September. Considerable variation in length is found for any given age of Spotted Mackerel. Females and males reach maturity at about 60 cm and 52 cm TL, respectively, at 1–2 years of age. They have been aged up to 7 years. Generation length is estimated to be between 3 and 4 years.

FISHERIES INTEREST: Spotted Mackerel is an important species for recreational and commercial fishers in Queensland, Australia. It was a major part of the set gill net and ring net commercial fisheries but since 2004 it has been predominantly a line-fished species. The commercial catch from the Great Barrier Reef in 2003 was 27.8 mt. The recreational catch in Queensland in 2010 was estimated at 45,681 individual fish. Spotted Mackerel is a secondary target species for the recreational land-based game fishery along the southeastern coast of Australia. It is caught by recreational fishermen in the Pilbara region of Western Australia. The highly aggregated, near-surface schooling behavior of the stock, coupled with its predictable seasonal movements along the east coast, allows ease of targeting by both the commercial and recreational sectors, thereby making the stock susceptible to overfishing. In 1999–2000, commercial catches of Spotted Mackerel increased significantly in response to the development of valuable overseas export markets. In the 2009/10 fishing season the estimated east coast harvest was 100 mt of commercial catch (12 mt net and 88 mt line), 11 mt of charter, and 305 mt recreational. The Gulf of Carpentaria commercial sector has not reported any harvest since 2008 (the catch between 2003 and 2008 was less than 400 mt).

THREATS: Early declines in catch and the lack of information on the stock structure and fisheries in the northern and western portion of this species' range indicate that the status of this species should be carefully watched, and it was listed as Near Threatened on the IUCN Red List, instead of Vulnerable. There is a need for more information on stock structure throughout its distribution; better information on biology, particularly in northern and western Australia; more information on recreational catch, particularly in northern and western Australia; and coordination among state jurisdictions so that monitoring and assessment can be conducted at a population level. Recent management measures need to be assessed in the future, and more prudent actions may be needed if fishing pressure increases in the recreational sector or the commercial catch quota is exceeded.

CONSERVATION: An early assessment based on data from 1960 to 2002 for the eastern Australia stock (northern Queensland to northern New South Wales) estimated biomass declines to be approximately 33–63% of pre-1960 unfished or virgin biomass levels from 1992 to 2002 (within three generation lengths). This assessment indicated that the estimated 2002 biomass was above BMSY. Model projections suggest that catches of greater than 300 to 350 mt in 2003 had a high risk of reducing the population in relation to MSY. Annual landings in 2009–10 reached 71% of the TAC. In the most recent (2009–10) stock assessment, Spotted Mackerel was recorded as sustainably fished, with total mortality estimates indicating fishing is occurring at upper levels. It was considered Fully Fished from 2009–12.

There are bag and size limits for this species. The Queensland Fishery is regulated under Queensland's Fisheries Regulations 1995. In response to concerns about sustainability, some major changes in management were made in 2002 and 2003 that apply to both commercial and recreational fishers. The minimum legal size was increased from 50 to 60 cm total length, the recreational bag limit was reduced from 30 to 5 fish per person, and the use of nets to target Spotted Mackerel was prohibited. Since these changes to management, line fishing has become the main commercial method to catch it.

REFERENCES: ASR 2011; Begg 1998; Begg and Hopper 1997; Begg and Sellin 1998; Begg et al. 1997, 1998a, 1998b, 2005; Collette and Nauen 1983; Collette and Russo 1980, 1985; Collette et al. 2011a, 2011b; Griffiths 2012; IGFA 2018; Kingsford and Welch 2007; Munro 1943; NSW 2014; SS 2011; Staunton-Smith et al. 2005; Taylor et al. 2012; Williamson et al. 2006; Zischke et al. 2012.

Japanese Spanish Mackerel

Scomberomorus niphonius (Cuvier, 1832)

COMMON NAMES: English – Japanese Spanish Mackerel
French – Thazard Oriental
Spanish – Carite Oriental
Japanese – Sawara

ETYMOLOGY: The distinguished French anatomist and ichthyologist Georges Cuvier (1832) named this species after its geographical origin, Japan, or Nippon.

SYNONYMS: *Cybium niphonium* Cuvier, 1832; *Cybium gracile* Günther, 1873; *Sawara niphonia* (Cuvier, 1832); *Scomberomorus niphonius* (Cuvier, 1832)

TAXONOMIC NOTE: Early records of this species from Australia refer to misidentified *Scomberomorus munroi* (Collette and Russo 1985).

FIELD MARKS: 1 The Japanese Spanish Mackerel is typically characterized by having seven or more rows of longitudinal spots on the sides of the body; some of the spots are connected, forming wavy lines. The spots are more numerous and smaller than in the Australian Spotted Mackerel, *S. munroi*, and about the size of the pupil.

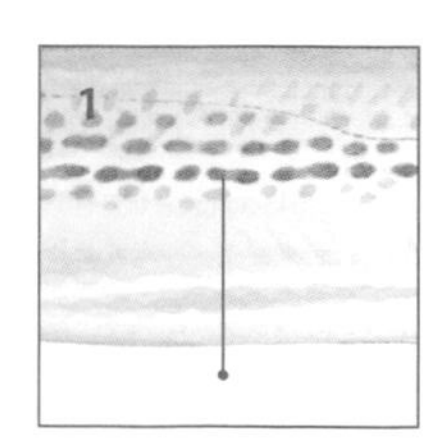

UNIQUE FEATURE: This is the only Spanish mackerel that has a straight intestine with no folds or limbs.

DIAGNOSTIC FEATURES: Gill rakers: There are a moderate number of gill rakers on the first arch: 2 or 3 on the upper limb and 9–12 on the lower limb, for a total of 11–15. **Fins:** The first dorsal fin has 19–21 spines; the second dorsal fin has 15–19 rays, followed by 7–9 finlets. The anal fin has 16–20 rays, followed by 6–9 finlets. There are 21–23 pectoral-fin rays. **Lateral line:** The lateral line gradually curves down toward the caudal peduncle. The lateral line system was described in detail by Nakae et al. (2013). **Vertebrae and intestine:** Vertebrae number 21–23 precaudal plus 27 or 28 caudal, for a total of 48–50, usually 49. This is the only species in the genus that has a straight intestine with no folds or limbs. Descriptions and illustrations of the osteology and soft anatomy of the Japanese Spanish Mackerel are included in the revision of the Spanish mackerels by Collette and Russo (1984). **Color:** The sides of the body are marked with 7 or more rows of longitudinal spots, some of which are connected. The spots are more numerous and smaller than in *S. munroi*, about the size of the pupil. Some lack spots but retain all other characteristics. The anterior quarter of the first dorsal fin and a narrow distal margin of the rest of the fin are black, while most of the basal membranes of the posterior three-quarters of the fin are jet black anteriorly, white posteriorly. The second dorsal fin is cream colored with some yellow anteriorly. The anal fin and anal finlets are transparent whitish.

Below: Unmarked individual.

GEOGRAPHIC RANGE: Japanese Spanish Mackerel live in the western North Pacific, confined to the subtropical and temperate waters of China, the Yellow Sea, and the Sea of Japan north to Vladivostok, Russia. They may occur in southern China, including Hainan Island. They have increased their abundance in Korean waters along with higher sea surface temperatures, and they may also be increasing their range in northern Japan.

SIZE: Japanese Spanish Mackerel reach a length of more than 100 cm FL and a weight of 9.4 kg. The IGFA all-tackle game fish record is a 20 lb, 10 oz (9.35 kg) fish caught at Shirasaki, Wakayama, Japan, in 2007. The length-weight relationship was published by Liu et al. Fish of age 1 grow rapidly, reaching 500 mm in length and 1,000 g in weight by autumn. Longevity is estimated to be 6 years, based on a growth study using scales in Japan. Generation length is estimated to be approximately 2 or 3 years.

HABITAT AND ECOLOGY: Japanese Spanish Mackerel are pelagic, oceanodromous, and found near shore (including semi-enclosed sea areas). The primary spawning ground is in the Seto Inland Sea, although some portion of the population may also spawn in the East China Sea. They undergo a spawning migration in spring (March to June) and a feeding migration in fall (September to November) in the Inland Sea of Japan.

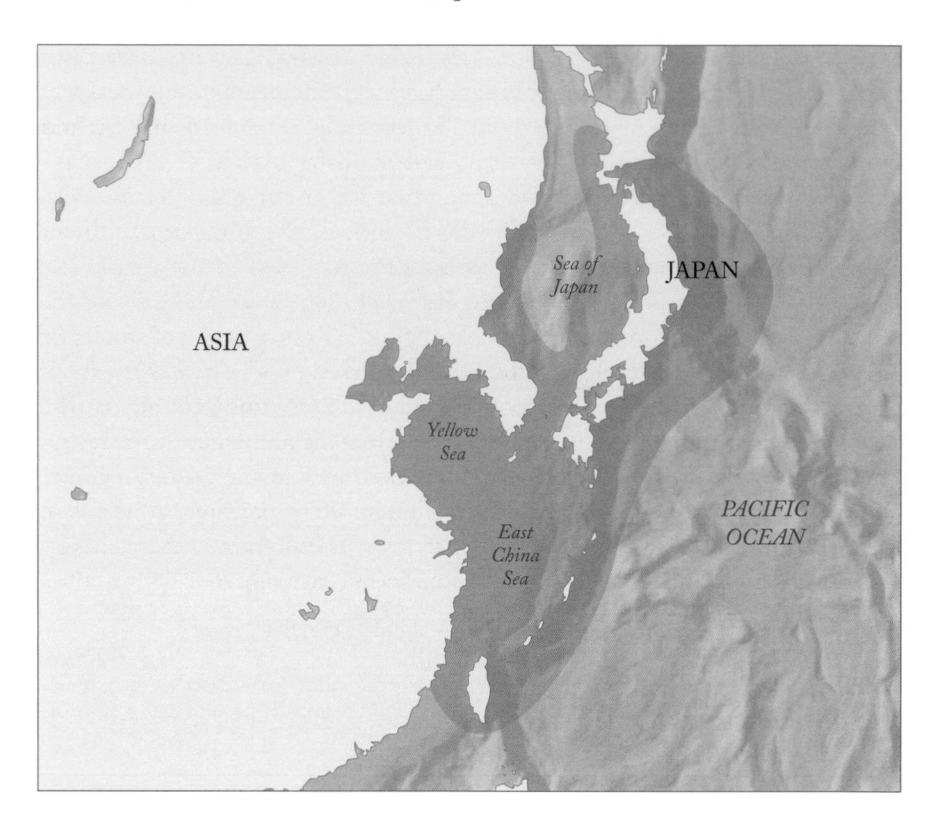

Right: Japanese Spanish Mackerel range map.

FOOD: Japanese Spanish Mackerel feed on pelagic and mesopelagic fishes and squids. The main prey of smaller individuals in the Seto Inland Sea were Japanese anchovy, sand lance, Japanese common squid, and cutlassfish. Medium- and larger-sized individuals fed on the same prey plus sardine, gizzard shad, round scad, and Pacific Chub Mackerel. Similarly, in the Southern Sea of Korea, the diet was mainly Pacific Chub Mackerel and anchovy, plus small quantities of shrimps, stomatopods, squids, crabs, polychaete worms, and copepods. Smaller individuals consumed small fishes such as anchovies and crustaceans. The portion of these prey items decreased with increasing size, and the decrease was paralleled with an increase in consumption of larger fishes such as Pacific Chub Mackerel and saury. Larvae exhibit almost exclusive piscivory from first feeding in tanks.

REPRODUCTION: There are two migrations in the Inland Sea of Japan, a spawning migration in the spring (March to June) and a feeding migration in the fall (September to November). Minimum length of maturity is 39.5 cm

FL for females and 26 cm FL for males in China. Length of 50% maturity is 60 cm FL for females and 40 cm FL for males in Japan. Early gonadal development is divisible into four stages: development from March to April, maturation from May to June, spawning from July to August, and resting from September until February the next year. Average annual fecundity is estimated at 550,000–870,000 eggs.

EARLY LIFE HISTORY: A few larvae of Japanese Spanish Mackerel have been reported from the waters around Taiwan. Eggs were collected in the plankton, and larvae were reared in the laboratory. Larvae hatched at 5.16 mm SL and were described and illustrated through 11 growth stages to a 35- to 38-day-old postlarva at 30.0 mm SL. Larvae feed exclusively on fish larvae from their first feeding stage and grow at rates of more than 1.0 mm per day under laboratory conditions, 1.47 mm per day in the field.

STOCK STRUCTURE: Morphometric and meristic data indicated that samples of Japanese Spanish Mackerel from the north, central, and southern part of the East China Sea were all from the same population. No significant population structure was found between samples in the East China Sea and the Yellow Sea. Analyses of mitochondrial DNA indicates that there is a single panmictic population that underwent significant population expansion in the late Pleistocene. Strong dispersal capacity of larvae and adults, coupled with long-distance migration and prevailing ocean currents, have apparently prevented significant population differentiation.

FISHERIES INTEREST: Japanese Spanish Mackerel play an important role in the commercial fisheries of China, Japan, and Korea. They are caught with gill nets, purse seines, and set nets. FAO worldwide reported landings show a gradual increase from 1900 mt in 1950 to 60,685 mt in 2006. They are caught throughout their range and are the most important Spanish mackerel in Japan, especially in the Seto Inland Sea. In the Inland Sea, the main fishing seasons are during their two migrations, from March to June and September to November. There are no major fisheries for this species on the eastern part of Japan.

There are also important fisheries in the Huanghai Sea (Yellow Sea) and Bohai Sea. In Taiwan, catches reported to FAO have declined from 15,000 mt in 2007 to 1,400 mt in 2008, but this needs to be confirmed, as catches are reported from distant-water longlines, which are not normally used to catch this species. In the East China Sea, Japanese Spanish Mackerel is caught by purse seiners. It is also an important fishery in South Korea, where 40,000 mt were recorded in 2007, accounting for 50% of the global catch.

THREATS: The catch in the western Seto Inland Sea, Japan, reached a maximum of 2,848 mt in the 1980s and gradually decreased. By 1994 it was only 14.7% of the 1980 level. The age composition changed from mostly 2-year-olds with some 3-, 4-, and 5-year-olds in 1986 to 73.9% of 1-year-olds by 1996. There was a slight increase to 5,000 mt in 2007, based on a recovery program that begin in 2002 with a restocking program. In the western part of Japan, catches are increasing, as this species is likely increasing its range, because it used to be rare in this region. It is listed as Data Deficient on the IUCN Red List. More information on catch and effort for this species in China and Korea is needed, as this species may qualify for a Threatened category.

CONSERVATION: Japanese Spanish Mackerel is heavily fished in many parts of its range, and there has been at least one occurrence of a localized collapse in the Inland Sea of Japan. A bioeconomic model was developed for this fishery in China, estimating maximum revenue, optimum economic effort, and optimum energy consumption. As part of a Japanese national project for recovery of fishery resources, release of juvenile Japanese Spanish Mackerel raised in the hatchery started in 1998. About 120,000 juveniles were released in the Inland Sea in 1999. Release of 100 mm juveniles reared in nursery facilities turned out to be more efficient than release of 40 mm juveniles. In addition, there are regulations to control effort (including regulating the number of boats and catch size) as well as seasonal closures in the Inland Sea. A population dynamics model that accounts for variation in natural recruitment was developed to simulate the effects of hatchery releases and fishing regulation in the eastern Seto Inland Sea. Although recovery efforts are underway in the Inland Sea, much less information is available on the harvest and population status of the population in China and Korea, where an estimated 50% or more of the global catch may occur.

REFERENCES: Chiu and Chen 1993; Collette and Nauen 1983; Collette and Russo 1985; Collette et al. 2011; Hamada and Iwai 1997; Huang and Xiong 1997; Huh et al. 2006; IGFA 2018; Inoue et al. 2007; Kishida 1986; Kishida et al. 1985; Kono et al. 1997; Lee et al. 2011; Liu 1981; Liu et al. 1982; Nakae et al. 2013; Nakajima et al. 2014; Obata et al. 2007, 2008; Qui and Ye 1996; Sha et al. 1966; Shoji and Tanaka 2001, 2004, 2005a, 2005b; Shoji et al. 2002, 2005; Shui et al. 2009; Wang 1982; Ye and Zhu 1984; Zhang et al. 2011.

Queen Mackerel

Scomberomorus plurilineatus Fourmanoir, 1966

COMMON NAMES: English – Queen Mackerel
French – Thazard Kanadi
Spanish – Carite Kanadi

ETYMOLOGY: Named *plurilineatus* by the French ichthyologist Pierre Fourmanoir (1966) in reference to the many broken lines on its sides.

SYNONYMS: *Scomberomorus lineolatus* (non Cuvier, 1829); *Scomberomorus guttatus* (non Bloch and Schneider, 1801); *Scomberomorus plurilineatus* Fourmanoir, 1966

TAXONOMIC NOTE: This species has been confused with *Scomberomorus lineolatus* and with *S. guttatus* (and also one of its junior synonyms, *Scomberomorus leopardus*). The primary source of published biological information is based on observations in Zanzibar as *S. lineolatus* (Williams 1960).

FIELD MARKS: 1 Sides of body silvery, with a series of about six to eight irregular rows of interrupted horizontal wavy black lines and spots.

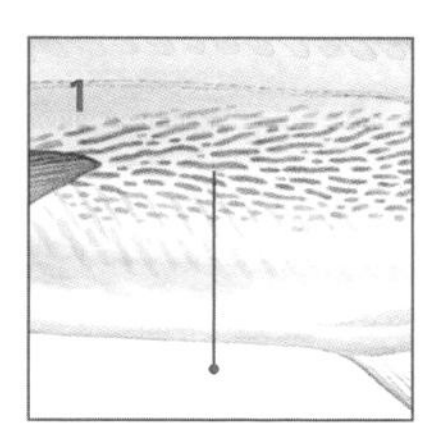

UNIQUE FEATURE: The only Spanish mackerel with short wavy lines and small spots on its sides.

DIAGNOSTIC FEATURES: Gill rakers: There are a moderate number of gill rakers on the first gill arch: 2 or 3 on the upper limb, 9–13 on the lower limb, usually 10–12, for a total of 12–15. **Fins:** The first dorsal fin has 15–17 spines; the second dorsal fin has 19–21 rays, followed by 8–10 finlets. The anal fin has 19–22 rays, followed by 7–10 finlets. There are many pectoral-fin rays, 21–26, modally 23. **Lateral line:** The lateral line lacks auxiliary branches anteriorly and gradually curves down toward the caudal peduncle. **Vertebrae and intestine:** Vertebrae number 19 or 20 precaudal plus 25–27 caudal, for a total of 43–46, usually 20 + 26 = 46. The intestine has 2 folds and 3 limbs. Descriptions and illustrations of the osteology of the Queen Mackerel are included in the revision of the Spanish mackerels in Collette and Russo (1984). **Color:** The sides of the body are silvery, with a series of about 6–8 rows of interrupted horizontal wavy black lines and spots that are much narrower than the space between them. Anteriorly, usually only one of these lines is above the lateral line and is replaced posteriorly by a number of short oblique black lines, becoming somewhat confused. Only 2 or 3 of the lines continue through to the caudal peduncle. The horizontal black lines on the body may be interrupted to varying degrees, beginning almost intact in places but broken up into a series of small rectangular "spots" in others. Juveniles have spots but develop the adult pattern of interrupted lines by the time they reach a length of 40 cm. The upper areas of the caudal peduncle and the median keel are black, the lower areas dusky. The first dorsal fin is black except for the lower areas of membranes, which may be pale posteriorly. The leading edge and tips of the rays of the second dorsal fin are dusky, the rest of the fin is silver to pale, and the finlets are dusky with a silver area at center. The anal fin and the leading edges and tips of the anal rays are dusky, while the rest of the fin is silvery. The anal finlets are white with a dusky central area. The inner surfaces of the pectoral fins are black, as is the pectoral axil, and dusky on the outside, with black edges. The pelvic fins are pale whitish, with the outer side of the mid-rays dusky.

GEOGRAPHIC RANGE: Queen Mackerel are confined to the western Indian Ocean along the coast of East Africa from the Seychelles, Kenya, and Zanzibar to Algoa Bay, South Africa, and along the west coast of Madagascar.

SIZE: Queen Mackerel reach a maximum length of 109 cm FL. The IGFA all-tackle game fish record is of a 12 lb, 4 oz (5.56 kg) fish caught off Benguerra Island, Mozambique, in May 2008. The mean length of fish landed by line fishermen in Natal during 1975–77 was 69.2 cm FL, with a range of 33.6–109 cm. The length-weight relationship of 171 specimens 39.0–113.5 cm TL from the Kenyan coastal artisanal fishery were presented by Mbaru et al. The maximum weight was 5.28 kg.

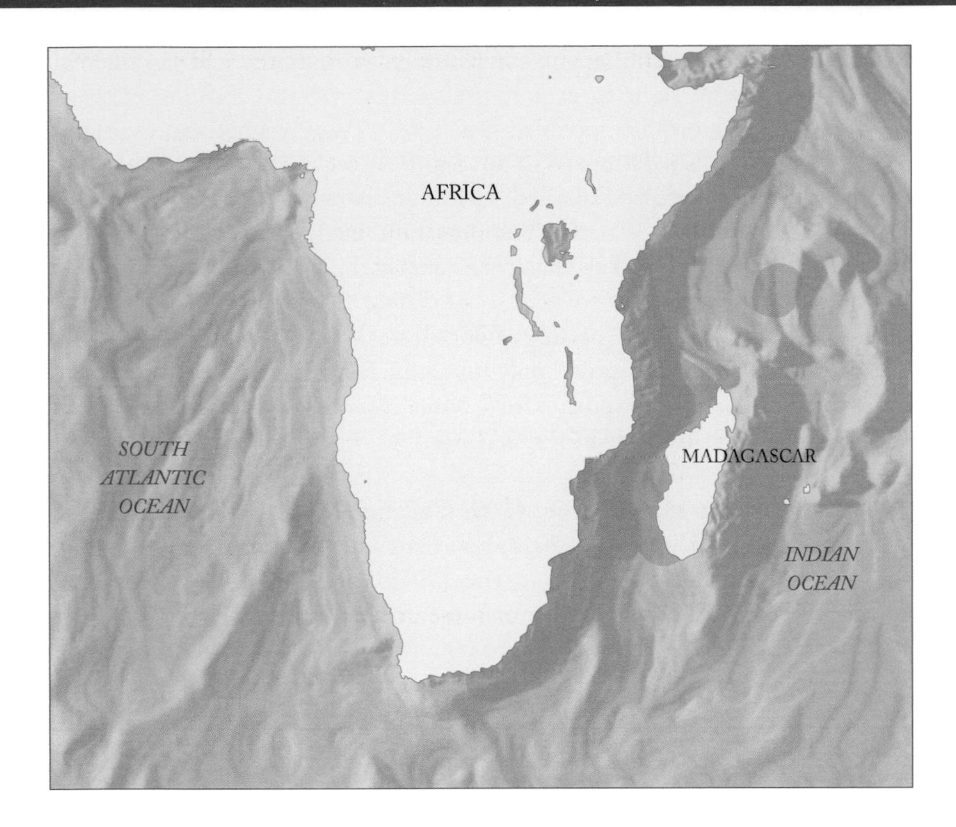

Left: Queen Mackerel range map.

HABITAT AND ECOLOGY: This is a pelagic, oceanodromous, and neritic species. It forms large schools in the Zanzibar Channel from March to September.

FOOD: Queen Mackerel feed mainly on anchovies (*Anchoviella* spp.), clupeids (*Amblygaster* spp., *Sardinella fimbriata*, and *S. perforata*), other small fishes, squids, and mantis shrimps.

REPRODUCTION: In South Africa, 50% maturity is attained at an age of 2 years and a length of 72–74 cm FL for males, 76–78 cm FL for females. Males appear to mature at a smaller size than females; the smallest male and female were 54.6 cm and 58.5 cm FL, respectively. Longevity is estimated to be 6 years.

FISHERIES INTEREST: Queen Mackerel is a commercial species taken with gill nets and by hook and line. It is among 39 major reef fish species harvested by the Kenyan coastal artisanal fishery. It is seasonally important in Tanzania and Kenya. In commercial fisheries in South Africa it is typically an incidental catch. There is a major recreational line fishery in Natal and Zululand, South Africa, for this species. Queen Mackerel is the highest ranked of the ten main species targeted by spear fishermen in KwaZulu-Natal. It is among the dozen fishes commonly caught in Durban Harbor by recreational anglers. However, landings of Queen Mackerel are not reported individually because it is mixed with other species of Spanish mackerels in some parts of its range.

THREATS: There are clashing interests between recreational and commercial fishermen over the right to exploit the species in Natal. Fishing effort in Madagascar has more than doubled since 1980. There has been an estimated 50% decline of unfished biomass in the South Africa portion of its range, and it is considered optimally exploited. However, there is little data from other regions within its range, where fishing pressure is likely to be at least as high as in South Africa. It is listed as Data Deficient on the IUCN Red List.

CONSERVATION: Beach seine netting of this species is prohibited in South Africa because netted shoals comprised only immature fish. There is a bag limit of 10 fish per fisher per day in South Africa. More information on species-specific landings and effort is needed for this region.

REFERENCES: Chale-Matsau et al. 1999; Collette and Nauen 1983; Collette and Russo 1985; Collette et al. 2011; Fourmanoir 1966; Guastella et al. 1994; IGFA 2018; Laroche and Ramananarivo 1995; Mann et al. 1997; Mbaru et al. 2010; Merrett and Thorp 1966; Penney et al. 1999; Van der Elst and Collette 1984; Williams 1960, 1964.

School Mackerel

Scomberomorus queenslandicus Munro, 1943

COMMON NAMES: English – School Mackerel, Queensland School Mackerel
French – Thazard du Queensland
Spanish – Carite de Queensland

ETYMOLOGY: Named *queenslandicus* by the Australian ichthyologist Ian S. R. Munro (1943) because of its occurrence in Queensland.

SYNONYMS: There are no synonyms for this species.

TAXONOMIC NOTE: Early records of School Mackerel before Munro described the species in 1943 were recorded under the name *Scomberomorus guttatus*, a species that does not occur in Australia.

FIELD MARKS:

1 School Mackerel have relatively few gill rakers (three to nine).
2 They have relatively few large spots on the posterior part of the body that are larger than the diameter of the eye.

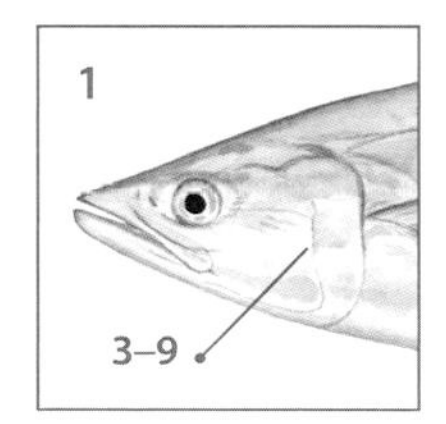

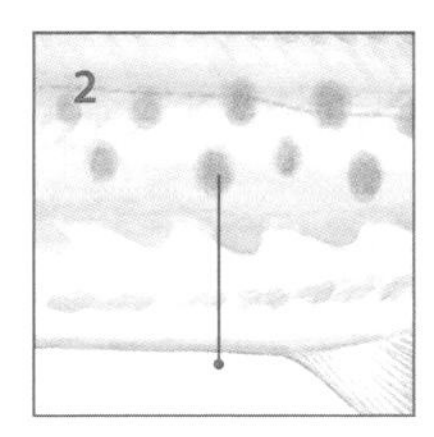

UNIQUE FEATURE: The Spanish mackerel with only a few large spots on the posterior part of the body.

DIAGNOSTIC FEATURES: Gill rakers: There are only a few gill rakers on the first arch: 0–2 on the upper limb, 3–8 on the lower limb for a total of 3–9, usually 7 or fewer. **Fins:** The first dorsal fin has 16–18 spines, usually 17; the second dorsal fin has 17–19 rays, followed by 9–11 finlets. The anal fin has 16–20 rays, usually 19, followed by 9–11 finlets. There are 21–23 pectoral fin rays. **Lateral line:** The lateral line gradually curves down toward the caudal peduncle. **Vertebrae and intestine:** Vertebrae number 19 or 20 precaudal plus 28 or 29 caudal, for a total of 48 or 49. The intestine has 2 folds and 3 limbs. The osteology and soft anatomy were described by Collette and Russo (1985). **Swim bladder:** No swim bladder is present. **Color:** The sides of adults are marked with about 3 rows of indistinct bronze-gray blotches, each a little larger than the diameter of the eye. The membrane of the first dorsal fin is jet black, with contrasting areas of intense white between the sixth and last spine. The second dorsal fin, finlets, and caudal fin are gray with darker margins. The pelvic fins, anal fin, and anal finlets are white. The caudal fin is pearly gray with darker margins. The pectoral fins are grayish, darkest on the inner surface. The characteristic body blotches are absent in juveniles.

GEOGRAPHIC RANGE: School Mackerel are confined to inshore coastal waters of southern Papua New Guinea and the northern three-quarters of Australia. The westernmost records are from Shark Bay and Onslow, Western Australia, and the eastern range extends south to Port Jackson and Botany Bay in the Sydney region of New South Wales.

SIZE: Female School Mackerel reach at least 78 cm FL, males at least 69 cm FL, although Lewis reported a maximum of 100 cm FL. School Mackerel attain 17 lb (7.7 kg) but are usually seen at about 5 lb (2.3 kg).

HABITAT AND ECOLOGY: School Mackerel are an epipelagic, neritic schooling species that often inhabit turbid coastal waters in embayments and estuaries. They form mixed schools with Narrow-barred King Mackerel over shallow reefs offshore of Queensland. Their depth profile has a lower limit of around 100 m, however, they often inhabit turbid coastal waters shallower than 30 m. School Mackerel are seasonally migratory and move into inshore waters, bays, and estuaries of Queensland during the southern midwinter and early spring.

FOOD: School Mackerel feed almost entirely on fishes. Most of the prey are pelagic herring-like fishes such as pilchards (*Sardinops*), sardines (*Sardinella*), and anchovies (*Engraulis*), but they capture other prey found closer to the bottom, such as triggerfishes and shrimps. School Mackerel feed on a wider variety of prey than Spotted Mackerel and eat more bottom-dwelling organisms than Spotted Mackerel, which are almost exclusively pelagic predators.

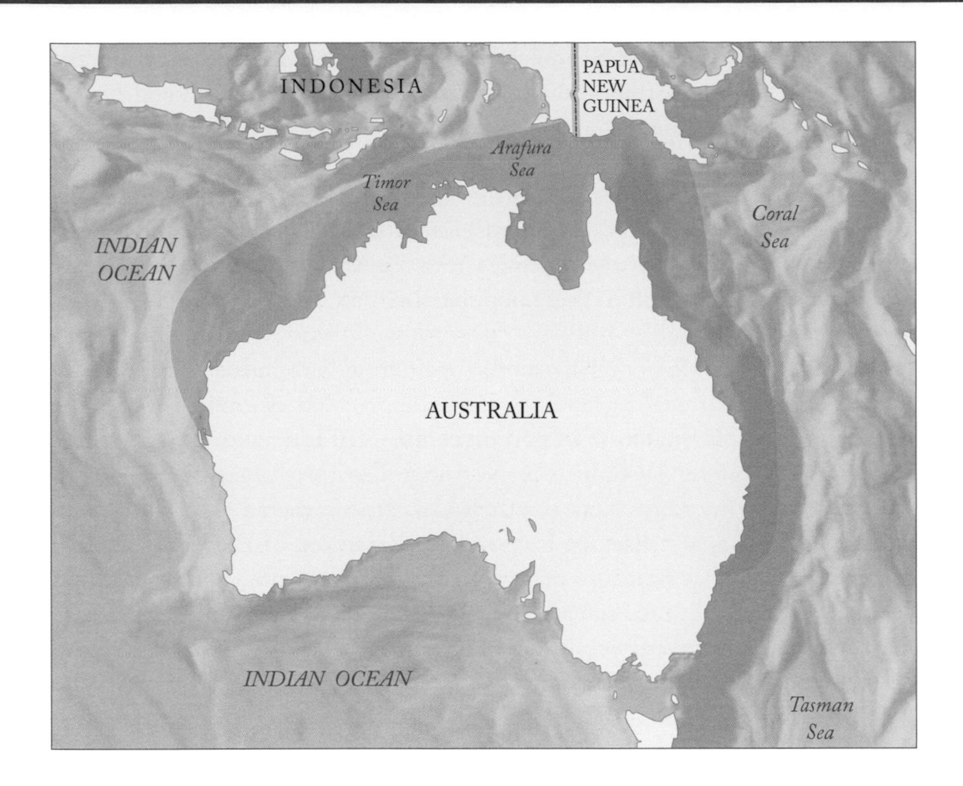

Left: School Mackerel range map.

REPRODUCTION: School Mackerel spawn between October and January throughout inshore coastal waters, bays, and estuaries along the east coast of Queensland. Age of first maturity is 1.5 years, the length of females at 50% maturity is 40–45 cm FL, and the length of males at 50% maturity is 35–40 cm FL. All School Mackerel longer than 70 cm FL were females. Batch fecundity has been estimated to be about 262,000 eggs per batch for females of 50 cm TL.

AGE AND GROWTH: Female School Mackerel live for up to 7 years, reaching a maximum size of about 78 cm FL. Males live for at least 10 years and reach 69 cm FL.

EARLY LIFE HISTORY: Larval School Mackerel were collected from October to February from coastal bays and the lagoon region of the Great Barrier Reef off Townsville, Queensland. Larvae were described from 193 specimens 3.5–9.8 mm SL. Larvae with 48 or 49 myomeres match the vertebral counts for School Mackerel. These larvae differ from Broadbarred Mackerel by the development of a third preopercular spine at about 5.0 mm in addition to the two rudimentary preopercular spines anterior to the main series of preopercular spines. Head, trunk, and fin pigmentation, morphometric data, and illustrations of six larvae 3.6–9.5 mm standard length were presented by Jenkins et al. Larval School Mackerel feed mostly on other larval fishes, particularly herring-like fishes, and larvaceans, with the proportion of larval fishes increasing as the proportion of larvaceans decreases with growth of the larva.

STOCK STRUCTURE: Tagged School Mackerel moved much shorter distances from their release site than did Spotted Mackerel. School Mackerel have a complex stock structure, with stocks being associated with large embayments. Reproductive patterns, tag and recapture information, otolith chemical analyses, and genetic studies all indicate that there are at least two stocks of School Mackerel in coastal waters of eastern Australia. Significant genetic differences were noted between samples from the Arafura Sea and Queensland, emphasizing the isolation of these populations.

FISHERIES INTEREST: School Mackerel, together with Spotted Mackerel (*S. munroi*), Broad-barred Mackerel (*S. semifasciatus*), and Narrow-barred King Mackerel (*S. commerson*) form important commercial and recreational fisheries throughout Western Australia, the Northern Territory, Queensland, and to a lesser extent, northern New South Wales. An estimated 1,917 fish were taken by recreational fishermen in the Gascoyne region of Western Australia in 1998–99, and they were also taken in the Pilbara region during 1999–2000. The estimated recreational catch in 2010 in Queensland was 60,453 individual fish. School Mackerel are a secondary target species for the recreational land-based game fishery along the southeastern coast of Australia. They are taken with others of their genus in a fishery in Queensland, with a reported annual take of 129–44 mt in 2006–10. School Mackerel were caught by commercial fishers with set mesh nets, drift nets, or ring nets up to 2002 and by recreational fishermen using hook and line.

THREATS: Surveys conducted across Queensland's recreational fisheries revealed that 40–65% of *S. queenslandicus* landed are caught by recreational fishers. While it is likely that the School Mackerel harvest is sustainable, its status will remain "uncertain" until there is greater confidence in commercial data and better quantification of the recreational harvest. Despite being commercially harvested as a food source, there are no data suggesting that this species is undergoing a significant population decline at present, so it was assessed as Least Concern on the IUCN Red List.

CONSERVATION: School Mackerel CPUE was stable from 1998 to 2003. The commercial catch from the Great Barrier Reef in 2003 was 29.6 mt. Commercial catches increased significantly from 1999 to 2000, while recreational catches decreased. This led to a change in management in 2002, where commercial net fishers are now prohibited from targeting this species. A formal stock assessment has been conducted. Total landings in 2006–10 ranged from 19–144 mt per year; in 2009–10, 136 mt were caught, of which 114 mt were taken with nets and 21 mt by line. Regulations include a bag limit of five individuals and minimal size limit of 50 cm that applies to both commercial and recreational fishers on the east coast of Queensland. Commercial fishers must also have a license with a fishery symbol from the Queensland Department of Agriculture and Fisheries, allowing them to fish for School Mackerel. These licenses also regulate fishing practices and gear. There are no size limits in the Northern Territory, and the size limits are 50 cm TL in Western Australia, with a mixed bag limit of 3 for other northern Spanish mackerels.

REFERENCES: ASR 2011; Begg 1998; Begg and Hopper 1997; Begg and Sellin 1998; Begg et al. 1997, 1998a, 1998b; Cameron and Begg 2002; CRC 2005; Collette 2001; Collette and Nauen 1983; Collette and Russo 1985; Collette et al. 2011; Donohue et al. 1982; Farmer and Wilson 2011; Griffiths 2012; Jenkins et al. 1984a, 1984b, 1985; Kailola 1991; Kailola et al. 1993; Kingsford and Welch 2007; Lewis 1981; McPherson 1985, 1987; Moran et al. 1983; Munro 1943; Rohan and Church 1979; Salini et al. 1994; SS 2011; Stevens and Davenport 1991; Sumner et al. 2002; Taylor et al. 2012; Torres Strait Protected Zone Joint Authority 2006; Ward and Rogers 2003; Williams 1997; Zischke et al. 2012.

Cero

Scomberomorus regalis (Bloch, 1793)

COMMON NAMES: English – Cero
French – Thazard Franc
Spanish – Carite Chinigua

ETYMOLOGY: Named *regalis*, Latin for "royal" or "kingly," by the German physician and naturalist Marcus Élieser Bloch (1797:31–34, with color plate 333), perhaps because this and other Spanish mackerels have been called "King Mackerel."

SYNONYMS: *Scomber regalis* Bloch, 1793; *Scomberomorus plumierii* Lacepède, 1803; *Cybium regale* Cuvier, 1829; *Scomberomorus regalis* (Bloch, 1793)

TAXONOMIC NOTE: Although Cero and Atlantic Spanish Mackerel can easily be distinguished morphologically or with nuclear molecular markers, individuals of the two species share a common mitochondrial genome, probably due to introgression from hybridization between the two species (Banford et al. 1999).

FIELD MARKS: 1 Pectoral fins are covered with scales.
2 First dorsal fin black anteriorly and distally, white at base.
3 Body with mid-lateral orange stripe bordered with orange spots above and below.

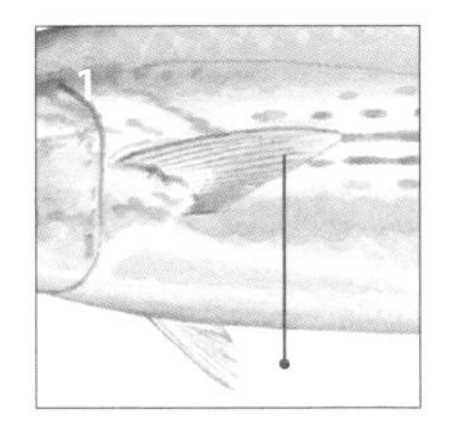

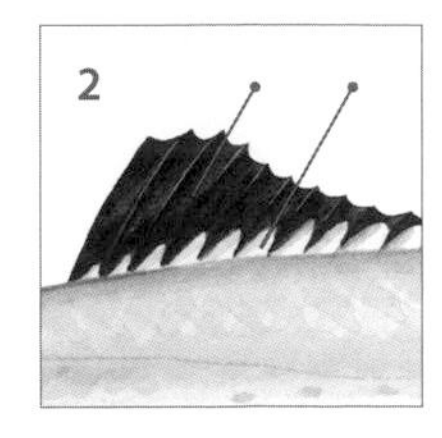

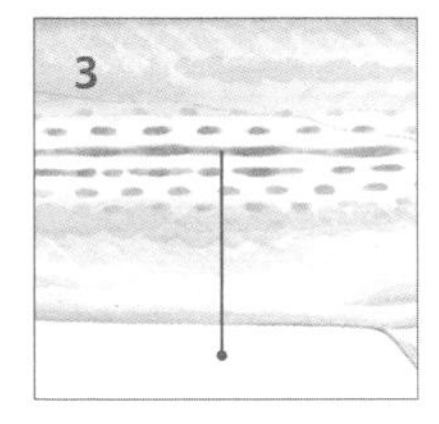

UNIQUE FEATURE: Unlike other species of Spanish mackerels, Cero prefer the clear waters around coral reefs as their habitat.

DIAGNOSTIC FEATURES: Gill rakers: There are a relatively moderate number of gill rakers on the first gill arch: 2 to 4 on the upper limb, 10 to 14 on the lower limb, total 12 to 18, usually 15 or 16. **Fins:** The first dorsal fin has 16–18 spines, usually 17; the second dorsal fin has 16–19 rays, followed by 7–9 finlets. The anal fin has 15–20 rays, usually 18 or 19, followed by 7–10 finlets, usually 8. The pectoral fin has 20–24 rays, usually 21 or 22. The pectoral fin is covered with small scales. The pelvic fins are relatively long, 4.4 to 6.3% of fork length, compared to *S. brasiliensis* (3.6–5.9%). **Lateral line:** The lateral line gradually curves down toward the caudal peduncle. **Vertebrae and intestine:** Vertebrae number 19 or 20 precaudal plus 28 or 29 caudal, for a total of 47 or 48. The intestine has 2 folds and 3 limbs. The osteology of Cero has been described and illustrated by Mago Leccia (1958) and Collette and Russo (1985). **Otoliths:** Sagittal otoliths were illustrated by Baremore and Bethea (2010). **Color:** The sides of the body are silvery, with one long mid-lateral stripe and several rows of yellow-orange streaks of variable length and small yellow spots above and below the stripe. The anterior third of the first dorsal fin is black, the posterior portion white.

GEOGRAPHIC RANGE: Cero are restricted to the western Atlantic, from Cape Cod, Massachusetts, to Rio de Janeiro, Brazil, but their distribution is concentrated in the West Indies.

SIZE: The maximum length in Florida is 83.5 cm FL, with a weight of 4.9 kg. The IGFA all-tackle game fish record is of an 18 lb (8.16 kg) fish caught off Bimini in the Bahama Islands in June 2013.

HABITAT AND ECOLOGY: Cero is epipelagic, but unlike most species of Spanish Mackerels, it is most abundant in the clear waters around coral reefs, occasionally travelling in small groups. It is found around Puerto Rico all year long.

FOOD: In the West Indies, 96% of the food of Cero consists of small schooling fishes, particularly herring-like fishes (*Harengula*, *Jenkinsia*, and *Opisthonema*) and silversides (*Allanetta*) but also includes squids and shrimps. Cero may jump free from the surface, perhaps attacking schools of forage fishes.

REPRODUCTION: Length at 50% maturity in southern Florida is 35 cm FL for males and 38 cm FL for females, with fecundity estimates of 161,000–2,234,000 eggs per female ranging between 38 and 80 cm FL. In Puerto Rico, length at 50% maturity is 35 cm FL for males and 41.3 cm FL for females. Spawning occurs year-round in southern Florida and Puerto Rico, with a peak in May in southern Florida and from April to September in Puerto Rico, but is restricted to the period from April to October on California Bank, south of Jamaica.

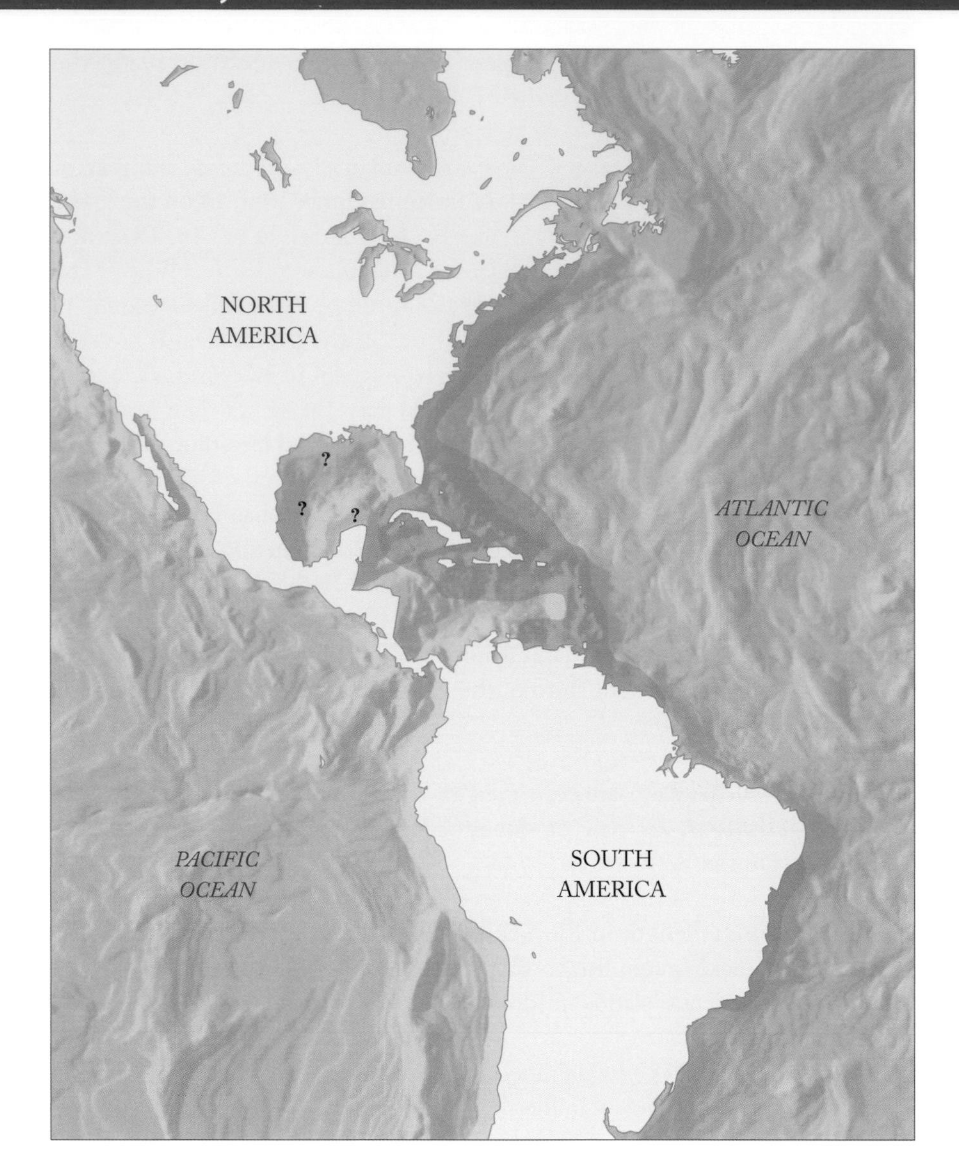

Right: Cero range map.

EARLY LIFE HISTORY: Cero eggs have one oil globule and range in size from 1.16–1.22 mm in diameter. Larvae are 3.4 mm notochord length at hatching. Six larvae ranging from 3.4 mm notochord length to 10.7 mm standard length were illustrated in Charles Mayo's PhD dissertation at the University of Miami and were reproduced in Richards' 2006 identification guide to larval western central Atlantic fishes. They differ from other western Atlantic Spanish mackerels by pigmentation in the gular region.

FISHERIES INTEREST: Cero are commercially caught with gill nets and on lines in the West Indies and the Bahamas and are also valued sport fish taken by trolling with cut bait in Florida. Cero were among the ten most abundant species caught trolling by charter boats in south Florida and the US Caribbean in 1984 and 1985. Fish Aggregating Devices (FADs) are used in fisheries for

Cero at St. Thomas and St. Croix in the US Virgin Islands. Catches of Cero are poorly reported, and the species is often included in landings of unidentified *Scomberomorus*. Some of the catches reported as *S. maculatus* by Cuba and the Dominican Republic are in fact attributable to Cero. Reported worldwide landings are relatively low and ranged between 100 and 800 mt from 1950–2005 and have been below 100 mt since that time.

THREATS: Cero support commercial and recreational fisheries throughout its range and is particularly abundant around Caribbean reefs. Catches are not well reported, are misreported, and are often combined with other members of the genus. It is listed as Least Concern on the IUCN Red List.

CONSERVATION: There are no species-specific conservation measures. Cero may be present in some marine protected areas within its range. Although taken by local fisheries, usually by hook and line, it is not a target species and is not usually fished with gill nets or purse seines.

REFERENCES: Banford et al. 1999; Baremore and Bethea 2010; Beardsley and Richards 1970; Bloch 1797; Brusher and Palko 1987; Collette and Nauen 1983; Collette and Russo 1985; Collette et al. 2011; Cooper 1982; Erdman 1956, 1977; Figuerola-Fernandez et al. 2007; Finucane and Collins 1984; Friedlander et al. 1994; IGFA 2018; Mago Leccia 1958; Manooch et al. 1978; Mayo 1973; Paine et al. 2007; Randall 1967, 1968; Richards 2006.

Broadbarred Mackerel

Scomberomorus semifasciatus (Macleay, 1883)

COMMON NAMES: English – Broadbarred Mackerel, Grey Mackerel. Although many recent references use the common name Grey Mackerel for this species, we will follow the International Game Fish Association (2016) and Grant's consistent use of Broadbarred Mackerel throughout his five editions of *Guide to Fishes* (1965–82), culminating in his expanded 1987 *Fishes of Australia*.
French – Thazard Tigré
Spanish – Carite Tigre

ETYMOLOGY: Named *semifasciatum* by the Australian ichthyologist William Macleay (1883) in allusion to the bars (*fascia*, Latin for "band") on the lower part of the body (*semi* is Latin for "partial").

SYNONYMS: *Cybium semifasciatum* Macleay, 1883; *Cybium tigris* De Vis, 1884; *Indocybium semifasciatum* (Macleay, 1883); *Scomberomorus semifasciatus* (Macleay, 1883)

TAXONOMIC NOTE: As noted in the Etymology section, this Spanish mackerel was given a good descriptive name easily translatable into its common name.

FIELD MARKS:

1. Small- and medium-sized individuals (less than 70 cm fork length) have 12–20 distinctive dark bars on the lower portion of their bodies, but these fade with growth.
2. The fleshy caudal keels are broader than those of the other Australian species of Spanish mackerels.

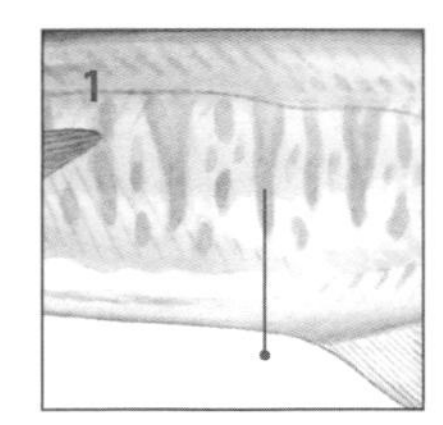

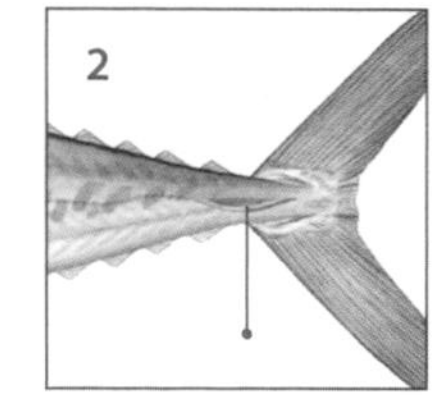

UNIQUE FEATURE: A Spanish mackerel with wide bars on its sides, much wider than in the Narrow-barred King Mackerel.

DIAGNOSTIC FEATURES: Gill rakers: There are a moderate number of gill rakers on the first arch: 1 or 2 on the upper limb, 5 to 11 on the lower limb, for a total of usually 11 or fewer. **Fins:** The first dorsal fin has 13–15 spines; the second dorsal fin has 19–22 rays, usually 20 or more, followed by 8–10 finlets. The anal fin has 19–22 rays, usually 21 or 22, followed by 7–10 finlets. There are many pectoral fin rays, 20–25, usually 23 or 24. **Lateral line:** The lateral line gradually curves down toward the caudal peduncle. **Vertebrae and intestine:** Vertebrae number 18 or 19 precaudal plus 25–27 caudal, for a total of 44–46, usually 45. The intestine has 2 folds and 3 limbs. The osteology and soft anatomy were described by Collette and Russo (1985). **Color:** Juveniles (less than 10 cm FL) are marked with 12–20 broad vertical dark-gray bands confined to the body above the lateral line, giving rise to its common name of Broadbarred Mackerel. The number of bands increases with age. The cheeks and belly are silvery white; the snout is dark gray with a patch of green above the eyes. The first dorsal fin is jet black anteriorly, white posteriorly. The second dorsal fin is cream colored with some yellow anteriorly. The anal fin and anal finlets are transparent white. The caudal fin is creamy white at the margins and dusky or blackish near the base. The pectoral fins are dusky. The vertical body bands are most marked in individuals less than 50 cm FL, and in larger fish there is a tendency for these markings to become less distinct, break into spots, or fade out more or less completely. Above 70 cm FL, dead fish assume a blotchy drab grayish yellow appearance with little or no evidence of markings, giving rise to the Australian common name of Grey Mackerel (Munro 1943).

GEOGRAPHIC RANGE: Broadbarred Mackerel are found along the northern coast of Australia from northern New South Wales north and west around the coast through the Gulf of Carpentaria to Shark Bay, Western Australia, and also in southern Papua New Guinea. Reports of this species from Thailand and Malaysia are based on misidentifications.

SIZE: Grows to 120 cm FL, 10 kg. The IGFA all-tackle angling record is of a 20 lb, 8 oz (9.3 kg) fish taken at The Patch, Dampier, Australia, in June 1997. Females grow faster than males and have a larger maximum asymptotic length. Five methods of determining length at age from otoliths were compared by Ballagh et al. and longevity was estimated to be 12 years.

HABITAT AND ECOLOGY: Broadbarred Mackerel are pelagic and oceanodromous. They are found most commonly around coastal headlands and rocky reefs but are also caught offshore. They tolerate turbid low-salinity waters and thus can inhabit nearshore areas such as river mouths and estuaries.

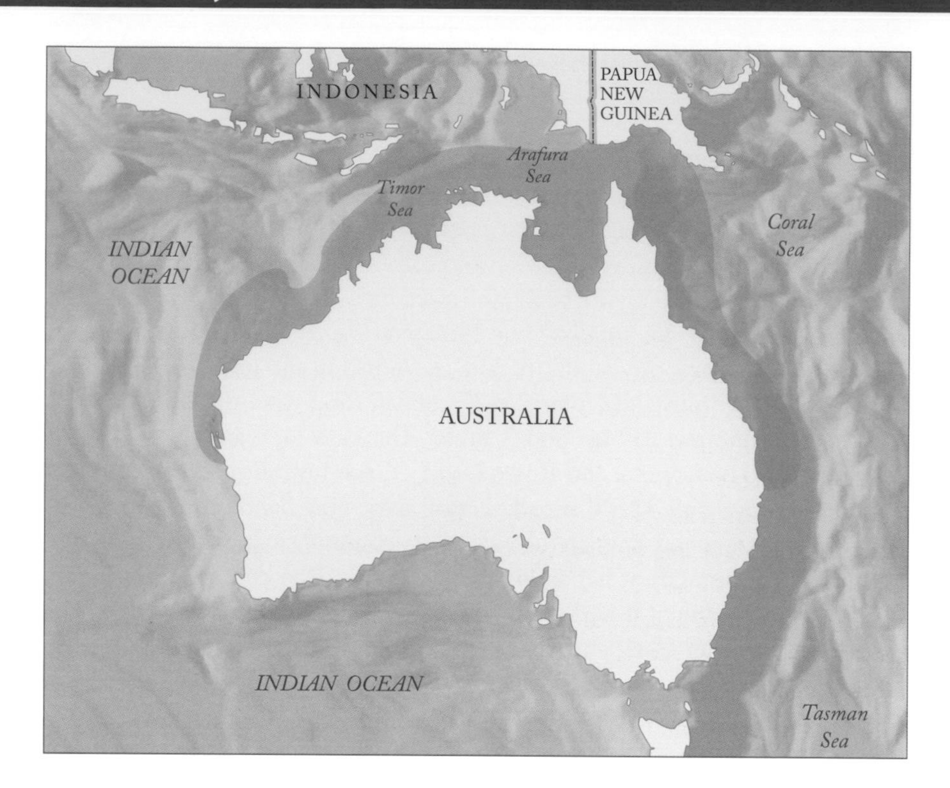

Right: Broadbarred Mackerel range map.

FOOD: Broadbarred Mackerel from the Dampier region of northwest Australia, estuaries along the coast of north Queensland, and the Norman River estuary of the Gulf of Carpentaria fed almost exclusively on pelagic fishes. They consumed prey fishes that were large relative to their size.

REPRODUCTION: Spawning of Broadbarred Mackerel apparently begins in September to October. Along the eastern Australian coast, length of 50% maturity was 67.45 cm FL for males and 81 cm FL for females. A later study found estimates of 50% maturity to be 57.1 cm for males and 60.2 cm for females. The smallest ripe male was 54 cm FL, and the smallest ripe female was 70 cm FL.

EARLY LIFE HISTORY: Larval Broadbarred Mackerel were collected from November to January from bays and the inside margin of the lagoon region of the Great Barrier Reef off Townsville, Queensland. Larvae were described from 101 specimens 3–10.5 mm standard length. Two types of larvae were found with 44–46 myomeres, which match the vertebral counts for Broadbarred Mackerel and Narrow-barred Spanish Mackerel. Broadbarred Mackerel larvae are distinguished by the presence of two rudimentary preopercular spines anterior to the main series of preopercular spines, instead of three. Head, trunk, and fin pigmentation, morphometric data, and illustrations of six larvae 3.8–10.5 mm SL were presented by Jenkins et al. Larger larvae, 12–44 mm SL, were taken inshore and up to 16–24 km from the coast with light traps. Larval Broadbarred

Mackerel feed mostly on other larval fishes, particularly herring-like fishes such as anchovies.

STOCK STRUCTURE: There are at least five biological stocks of Broadbarred Mackerel across northern Australia: Western Australia, northwest Northern Territory, Gulf of Carpentaria, northeast Queensland, and central east Queensland, complicating management.

FISHERIES INTEREST: Broadbarred Mackerel are an important commercial species taken across northern Australia predominantly by commercial offshore gill netters. They are also a highly prized game fish in the recreational fishing sector, especially in the Gulf of Carpentaria. During the late 1990s, most of the catch was taken in the Gulf of Carpentaria by Queensland and Northern Territory commercial gill net fishers, with a national harvest of about 800 mt. Broadbarred Mackerel are the second most important species in the Northern Territory Spanish mackerel troll fishery. Four species of *Scomberomorus*, including *S. semifasciatus*, along with *Grammatorcynus*, form Queensland's second most important fin-fishery. They are captured in sheltered waters by set netting and are an important target for sport fishermen who fish by trolling. Fish of 60 to 90 cm FL are caught on fishing grounds north of Yeppoon (Queensland) in November, while small groups of fish are taken in estuaries north of Moreton Bay. The commercial catch from the Great Barrier Reef in 2003 was 181 mt. They are caught by hook and line throughout the year in Western Australia, with the majority of boat-based recreational catch occurring on the North Coast, with some catches on the Gascoyne and west coasts. The reported annual catch has varied between 193 and 444 mt during 2006–10, with the most recent catch (2009–10) 193 mt, of which 181 mt were caught with nets and 12 mt by line. Broadbarred Mackerel are also taken by trawlers in the Gulf of Papua.

THREATS: The northwest Northern Territory stock declined substantially as a result of high Taiwanese catches in the 1970s and 1980s but has since recovered with cessation of foreign fishing. Despite being commercially harvested as a food source, there are no data suggesting that any of the stocks of Broadbarred Mackerel are undergoing a significant population decline at present, but some may be threatened by targeted fishing in spawning sites. This species was assessed as Least Concern on the IUCN Red List.

CONSERVATION: The five recognized stocks of Broadbarred Mackeral are currently considered sustainable. The Western Australia stock is the most different and is managed as a component of the Mackerel Managed Fishery of Western Australia, which has the Narrowbarred Spanish Mackerel as its major component. The only regulations for the recreational harvest of the two Northern Territory stocks of Broadbarred Mackerel are the general personal possession limit of 30 fish. Management of the fishery in the Gulf of Carpentaria is more complicated, with management of the shared resource by the Queensland Fisheries Joint Authority. Queensland introduced changes in the net fishery at the beginning of the 2012 season, which decreased the total length of available net from 27 km to 9 km in the offshore component of the fishery. Regulations include a minimum size limit of 50 cm FL that applies to both commercial and recreational fishers on the east coast of Queensland. Recreational fishers are limited to 10 Broadbarred Mackerel per person. Licenses also regulate fishing practices and gear.

REFERENCES: ASR 2011; Baker and Sheaves 2005; Blaber 1986; Broderick et al. 2011; Cameron and Begg 2002; Charters et al. 2010; Collette 2001; Collette and Nauen 1983; Collette and Russo 1985; Collette et al. 2011; Grant 1987; Grubert et al. 2013; IGFA 2018; Jenkins et al. 1984a, 1984b, 1985; Macleay 1883; Munro 1943; Newman et al. 2010; Roelofs et al. 2014; Ryan et al. 2013, 2915; Salini et al. 1990, 1998; Stapley and Welch 2009; Thorrold 1993; Ward and Rogers 2003; Welch and Ballagh 2009; Welch et al. 2009a, 2009b; Williamson et al. 2006.

Pacific Sierra

Scomberomorus sierra Jordan & Starks, 1895

COMMON NAMES: English – Pacific Sierra
French – Thazard Sierra
Spanish – Carite Sierra

ETYMOLOGY: The American ichthyologists David Starr Jordan and Edwin Starks (1895) used the common name *sierra* for the scientific name.

SYNONYMS: *Scomberomorus maculatus* non Mitchill, 1815; *Scomberomorus sierra* Jordan and Starks, 1895

TAXONOMIC NOTE: Sometimes this species has erroneously been considered a synonym of the western Atlantic *Scomberomorus maculatus*, although it is closely related to that species as well as the other Gulf of California Spanish mackerel, *S. concolor*, and the eastern Atlantic *S. tritor* (Collette and Russo 1985, Banford et al. 1999).

FIELD MARKS:

1 Pacific Sierra have 12–17 gill rakers on the first gill arch, compared to the Monterey Spanish Mackerel, which has 21–27.
2 Females of both species have golden spots on their sides. Pacific Sierra have two to four rows, with up to seven or eight rows in large speciemes. Female Monterey Spanish Mackerel have only two rows of spots. Male Pacific Sierra also have spots on their sides, while male Monterey Spanish Mackerel lack spots.

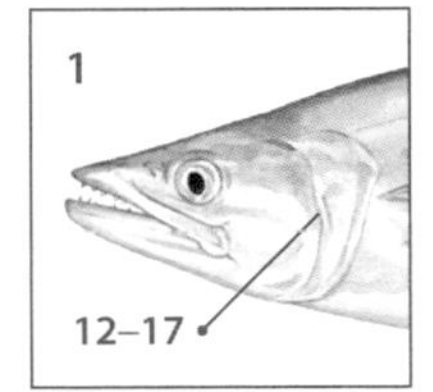

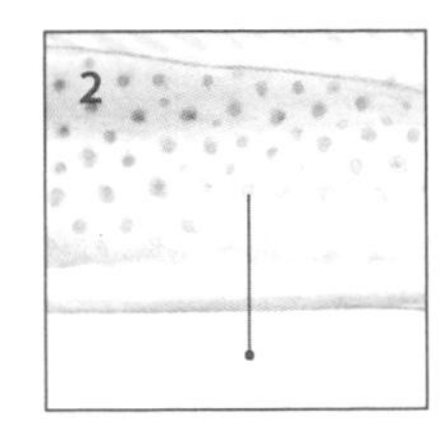

UNIQUE FEATURE: The Sierra Mackerel is the only Pacific species of the three species of Spanish mackerels that has numerous small round orange-colored spots on its sides.

DIAGNOSTIC FEATURES: Gill rakers: Gill rakers on first arch moderate, 2–4 on the upper limb, 9–14 on the lower limb, usually 12 or 13, for a total of 12–17, usually 14–16. **Fins:** The first dorsal fin has 15–18 spines; the second dorsal fin has 16–19 rays, followed by 7–10 finlets. The anal fin has 16–21 rays, followed by 7–10 finlets. The pectoral fin has 20–24 rays, modally 21. **Lateral line:** The lateral line gradually curves down toward the caudal peduncle. **Vertebrae and intestine:** Vertebrae number 19–21 precaudal plus 26–29 caudal, for a total of 46–49, usually 48. The intestine has 2 folds and 3 limbs. Descriptions and illustrations of the osteology and soft anatomy of the Pacific Sierra are included in the revision of the Spanish mackerels (Collette and Russo 1984). **Color:** The sides of the body are silvery and marked with numerous brownish (orange in life) spots, three rows of spots below the lateral line, one above, and up to eight in large specimens. The anterior half of the first dorsal fin and the margin of the posterior half are black. The base of the posterior half is white. The second dorsal fin has black margins and is tinged with yellow. The anal fin is white.

GEOGRAPHIC RANGE: The Pacific Sierra is endemic to the eastern Pacific and is found from southern California to Antofagasta, Chile, including the Gulf of California and the Galapagos, Cocos, Malpelo, and Tres Marias islands.

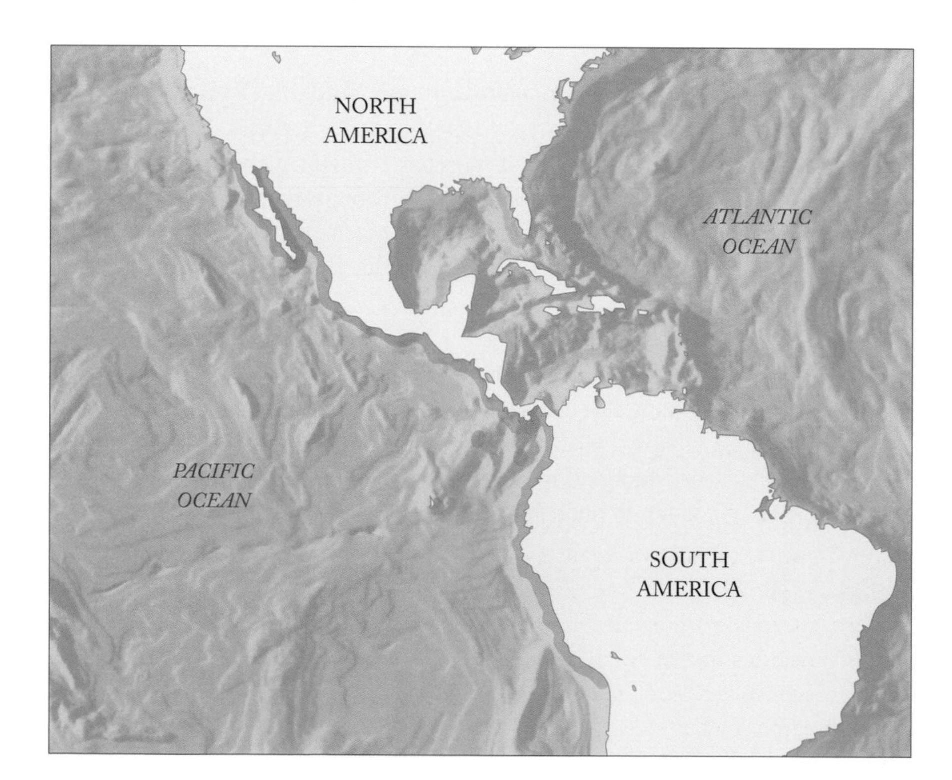

Right: Pacific Sierra range map.

SIZE: Pacific Sierra grow to 97 cm FL, 8.2 kg. The IGFA all-tackle game fish record is shared by two 18 lb (8.16 kg) fish from Ecuador caught in 1990, one from Salinas and the other from Isla de la Plata. Females attain a larger mean size (44.9 cm FL) than males (41.1 cm FL) on the Pacific coast of Mexico.

HABITAT AND ECOLOGY: It occurs near the surface of coastal waters.

FOOD: Adult Pacific Sierra mainly feed on small fishes. In Colombia and Mexico, they feed on anchovies (*Anchoa* and *Cetengraulis*) and other herring-like fishes (*Odontognathus* and *Opisthonema*).

REPRODUCTION: The size of first maturity for Pacific Sierra is 26–32 cm FL in Colombia and 44.3 cm FL in the Gulf of California, corresponding to an age of 3 years. Spawning takes place near the coast over most of its range. Spawning occurs off Mexico from July through September, in the Gulf of Nicoya, Costa Rica, between August and the end of November, and in Colombia between November and April. Based on an estimated average age of first maturity and a longevity of 8 years, a generation length of 2–4 years is estimated.

EARLY LIFE HISTORY: Twelve juveniles, 21–71 mm, from Costa Rica and Ballenas Bay, Baja California, were described, and the smallest and largest of these were illustrated by Eckles.

FISHERIES INTEREST: Pacific Sierra is an abundant game fish along the Pacific coasts of Mexico and Central America, and it is important in commercial fisheries. It is caught by gill nets in artisanal fisheries throughout Mexico, Nicaragua, Panama, and Colombia. The Pacific Sierra has become one of the most important reef fish fisheries in the Gulf of California in recent years, with the annual landings increasing from 433 mt in 1956–61 to 2,583 mt in 2000–2005. Total reported landings show a gradual, but variable, increase from 500 mt in 1950 to 13,649 mt in 2014 and have been fairly stable, 8,000–13,000 mt over the last 10 years (2004–14).

THREATS: Pacific Sierra are widespread and, although population levels appear to fluctuate in Peru, they are relatively stable at present despite an active fishery. Pacific Sierra were listed as Least Concern on the IUCN Red List.

CONSERVATION: There are no species-specific conservation measures for Pacific Sierra. However, in Mexico, there is a sport fishing limit of a total of 10 fish per day per person and no more than 5 of a single species for all sport fisheries. In Peru, there is a minimum catch size of 60 cm and a maximum tolerance of 10% juveniles in the catch. Increasing the age of first capture to 5.5 years would result in greater yield without negative consequences to the stock. More research is needed on this species' biology, particularly on age, growth, reproductive biology, and natural mortality rates.

REFERENCES: Aguirre-Villaseñor et al. 2006; Artunduaga Pastrana 1976; Banford et al. 1999; Collette 1995; Collette and Nauen 1983; Collette and Russo 1985; Collette et al. 1963, 2011; Domínguez López et al. 2010; Eckles 1949; Erdman 1971; Erisman et al. 2010, 2011; Klawe 1966; Kong 1978; Moreno-Sánchez et al. 2011; Ramírez-Pėrez 2010; Ramírez-Pérez et al. 2015; Robles-Cota 2011; Schaefer 2001; Vega 2004; Williams et al. 2011.

Chinese Seerfish

Scomberomorus sinensis (Lacepède 1800)

COMMON NAMES: English – Chinese Seerfish
French – Thazard Nébuleux
Spanish – Carite Indochino

ETYMOLOGY: The Chinese Seerfish was described by the French naturalist Bernard Germaine Etienne de la Ville, Comte de Lacepède (1800:599), based on a Chinese drawing with no etymology for the name he chose, but the name *sinensis* clearly refers to China.

SYNONYMS: *Cybium chinense* Cuvier and Valenciennes, 1831; *Cybium cambodgiense* Durand, 1940

TAXONOMIC NOTE: This species may be misidentified as *Scomberomorus niphonius* in southern China.

FIELD MARKS:

1 The Chinese Seerfish is the only species of Spanish mackerel that has a deep dip in the lateral line beginning under the first dorsal fin. The Atlantic King Mackerel and the Indo-Pacific Narrow-barred Spanish Mackerel also have a deep dip in the lateral line but the dip is much further back, under the dorsal finlets.
2 Adults have two poorly defined rows of large spots on the sides that are mostly larger than eye diameter.
3 The fins are mostly blackish.

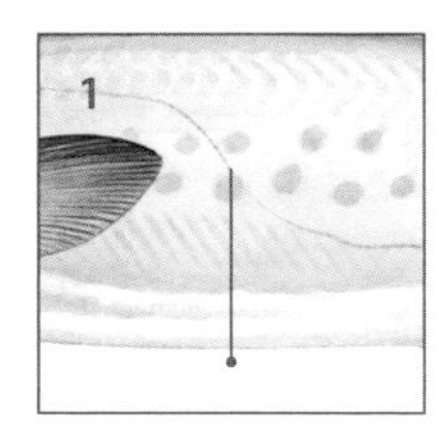

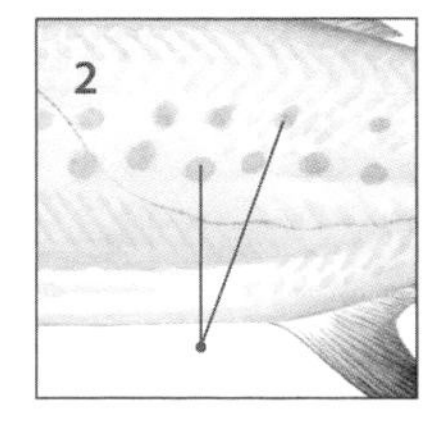

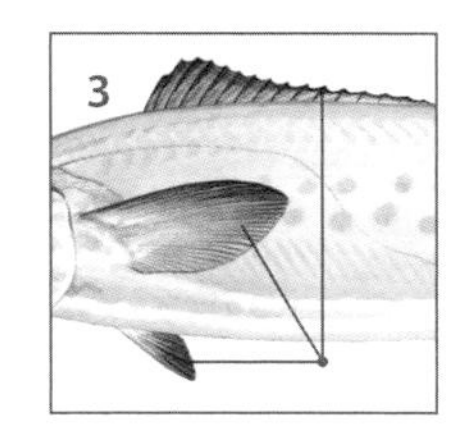

UNIQUE FEATURE: This is the largest species of Spanish Mackerel, the only one that has a swim bladder, and the only one that moves long distances into fresh water.

DIAGNOSTIC FEATURES: Gill rakers: There are a moderate number of gill rakers on the first arch: 1–3 on the upper limb and 10–12 on the lower limb for a total of 11–15. **Fins:** The first dorsal fin has 15–17 spines; the second dorsal has 15–17 rays, followed by 6–7 finlets. The anal fin has 16–19 rays, followed by 5–7 finlets. There are 21–23 pectoral-fin rays. **Lateral line:** The lateral line curves down abruptly beneath the first dorsal fin. **Swim bladder:** This is the only species of Spanish mackerel that has a swim bladder. **Vertebrae and intestine:** Vertebrae number 19 or 20 precaudal plus 21 or 22 caudal, for a total of 41 or 42. The intestine has 2 folds and 3 limbs. Descriptions and illustrations of the osteology and soft anatomy of the Chinese Seerfish are included in the revision of the Spanish mackerels, Collette and Russo (1984). **Color:** The back is greenish blue, the belly silvery, and the fins mostly blackish. The pelvic and anal fins have blackish margins. The anal finlets are colorless. There are two rows of large (larger than the diameter of the eye) indistinct round spots on the sides in adults. Juveniles have saddle-like blotches extending down to about the middle of the body.

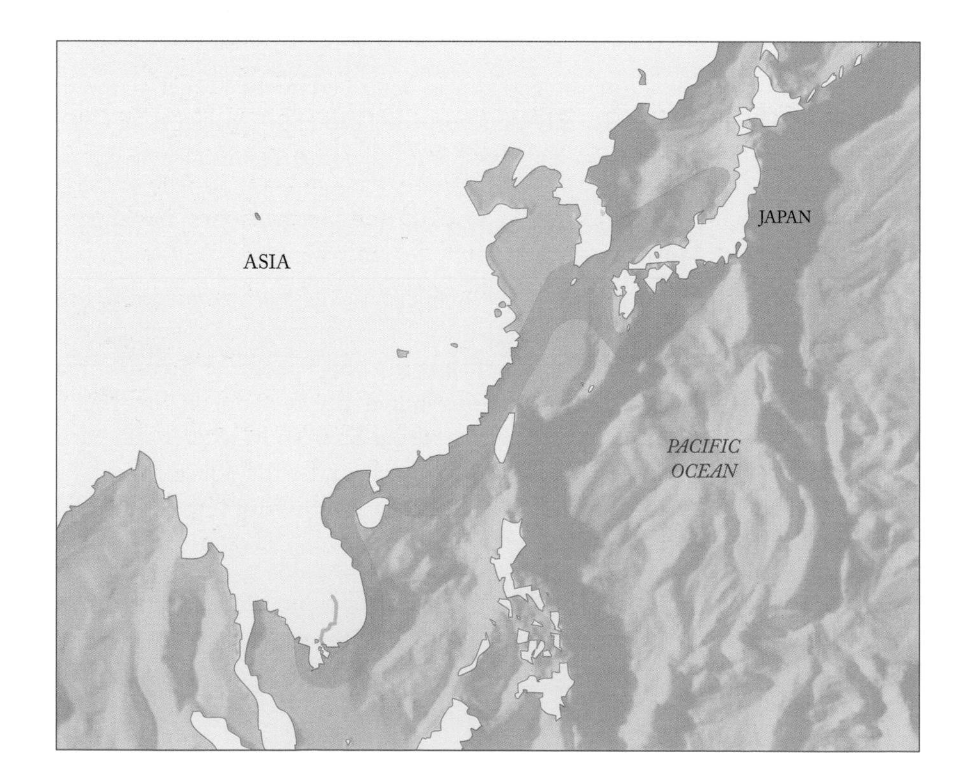

Right: Chinese Seerfish range map.

GEOGRAPHIC RANGE: Chinese Seerfish are restricted to the western Pacific, from Akita, Honshu, Sea of Japan, the Yellow Sea, and China south to Vietnam and Cambodia, where it enters the Mekong River. This is the only scombrid species that moves long distances into fresh water.

SIZE: Chinese Seerfish grow to at least 218 cm FL and 131 kg. The IGFA all-tackle game fish record is of a 288 lb, 12 oz (131 kg) fish caught off Cheju-Do, Korea, in October 1982. There is a length-weight graph for fish up to 120 cm FL and 18 kg in D'Aubenton and Blanc.

HABITAT AND ECOLOGY: The Chinese Seerfish is an epipelagic, neritic, and estuarine species that can travel more than 300 km up the Mekong River, above Phnom Penh, Cambodia. Very little information is available about its biology.

FOOD: The only published report of the diet of Chinese Seerfish is an old statement that they feed on "sardines and carangids" in Japan.

REPRODUCTION: The Chinese Seerfish is presumed to reproduce in marine waters, but juveniles as small as 165 mm FL have been reported from a large freshwater lake, Tonle Sap, which is hundreds of kilometers up the Mekong River.

FISHERIES INTEREST: Chinese Seerfish are a prized food fish in Japan and probably in China as well. They are caught in the Mekong River of Cambodia and commanded a high price in the Phnom Penh market in 1964. They are utilized fresh, dried, or salted and smoked and are consumed pan-fried, broiled, and baked. However, they are not currently common in markets in Japan. There are no landings data available for this species.

CONSERVATION: There are no species-specific conservation measures. Research is needed to determine the status of this species in addition to more information on its biology. This species is very poorly known and may be confused with *Scomberomorus niphonius* in some parts of its range. Given its large size, it is likely to be vulnerable to fishing pressure throughout its range, but no landings data are available. It is listed as Data Deficient on the IUCN Red List.

REFERENCES: Blanc et al. 1965; Collette 2001; Collette and Nauen 1983; Collette and Russo 1985; Collette et al. 2011; D'Aubenton and Blanc 1965; Kishinouye 1915, 1923; Lacepède 1800.

West African Spanish Mackerel

Scomberomorus tritor (Cuvier, 1832)

COMMON NAMES: English – West African Spanish Mackerel
French – Thazard Blanc
Spanish – Carite Lusitánico

ETYMOLOGY: Named *tritor* by the great French anatomist and ichthyologist Georges Cuvier (in Cuvier and Valenciennes 1832), with no indication of the meaning of the name.

SYNONYMS: *Cybium tritor* Cuvier, 1832; *Apolectus immunis* Bennett, 1831; *Scomberomorus argyreus* Fowler, 1905; *Scomberomorus maculatus* (non Mitchill, 1815); *Scomberomorus tritor* (Cuvier, 1832).

TAXONOMIC NOTE: This species has been erroneously considered as a synonym of the western Atlantic *Scomberomorus maculatus* by many authors. It is closely related to several other western and eastern Atlantic Spanish mackerels that share the strange character of having teeth in their nasal capsules (Collette and Russo 1985, Banford et al. 1999).

FIELD MARKS: 1 This is the only species of Spanish mackerel in the eastern Atlantic, and it differs from other species of Spanish mackerels in lacking any bold bars or spots, having instead a pattern of irregular, less-distinct elongate spots, which tend to form faint crossbars in large adults.

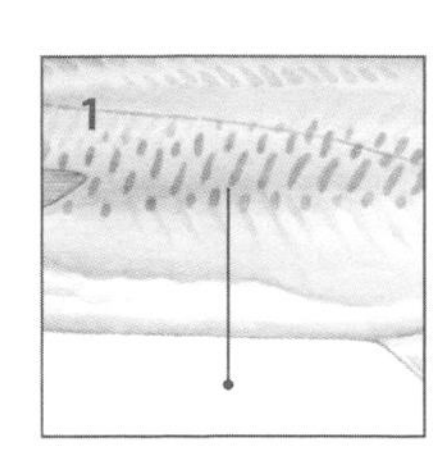

UNIQUE FEATURE: The only species of Spanish Mackerel that occurs natively in the eastern Atlantic.

DIAGNOSTIC FEATURES: Teeth: Teeth on lower jaw 3–21, mostly 12–15, upper jaw 13–27, mostly 17–19. **Gill rakers:** Gill rakers on first arch moderate, 1–3 on the upper limb, usually 2; 10–13 on the lower limb, for a total of 12–15. **Fins:** The first dorsal fin has 15–18 spines; the second dorsal fin has 16–19 rays, usually 17, followed by 7–9 finlets. The anal fin has 17–20 rays, followed by 7–9 finlets. The pectoral fin has 20–22 rays. **Lateral line:** The lateral line gradually curves down toward the caudal peduncle. **Vertebrae and intestine:** Vertebrae number 18 or 19 precaudal plus 27 or 28 caudal, for a total of 46 or 47, usually 19 plus 27, total 46. The intestine has 2 folds and 3 limbs. The osteology and soft anatomy were described and illustrated by Collette and Russo (1985). **Color:** The sides of the body are silvery and marked with about three rows of vertically elongate dark spots; some large individuals have large vertical bars. The anterior half of the first dorsal fin and the margin of the posterior half are black; the base of the posterior half is white.

GEOGRAPHIC RANGE: The West African Spanish Mackerel is present in the eastern Atlantic from the Canary Islands and Senegal south through the Gulf of Guinea to Baía dos Tigres in southern Angola. It is rare in the Mediterranean Sea, with several verified records from Nice, Villefranche, and Palermo along the coasts of France and Italy.

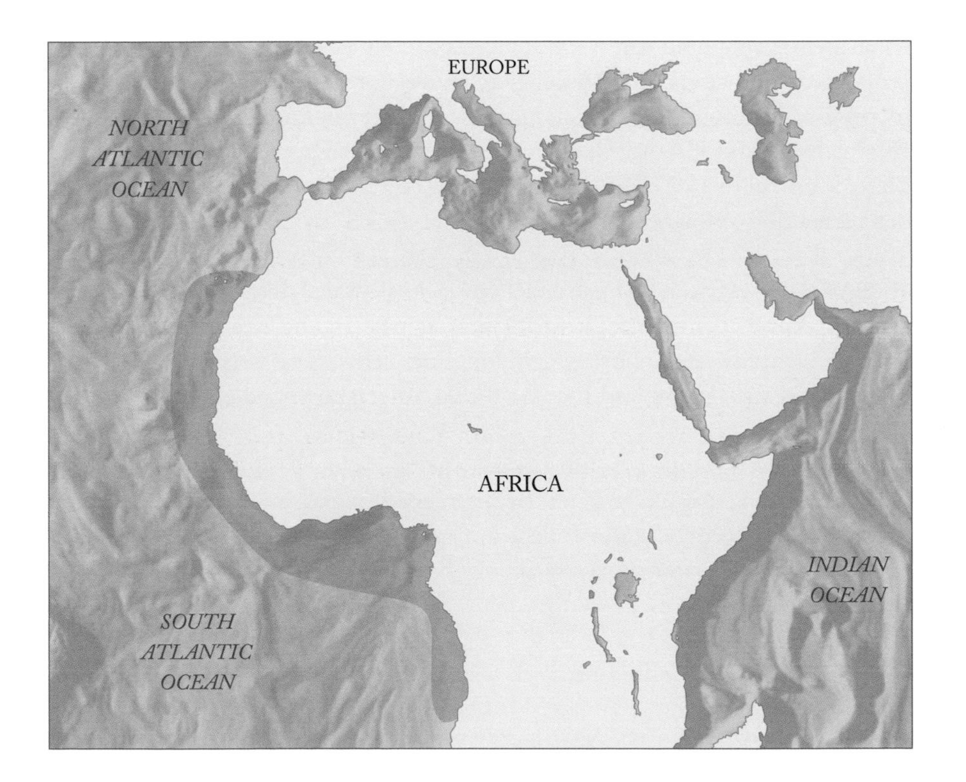

Left: West African Spanish Mackerel range map.

SIZE: West African Spanish Mackerel grow to 100 cm FL. The IGFA all-tackle game fish record is of a 13 lb, 3 oz (6 kg) fish caught off Grand Bereby, Ivory Coast, in December 1998. Length-weight relationships of samples 58–468 cm from several estuaries were presented by Écoutin and Albaret.

HABITAT AND ECOLOGY: The West African Spanish Mackerel is a coastal pelagic species that inhabits warm waters and enters coastal lagoons such as the Ebrié Lagoon in the Ivory Coast and the Lagos Lagoon in Nigeria.

FOOD: This species feeds on herring-like fishes, particularly *Ethmalosa fimbriata*, in coastal lagoons and on sardines, anchovies, and other small fishes.

REPRODUCTION: West African Spanish Mackerel reproduce in July to August in Mauritania, April to October in Senegal, and February to September in Guinea-Bissau. Length at 50% maturity was estimated in Senegal for males as 33.1 cm FL and 34.1 cm FL for females. In Guinea-Bissau, length at 50% maturity was 33.5 cm FL for females and 32.2 cm FL for males. A 95 cm FL female contained about 1 million eggs.

EARLY LIFE HISTORY: Juveniles have been seined along the shore near Dakar in July. Seven larvae 3.5–8.1 mm were caught south of the Ivory Coast at water temperatures of 23.2–26°C and salinities of 34.38–35.45 ppt.

FISHERIES INTEREST: West African Spanish Mackerel are taken by artisanal fishermen throughout the Gulf of Guinea: Senegal, Guinea, Senegambia, Ivory Coast, Ghana, and Nigeria. They are one of the species frequently used to prepare "lanhouin," a traditional processed fermented fish condiment in Benin, Togo, and Ghana. Catches are likely underreported and have ranged up to 5,389 mt in 1990. For the past 10 years catches have generally fluctuated between 1,500 and 4,500 mt without trend.

THREATS: West African Spanish Mackerel is mostly taken by artisanal fishermen and sometimes commercially with purse seines. In the Mediterranean it is only incidentally caught by pelagic longlines. This species is listed as Least Concern on the IUCN Red List for the Mediterranean and globally.

CONSERVATION: There are no species-specific conservation measures for West African Spanish Mackerel. Better reporting and more catch and effort information is needed for most species of small tunas in the Atlantic. Although catch landings are not regularly reported, there is no current indication of decline.

SELECTED REFERENCES: Banford et al. 1999; Collette and Nauen 1983; Collette and Russo 1985; Collette et al. 2011; Diaha et al. 2010; Fagade and Olaniyan 1973; E. Postel 1954, 1955; Tortonese 1975.

Albacore

Thunnus alalunga (Bonnaterre, 1788)

COMMON NAMES: English – Albacore
French – Germon
Spanish – Atún Blanco
Japanese – Binchô, Binnaga

ETYMOLOGY: Named *alalunga*, a combination of the Latin *ala*, "wing," and *lunga*, "long," in reference to the diagnostically long pectoral fin, by the French naturalist Pierre Joseph Bonnaterre (1788:139), based on an account in Cetti (1788) from the Mediterranean Sea.

SYNONYMS: *Scomber alalunga* Bonnaterre, 1788; *Scomber alatunga* Gmelin, 1789; *Scomber albicans* Walbaum, 1792; *Scomber germon* Lacepède, 1800; *Scomber germo* Lacepède, 1801; *Thynnus pacificus* Cuvier, 1832; *Scomber germo* Bennett, 1840; *Orcynus pacificus* Cooper, 1863; *Germo germon steadi* Whitley, 1933; *Thunnus alalunga* (Bonnaterre, 1788)

TAXONOMIC NOTE: Albacore are often confused with juvenile Bigeye Tuna, which also have long pectoral fins, especially at sizes under 50 cm fork length, but the pectoral fins of the Albacore have rounded rather than pointed tips.

FIELD MARKS:
1 In individuals greater than 50 cm fork length, the pectoral fins are longer than in other tunas, usually 30% of fork length.
2 The narrow white trailing margin of the caudal fin distinguishes the Albacore from the other members of the genus.

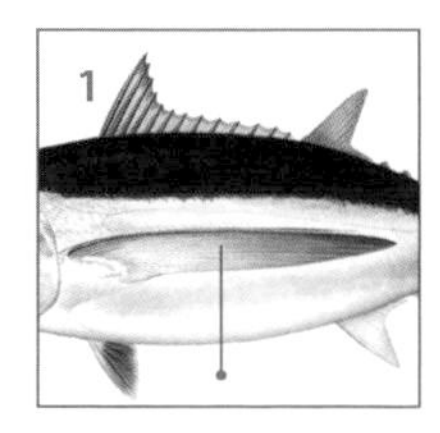

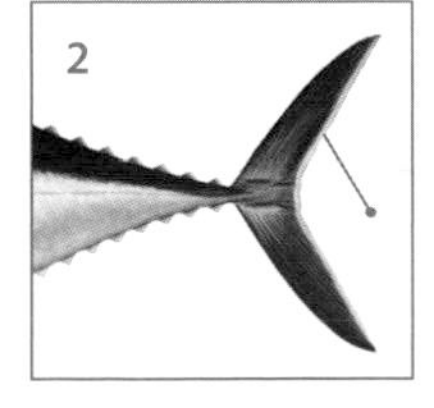

UNIQUE FEATURE: Albacore have the lightest flesh of all the tunas and are mainly canned as white-meat tuna.

DIAGNOSTIC FEATURES: Albacore are medium-sized tunas with the greatest body depth at a more posterior point than in other tunas, at or slightly anterior to the origin of the second dorsal fin, rather than at the middle of the first dorsal-fin base. **Gill rakers:** Gill rakers number 25 to 31 on the first arch and average a little more than 28. **Fins:** There are 12–14 spines in the first dorsal fin, 13–16 rays in the second dorsal fin, and 7–3 dorsal finlets. There are 11–15 rays in the anal fin, followed by 7–9 anal finlets. The pectoral fins have 31–36 rays and are remarkably long, usually 30% of fork length in fish 50 cm or longer, reaching well beyond the origin of the second dorsal fin. Smaller individuals have relatively shorter pectoral fins, overlapping in length with those of small Bigeye Tuna. **Liver:** Striations cover the ventral surface of the liver, and the right lobe is not significantly longer than the central and left lobes. **Swim bladder:** A swim bladder is present but poorly developed. **Spleen:** The spleen is on the left side of the body cavity and the stomach is on the right, reversed from the positions in the other species of *Thunnus*. **Vertebrae:** Vertebrae number 18 precaudal plus 21 caudal, for a total of 39 vertebrae. Descriptions and illustrations of the osteology and soft anatomy are contained in the revision of *Thunnus* by Gibbs and Collette (1967). **Color:** Back metallic dark blue changing to whitish on lower sides and belly. A lateral iridescent blue band runs along the sides in live individuals. The first dorsal fin is deep yellow, the second dorsal and anal fins are light yellow, the anal finlets are dark, and the posterior margin of the caudal fin is white.

GEOGRAPHIC RANGE: Albacore are cosmopolitan in tropical and temperate waters of all the oceans, including the Mediterranean Sea, but avoid surface waters between 10° N and 10° S. In the Atlantic, it is widely distributed between 60° N and 50° S. Albacore distribution in the western Pacific extends in a broad band between 40° N and 40° S and continues across to the eastern Pacific, where it separates into two branches. The northern branch extends from British Columbia to the tip of Baja California, and the southern branch extends from southern Peru to southern Chile.

SIZE: Albacore can reach up to 140 cm FL and 60 kg. The IGFA all-tackle game fish record is of an 88 lb, 2 oz (39.97 kg) fish taken off of Gran Canaria, Canary Islands, in November 1977. A 128 cm FL fish from the North Pacific was determined to be 15 years old, slightly older than estimates obtained from other Albacore age and growth studies.

HABITAT AND ECOLOGY: Albacore are a highly migratory, epipelagic and mesopelagic, oceanic species that is abundant in surface waters of 15.6°C to 19.4°C. Deeper swimming, large albacore are found in waters of 13.5°C to 25.2°C. Temperatures as low as 9.5°C may be tolerated for short periods. Vertical behavior differed substantially between tropical and temperate latitudes. At tropical latitudes, Albacore showed a distinct diel pattern in vertical habitat use, occupying shallower warmer waters above the mixed layer at night and deeper cooler waters below the mixed layer during the day. In contrast, there is little evidence of a diel pattern of vertical behavior in Albacore at temperate latitudes, with fish limited to shallow waters above the mixed layer almost all of the time. Albacore are known to concentrate along thermal discontinuities. They occasionally form mixed schools with Skipjack Tuna and Yellowfin Tuna. Use of combined Japanese and US tagging data confirm the frequent westward movement of young Albacore and less frequent eastward movements in the North Pacific. This corresponds to Albacore life history, where immature fish recruit into surface fisheries in the western and eastern Pacific and before maturing, gradually move near their spawning grounds in the central and western Pacific, where they are vulnerable to more deeply set longline gear. Details of migration in the eastern North Pacific remain unclear, but juvenile fish (2- to 5-year-olds) are believed to move into the eastern Pacific Ocean in the spring and early summer and return to the western and central Pacific perhaps annually, in the late fall and winter, where they tend to remain as they mature.

Below Albacore range map.

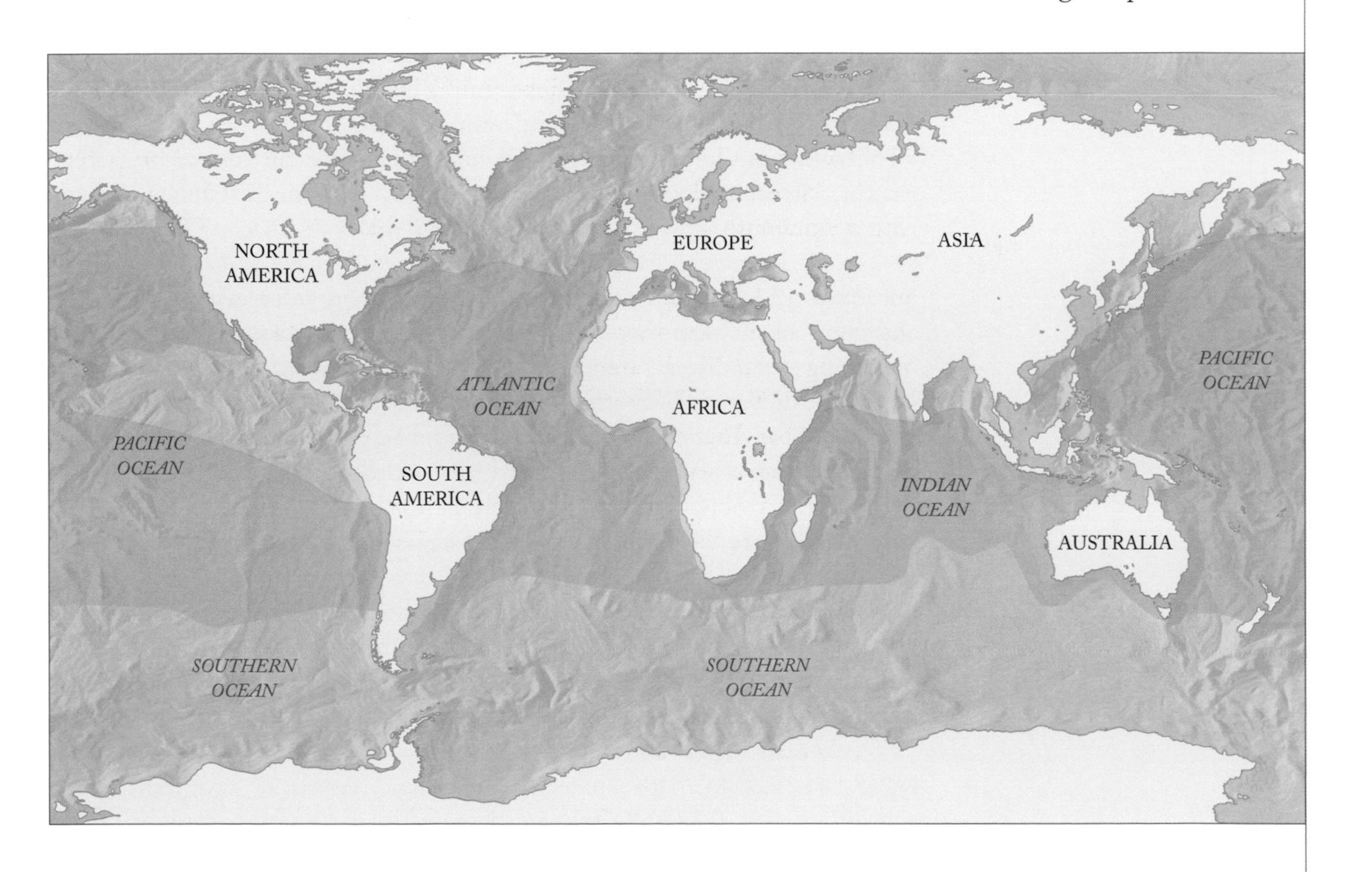

FOOD: Like other tunas, Albacore feed on a wide variety of fishes, squids, and crustaceans. The relative proportions of the three major types of food items vary both geographically and seasonally. Mediterranean Albacore consume a particularly high proportion of squids, 95% of stomach contents in one study, but in the central Mediterranean, Albacore diets were dominated by paralepidid fishes, hyperidean amphipods, and squids. In the eastern North Atlantic, Albacore consumed a higher proportion of crustaceans and a low proportion of fishes in the most offshore sampling zones. In the western South Atlantic, Albacore feed mainly on squids, followed by fishes and crustaceans. Diversity of prey in Albacore diets was low in the western South Pacific, with 21–31 taxa identified in stomach contents of fish off the coast of New Zealand, where Albacore inhabit surface waters. They fed mostly on crustaceans, with smaller amounts of fishes, squids, and gelatinous plankton. In contrast, prey diversity was much higher, 72–163 prey taxa identified in stomach contents, for Albacore caught in tropical waters off New Caledonia and Tonga, where they move into deeper, cooler waters during the day. Important prey items included several families of primarily deep-sea fishes.

REPRODUCTION: Albacore mature relatively late and at a larger size compared to other tunas, with estimates of length at 50% maturity around 90 cm FL in all oceans except the Mediterranean. In the South Pacific the length at 50% maturity is about 87 cm FL at an estimated age of 4.5 years, while in the western Indian Ocean length at 50% maturity for female albacore was 85.3 cm FL. Maturity appears to occur earlier for Albacore in the Mediterranean, with a minimum size at first maturity for females of 56 cm FL at an age of 2 years. Immature Albacore (<80 cm) generally have a sex ratio of 1:1, but males predominate in catches of mature fish. Spawning occurs in tropical and subtropical waters when sea surface temperatures are 24°C or higher. Spawning occurs over 7 months, March to September, in tropical oceanic waters off Taiwan and the Philippines but is reduced to only 3 months in the Mediterranean, June to August. Like other tunas, Albacore are batch spawners with asynchronous egg development and indeterminate annual fecundity. Fecundity increases with size, but there is no clear correlation between fork length and ovary weight and number of eggs. In the western Indian Ocean, Albacore spawn on average every 1.7–2.2 days during the spawning season from November to January. They spawn in the South Pacific every 1.3 days during peak spawning from October to December, but they may spawn daily in the Mediterranean. Generation length is not well known for Albacore but is conservatively estimated to be 8–10 years.

EARLY LIFE HISTORY: Albacore eggs are pelagic, spherical, and transparent and are smaller (0.84–0.94 mm in diameter) than the eggs of other tunas. They contain an oil globule. Seven Albacore larvae 3.4–16.5 mm TL from the Indo-Pacific were illustrated by Ueyanagi. Albacore larvae resemble those of Yellowfin Tuna but differ in lacking pigmentation on the tip of the snout in larvae smaller than 9 mm. Albacore larvae seem restricted to warm

(24–25°C surface waters), above 60 m. Larval Albacore in the Mediterranean start feeding on copepodites and nauplii, switch to cladocerans and calanoid copepodites, and begin piscivory after notochord flexion.

STOCK STRUCTURE: Six major stocks of Albacore are recognized: North and South Atlantic, Mediterranean Sea, North and South Pacific, and Indian Ocean. There is substantial genetic evidence supporting the six stocks, and each is managed separately. However, alternate hypotheses of stock structure have been suggested, including four genetically homogenous populations (Mediterranean and Atlantic, Indian, and Pacific oceans). Analysis of molecular data is consistent with a lack of migration or gene flow from the North Atlantic to the Mediterranean Albacore population. Genetic heterogeneity has been reported within the Mediterranean even though there are no pronounced oceanographic barriers that might limit gene flow.

FISHERIES INTEREST: Albacore is one of the principal market species of tuna fisheries worldwide, representing 5% of the total catch in 2015. Albacore is mainly marketed as canned white-meat tuna, the most expensive canned tuna. It is caught with six main types of gear: pole and line, longline, gill net, troll, purse seine, and midwater trawl. Small juveniles (50–90 cm FL) are generally taken with surface gear (pole-and-line and trolling) and are based in temperate latitudes. Large adult fish (90–140 cm FL) are generally taken with gill net and longline in tropical and subtropical waters. FAO worldwide reported landings show a steady increase from 103,676 mt in 1950 to over 200,000 mt in 1971. Catches have generally fluctuated between 200,000 and 250,000 mt since that time.

Atlantic: Atlantic-wide Albacore landings increased from less than 40,000 mt in 1950 to a peak catch of more than 90,000 mt in 1965 and have ranged between 40,000 and 50,000 mt since 2007. The Atlantic population of Albacore is represented by three stocks: North and South Atlantic and Mediterranean.

North Atlantic: The main nations fishing in the North Atlantic are Spain, France, and Taiwan. Catches of North Atlantic Albacore are primarily made by pole and line (35%), trolling (28%), trawling (17%), and longline (17%). Catches of northern Albacore peaked at 65,000 mt in the mid-1960s, declined to a low of 20,000 mt in 2008, and then climbed back to 30,100 mt in 2016. Surface fisheries concentrate mainly in the Bay of Biscay, the Azores, and the Canary Islands during summer and fall, taking young fish, whereas longline vessels operate throughout the Atlantic year-round and target larger fish.

South Atlantic: Currently, the main nations fishing in this area are longliners from Taiwan (56%) and pole-and-line fishers from South Africa (18%) and Namibia (13%). Surface fisheries operate mainly between October and May, capturing juvenile and subadult fish. Catches have varied from a high of 41,000 mt in 1987 to a low of about 13,700 mt in 2016, a 10% decrease from 2015.

Mediterranean: Albacore are caught by a number of Mediterranean fleets, including those of Cyprus, Italy, Greece, Spain, and Malta. Longline is the primary gear, but catches are also made using gill nets and pole and line. Reported catches have been increasing, but there is a lack of reporting from some fisheries. Reported Albacore catches in the Mediterranean since 1982 have fluctuated between 1,235 mt in 1983 and 7,898 mt in 2003. Estimated catches in 2016 were 3,519 mt.

North Pacific: North Pacific Albacore support one of the most valuable tuna fisheries in the eastern Pacific. Currently, the main nations fishing in this area are Japan and the United States. The Japanese fishery has historically harvested the the largest amount, almost 75%, while the US fishery harvests most of the rest. The North Pacific Albacore catch in 2016 was about 51,200 mt, a 21% decrease from 2015. The main fishing gears are longline (39%) and pole and line (37%), followed by trolling (22%). Catches by longlining have shown a decreasing trend since 1997.

South Pacific: Taiwan, China, and Fiji are the major nations fishing for Albacore in the South Pacific. Albacore catches in this region increased from approximately 40,000 mt in the late 1960s / early 1970s to 80,000 mt by 2009. The catch of 68,600 mt in 2016 is similar to that of 2015. The main fishing gear is longline, which accounts for 96% of the catch.

Indian Ocean: Currently, Taiwan is the main nation fishing for Albacore in the Indian Ocean. Catches of Albacore in the Indian Ocean gradually increased from 20,000 mt in the early 1990s and have fluctuated between 25,000 and 40,000 mt for the past 30 years. Catches have been made almost exclusively by longline except for a period from 1980–95, when drift gill nets accounted for as much as a third of the catches.

THREATS: Currently, Albacore stocks are not overfished or experiencing overfishing. The drift net fishery for Albacore has been banned since 2002 in the EC countries and since 2004 in all the ICCAT Mediterranean countries, but it is known that illegal fishing activity still occurs in some areas. Data reporting is deficient for many fisheries that take Mediterranean Albacore, which adds considerable uncertainty to the assessment. Albacore are listed on the IUCN Red List as Near Threatened.

CONSERVATION: Albacore are listed as a highly migratory species in Annex I of the 1982 Convention on the Law of the Sea and are managed by the member nations of four tuna RFMOs.

North Atlantic: North Atlantic Albacore were overfished in the early 2000s, but ICCAT adopted a rebuilding plan that reduced TACs for several years, allowing the spawning stock biomass to recover. The 2016 assessment indicated the stock was not overfished and overfishing was not occurring. The stock is currently managed with quota allocations for the major harvesters, and a landings cap of 200 mt for minor harvesters, limitations on vessel numbers, and harvest control rules are being developed.

South Atlantic: The South Atlantic Albacore stock was considered to be overfished in the late 1990s, but stock status has been improving, and at the time of the 2016 assessment, the stock was not overfished and overfishing was not occurring. Recent catches have been well below the TACs. Current management measures include catch limits for major harvesters, with others restricted to catches of no more than 25 mt.

Mediterranean: The Mediterranean stock of Albacore was assessed in 2017, and it was estimated that current biomass is close to BMSY and fishing mortality is below FMSY. However, there was a lack of data for the assessment, resulting in considerable uncertainty. Management measures adopted by ICCAT include limits on vessel numbers and a two-month closed period for the fishery that coincides with a seasonal closure of the Mediterranean swordfish fishery.

North Pacific: The North Pacific Albacore stock was considered to be fully exploited in 2007, but the 2017 stock assessment indicated that North Pacific Albacore is currently not in an overfished state and is not being overfished. Catches over the past five years have been well below the estimated MSY of 132,000 mt. Management measures of the WCPFC and IATTC adopted in the mid-2000s called for members not to increase directed fishing effort.

South Pacific: The last full stock assessment was conducted by the WCPFC in 2015 and only covered the WCPFC Convention area south of the equator. The assessment indicated that the stock is not overfished and overfishing is not occurring. Analyses suggested further increases in effort will probably yield little or no increase in long-term catches and may result in further reduced catches. Recent catches have been below the estimated MSY of 76,800 mt. The WCPFC adopted a management measure to limit vessels fishing for South Pacific Albacore to the number fishing in 2005 or the 2000–2004 average.

Indian Ocean: The Indian Ocean Albacore stock was assessed in 2016 and at that time was not overfished and overfishing was not occurring. The stock had been considered to be fully exploited in 2007. Average catches over the past five years have been slightly below the estimated MSY of 38,800 mt. The IOTC has not adopted conservation and management measures for this stock.

SELECTED REFERENCES: Catalán et al. 2007; Chen et al. 2010, 2012; Collette and Nauen 1963; Collette and Nauen 1983; Collette et al. 2007; Consoli et al. 2008; Davies et al. 2011; Dragovich 1969; Evano and Bourjea 2012; Farley et al. 2013a, 2013b, 2014; Gibbs and Collette 1967; Ichinokawa et al. 2008; Iverson 1962; Montes et al. 2012; Nakadate et al. 2005; Nikolic et al. 2016; Pusineri et al. 2004; Schaefer 2001; Ueyanagi 1969; Viñas et al. 2004.

Yellowfin Tuna

Thunnus albacares (Bonnaterre, 1788)

COMMON NAMES: English – Yellowfin Tuna
French – Thon Jaune
Spanish – Aleta Amarilla

ETYMOLOGY: Named *albacares* by the French naturalist Pierre Joseph Bonnaterre (1788:140) based on an account in Sloane (1707) from Jamaica.

SYNONYMS: *Scomber albacares* Bonnaterre, 1788; *Scomber albacorus* Lacepède, 1800; *Thynnus argentivittatus* Cuvier in Cuvier and Valenciennes, 1832; *Scomber Sloanei* Cuvier in Cuvier and Valenciennes, 1832; *Thynnus Albacora* Lowe, 1839; *Thynnus macropterus* Temminck and Schlegel, 1844; *Orcynus subulatus* Poey, 1875; *Thunnus allisoni* Mowbray, 1920; *Neothunnus catalinae* Jordan and Evermann, 1926; *Neothunnus itosibi* Jordan and Evermann, 1926; *Kishinoella zacalles* Jordan and Evermann, 1926; *Semathunnus guildi* Fowler, 1933; *Neothunnus albacora* forma *longipinna* Bellón and Bardán de Bellón, 1949; *Neothunnus albacora* forma *brevipinna* Bellón and Bardán de Bellón, 1949

TAXONOMIC NOTE: Although several populations have been described, morphological and genetic data support a single, circumglobal species (Gibbs and Collette 1967, Scoles and Graves 1993).

FIELD MARKS:
1 Pectoral fins extend just beyond the second dorsal-fin origin.
2 Live animals have about 20 broken vertical lines of light spots below the lateral line.
3 Large individuals have elongate second dorsal and anal fins.

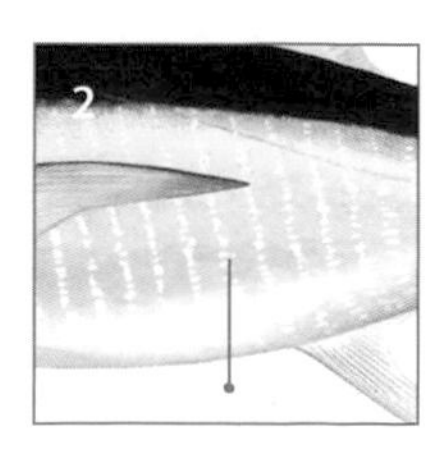

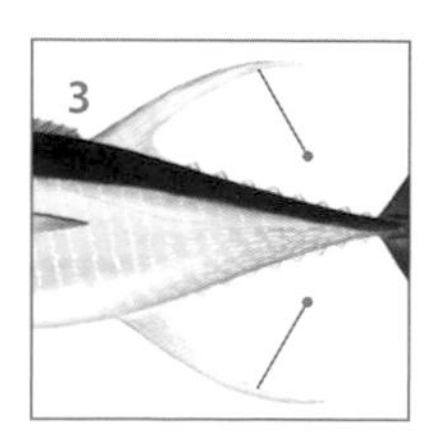

UNIQUE FEATURE: The Yellowfin Tuna is the only tuna in which large individuals have greatly elongate second dorsal and anal fins.

DIAGNOSTIC FEATURES: Yellowfin Tuna are large, with the body deepest near the middle of the first dorsal-fin base. **Gill rakers:** Gill rakers number 26 to 34 on the first arch. **Fins:** Large individuals develop elongate second dorsal and anal fins, which can become well over 20% of fork length. The pectoral fins are moderately long, usually reaching beyond the origin of the second dorsal fin but not beyond the end of its base, usually 22 to 31% of FL. **Liver:** There are no striations on the ventral surface of the liver and the right lobe is significantly longer than the central and left lobes. **Swim bladder:** A swim bladder is present. **Vertebrae:** Vertebrae number 18 precaudal plus 21 caudal. Descriptions and illustrations of the osteology and soft anatomy are contained in the revision of the genus by Gibbs and Collette (1967). Illustrations of the sagittal otoliths were published by Baremore and Bethea (2010). **Color:** The back is metallic dark blue changing through yellow to silver on belly; belly frequently crossed by about 20 broken, nearly vertical lines; dorsal and anal fins bright yellow; finlets bright yellow with a narrow black border. A recently collected specimen had black finlets instead of yellow.

GEOGRAPHIC RANGE: Yellowfin Tuna are found worldwide in tropical and subtropical seas but are absent from the Mediterranean Sea. They primarily occur between 20° N to 20° S, but during summer months, they have been caught at latitudes as high as 45–50° in both the northern and southern hemispheres. In the eastern Pacific, they range from southern California and the southwest and central eastern parts of the Gulf of California to Peru, including all of the oceanic islands. They may occur off Oregon and Washington during EI Niño years. During the austral summer, Yellowfin Tuna have been caught in waters south of the Cape of Good Hope.

SIZE: Maximum size is about 200 cm FL. The IGFA all-tackle game fish record is of a 427 lb (193.68 kg) fish caught off Cabo San Lucas, Baja California, Mexico. In the Indian Ocean, longevity is at least 7 years, although very few individuals live past 4 years. The apparent longevity in the eastern Pacific is 8 years, 6.5 years in the western Pacific, and 8 years in the Atlantic. The smallest mature individuals in the Pacific off the Philippines and Central America are in thc 50–60 cm FL size group at an age of 12–15 months. Based on age-structure data across all stocks, generation length is estimated to be between 2.2 and 3.5 years.

HABITAT AND ECOLOGY: Yellowfin Tuna are epipelagic and oceanic, occupying surface waters ranging from 18°C to 31°C. They spend most of their time in the mixed layer above the thermocline but routinely make repetitive bounce-dives during the day into cooler waters at depths of 200–300 m to forage on deep scattering-layer prey organisms. Around the Hawaiian Islands, large adults spent 60–80% of their time in or right below the uniform-temperature surface layer, above 100 m. Electronic archival tag data indicate that Yellowfin Tuna can dive to depths of at least 1,602 m, and deep dives in excess of 500 m are relatively common. Vertical distribution and time at depth is restricted by their physiological tolerance to oxygen and temperature. There is a close correlation between the vulnerability of the fish to purse seine capture, the depth of the mixed layer, and the strength of the temperature gradient within the thermocline. Yellowfin tuna tend to avoid waters with oxygen concentrations less than 2 ml/l. They are essentially confined to the upper 100 m of the water column in areas with marked oxyclines (upwelling zones) and are not usually caught below 250 m in the tropics. Yellowfin Tuna spend more time nearer the surface during darkness than during daylight hours. They school in surface waters primarily by size, either in monospecific or multi-species groups, and are often associated with floating objects. In some areas with shallow thermoclines, such as the eastern Pacific, larger fish (greater than about 85 cm FL) are frequently associated with porpoises.

Below: Yellowfin Tuna range map.

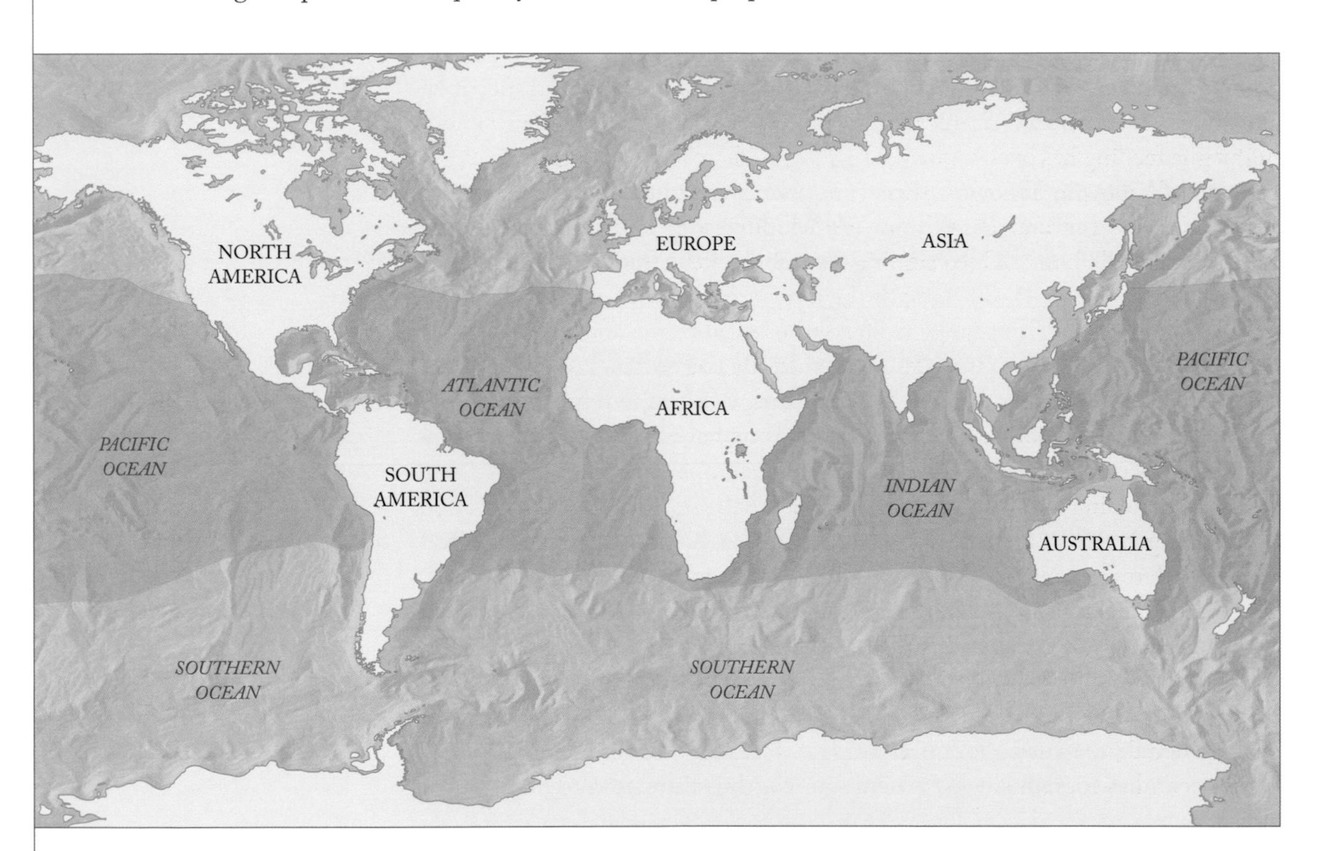

MOVEMENTS: Yellowfin Tuna are not highly migratory as are some tunas; results from conventional and electronic tagging studies suggest that net movements vary regionally. Within the Atlantic Ocean, conventional tagging results indicate limited dispersal of age-1 fish, but trans-Atlantic movements for older fish have been documented. In particular, juveniles tagged in the eastern Atlantic have been recovered in the western Atlantic, while juveniles and adults tagged in the western Atlantic have been recovered in the eastern Atlantic. In contrast, tagging studies in the western and eastern Pacific demonstrated movements that are more restricted. In the eastern North Pacific, archival tagging demonstrated movements of Yellowfin Tuna north and south with the seasons, but there was limited dispersal to the west, outside the region of high productivity.

FOOD: Yellowfin Tuna are opportunistic visual predators that consume a great diversity of prey types. By volume, fishes, squids, and crustaceans are the most important diet items. Fish prey can include both pelagic and mesopelagic species, and at times coastal fishes constitute a major food source. Stomach contents of Yellowfin Tuna swimming in mixed-species aggregations with two species of porpoises in the eastern tropical Pacific indicate that an ommastrephid squid (*Dosidicus gigas*) was the most important food item in all three species. Frigate tunas (*Auxis* species) and the pelagic halfbeak *Oxyporhamphus* were also important food items for both the Yellowfin Tuna and the Spotted Porpoise, and a portunid crab was also important for the tuna. Yellowfin Tuna and Blue Marlin fed mainly on squids in the eastern Arabian Sea. Yellowfin Tuna are known to alter their habitat utilization to take advantage of local prey concentrations. They feed primarily during the day, but nighttime feeding has been reported. In general, there is a relationship between the size of Yellowfin Tuna and their prey, and ontogenetic shifts in diet have been observed in several areas. Long-term Yellowfin Tuna diet analysis in the eastern Pacific demonstrated a significant shift in diet composition over a 10-year period.

REPRODUCTION: Spawning is widespread spatially and temporally throughout the world's oceans at sea surface temperatures of 24°C or higher, with the majority of spawning occurring at temperatures greater than 26°C. Spawning seasonality occurs in the northern and southern summer months in subtropical regions. Spawning occurs almost entirely at night between 2200 and 0600 hours. Length at 50% maturity in the eastern Pacific was 69 cm for males and 92 cm for females, corresponding to an age of about 2.1 years for females. In the western Indian Ocean, length at 50% maturity was 75 cm FL and in the Sea of Oman, 77.2 cm. The estimated batch fecundity in the eastern Pacific was 2.5 million oocytes, or 67.3 oocytes/gram of body weight. The fraction of mature females with postovulatory follicles was 0.66, indicating that the average female spawned about every 1.52 days. In the western Indian Ocean, batch fecundity for females of 79–147 cm FL was estimated to be 3.1 million eggs.

EARLY LIFE HISTORY: Fertilized eggs of captive Yellowfin Tuna brood stock ranged from 0.85 to 1.13 mm in diameter (average 0.97) and contained a single oil globule averaging 0.22 mm. Five larvae ranging from 3.2 mm notochord length to 8.5 mm SL were described and illustrated in Richards' 2006 identification guide to western central Atlantic fishes. Pigmentation is present on the midbrain, gut, jaw tips, and first dorsal fin after reaching a length of about 5 mm. Larval distribution in equatorial waters is transoceanic the year round, but there are seasonal changes in larval density in subtropical waters. It is believed that the larvae occur exclusively in the warm water above the thermocline. Larval Yellowfin Tuna reared at decreasing pH levels (pH 8.1, 7.6, 7.3, and 6.9) showed increasing organ damage in the kidney, liver, pancreas, eye, and muscle, correlating with decreased growth and survival.

Above: Juvenile Yellowfin Tuna, approx. 30 cm FL.

STOCK STRUCTURE: Yellowfin Tuna were described as distinct species in several geographic areas based on morphological characters. Although conventional and electronic tagging indicate that some individuals undertake long movements, 95% of linear displacements indicate restricted movements and regional fidelity. An evaluation of all data, including morphometrics and meristics, length at maturity, and tagging data, suggest there are multiple stocks of Yellowfin in the eastern Pacific. Genetic studies reported little, if any, genetic differentiation between Yellowfin Tuna from different ocean basins. Connectivity between oceans was believed to be facilitated by the presence of Yellowfin Tuna south of the Cape of Good Hope during the austral summer. Recent studies using next-generation sequencing protocols to survey literally thousands of nuclear loci have reported small but statistically significant heterogeneity between collections of Yellowfin Tuna from different ocean basins, as well as between collections across the tropical Pacific Ocean. Currently, Yellowfin Tuna are pragmatically managed as four stocks, one by each of the relevant tuna RFMOs.

FISHERIES INTEREST: Yellowfin Tuna are fast growing, widely distributed, and highly productive. They are caught by a variety of gears, including purse seine, pelagic longline, gill net, and pole and line. The species also supports recreational fisheries in many locations. Purse seines account for the largest catches of Yellowfin Tuna in all oceans. Worldwide landings show a gradual but variable increase from 111,000 mt in 1950 to 1,240,000 mt in 2000. Landings have been relatively stable for the past 15 years, peaking at 1,467,000 mt in 2014. The vast majority of purse seine catches are destined for canneries, while larger fish from longline fisheries are sold fresh or frozen, with some individuals entering the sashimi markets.

Atlantic Ocean: Yearly catch levels of Yellowfin Tuna in the Atlantic Ocean increased from slightly over 50,000 mt in the late 1950s to more than 170,000 mt in the mid-1990s and have fluctuated around 100,000 mt for the past 10 years. The catch in 2016 of 127,800 mt was above MSY (126,000 mt) and the TAC (110,000 mt). The fishery is concentrated in the eastern Atlantic, especially in the region around the Gulf of Guinea. Purse seine catches have increased over time and now account for almost 75% of the catch. Increasing fractions of purse seine sets are associated with FADs, as opposed to sets on free-swimming schools. The average size of Yellowfin Tuna is smaller in FAD-associated sets than free-swimming school sets. Landings of the pelagic longline fishery, which catch the largest average-size Yellowfin Tuna, have decreased by more than 50% over the past 20 years. Due to changes in fishery selectivity, the average size of the fish harvested has decreased, and this has resulted in a lower yield per recruit. Consequently, the maximum sustainable yield for the Atlantic Ocean has decreased from over 170,000 mt in the late 1970s to just over 125,000 mt today.

Eastern Pacific Ocean: Prior to 1960, bait boats were responsible for the majority of catches of Yellowfin Tuna in the eastern Pacific, but since that time purse seine catches have dominated the fishery, accounting for 95% of the catch in 2016. Purse seines have targeted Yellowfin Tuna associated with dolphins, free schools, and increasingly, those associated with FADs. The total catch in the eastern Pacific has gradually increased from under 100,000 mt per year in the 1960s, peaking at 443,000 mt in 2005. Since that time, catch levels have decreased, averaging 231,000 mt over the past five years (2012–16). Variations in catches reflect changes in fishing effort and selectivity, as well as regime shifts in productivity. Since the mid-1970s, there appear to have been three different productivity regimes that have affected recruitment.

Western and Central Pacific Ocean: Catches of Yellowfin Tuna in the western and central Pacific Ocean (WCPO) were less than 100,000 mt per year in the 1960s but began to increase dramatically in the late 1970s and have averaged close to 575,000 mt over the past five years. Prior to the early 1980s, pelagic longlines were responsible for the majority of the catch of Yellowfin Tuna in the WCPO. Since that time, purse seines have been the dominant gear type, and now account for about 61% of the catch. Longline catches in recent years (70,000–80,000 mt) are well below levels in the late 1970s to early 1980s, which peaked at about 110,000 mt, presumably related to changes in targeting practices by some of the larger fleets.

Indian Ocean: Catches of Yellowfin Tuna in the Indian Ocean were less than 100,000 mt per year through the mid-1980s and then increased rapidly, peaking at 530,000 mt in 2004. Over the past five years, catches have averaged just under 400,000 mt per year. Pelagic longlines were the dominant gear type through the 1980s, but since that time purse seine catches have increased in importance. Unlike in the other oceans, gill nets and other miscellaneous gear types account for almost 40% of the catch. Purse seine and pelagic longline fishing effort and catches declined substantially during the mid-to-late 2000s due to concerns over Somali-based piracy in the western Indian Ocean. Not only was there a decrease in the nominal effort (number of boats, total carrying capacity, number of fishing and searching days, total number of sets), but there were also changes in the fishing behavior due to the new security measures in place (boats working in pairs with military personnel on board, restriction on fishing areas, etc.).

THREATS: Yellowfin Tuna are close to fully fished or are overfished throughout their range, with overfishing occurring in the eastern Pacific and Indian oceans and likely in the Atlantic Ocean as well. Changes in selectivity in many areas due to increased catches from purse seine fleets fishing with FADs have resulted in increased catches of small undesirable-size fish and reduced maximum sustainable yields. Increases in purse seine capacity and efficiency, in combination with a lack of meaningful conservation measures to limit fishing effort, is of concern in many areas. Yellowfin Tuna was listed as Near Threatened on the IUCN Red List primarily because population declines would be much greater if it were not for the catch quotas that have been implemented.

CONSERVATION:

Atlantic Ocean: Yellowfin Tuna in the Atlantic Ocean were assessed in 2016. Fishing mortality was estimated to be 0.77 of FMSY and the current spawning stock biomass was estimated to be 0.95 of that necessary for MSY. The stock is considered to be slightly overfished, but overfishing is not occurring. However, catches in 2016 exceeded MSY. Current management measures include a total allowable catch for the Atlantic (but not individual country quotas), a two-month annual time/area closure in the Gulf of Guinea for fishing in association with FADs, and capacity limits on the number of purse seine and longline vessels for some countries.

Eastern Pacific Ocean: As of the 2018 assessment update, the recent fishing mortality is slightly above the FMSY. The current spawning biomass is estimated to be above the level necessary to produce MSY. The MSY is estimated to be 264,000 mt, which is above the mean annual catches over the past five years. The IATTC has adopted a harvest control rule for tropical tunas. Current conservation measures for Yellowfin Tuna include a 72-day annual closure for all large purse seine vessels (>182 mt capacity), a 1-month closure west of the Galapagos Islands, and limits on the number of FADs a vessel can deploy.

Western and Central Pacific Ocean: The most recent assessment of Yellowfin Tuna in the WCPO (2017) indicated that the stock is not overfished, with the current SSB at 1.39 of BMSY. Overfishing is not occurring, with current fishing mortality at 0.74 of FMSY. The estimated MSY is 664,200 mt, well above average catches of the past five years (573,000 mt). Active conservation measures include a limit on Yellowfin Tuna catches and vessel freezing capacity (not to increase), no discards of tropical tunas, and a three-month closure on FAD fishing in tropical waters, limits on the total number of vessel days, and an FAD management plan.

Indian Ocean: The 2016 assessment update indicated Yellowfin Tuna in the Indian Ocean are overfished, with the current SSB at 89% of BMSY. Overfishing is occurring, with current fishing effort at 1.11 FMSY. MSY is estimated to be 422,000 mt, and while the average catches over the past five years have been below this value, they have been increasing in recent years. A rebuilding plan is being developed for the stock, with current conservation measures including catch limits by gear type, an FAD management plan, a ban on discards of tropical tunas, and a mandate to develop and implement a quota allocation system.

REFERENCES: Aguila et al. 2015; Alverson 1963; Appleyard et al. 2001; Baremore and Bethea 2010; Bellón and Bardán de Bellón 1949; Bonnaterre 1788; Brill et al. 1999; Cole 1980; Collette and Nauen 1983; Cuvier and Valenciennes 1832; Dragovich 1970; Dragovich and Potthoff 1972; Fowler 1933; Frommel et al. 2016; Gibbs and Collette 1967; Grewe et al. 2015; Hoolihan et al. 2014; Hosseini and Kaymaram 2016; IGFA 2018; Jordan and Evermann 1926; Lacepède 1800; Langley et al. 2007, 2009a, 2009b; Manooch and Mason 1983; Margulies et al. 2007, 2016; Matthews et al. 1977; Minte-Vera et al. 2018; Mowbray 1920; Olson and Boggs 1986; Pecoraro et al. 2016a, 2016b; Perrin et al. 1973; Poey 1875; Richards 2006; Schaefer 1998, 2001, 2008; Schaefer et al. 2007, 2009, 2011, 2013; Scoles and Graves 1983; Sloane 1707; Temminck and Schlegel 1844; Varghese et al. 2014; Wild 1986; Zhu et al. 2011; Zudaire et al. 2013.

Blackfin Tuna

Thunnus atlanticus (Lesson, 1831)

COMMON NAMES: English – Blackfin Tuna
French – Thon a Nageoires noires
Spanish – Atun aleta negra

ETYMOLOGY: The Blackfin Tuna was appropriately named *atlanticus*, the "bonite de l'ocean Atlantique," by the French naturalist René Primevère Lesson (1831:165–166), based on specimens that he collected in the West Indies while employed as a botanist on the world cruise of the *Coquille*.

SYNONYMS: *Thynnus atlanticus* Lesson, 1831; *Orcynus balteatus* (Cuvier, 1832); *Parathunnus ambiguus* Mowbray, 1935; *Thunnus balteatus* (Cuvier, 1832); *Thunnus coretta* (Cuvier, 1829); *Thynnus balteatus* Cuvier, 1832; *Parathunnus atlanticus* (Lesson, 1831); *Parathunnus rosengarteni* Fowler, 1934; *Thunnus atlanticus* (Lesson, 1831)

TAXONOMIC NOTE: The Blackfin Tuna is one of the three tropical species of tunas placed in the subgenus *Neothunnus*.

FIELD MARKS: 1 The Blackfin Tuna has very few gill rakers on first gill arch, only 19–25.
2 The pectoral fins are moderately long.
3 The finlets are dusky, with a trace of yellow.

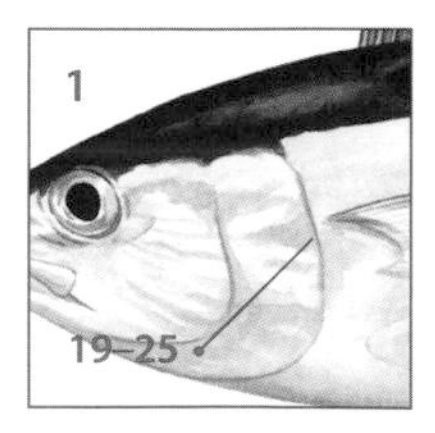

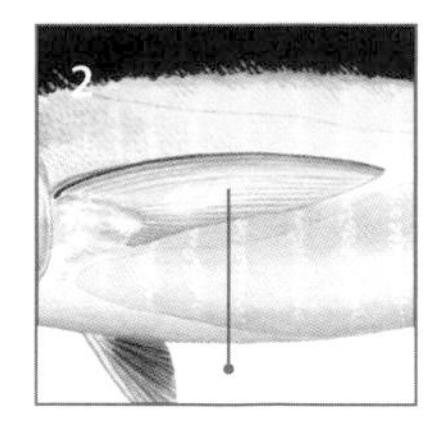

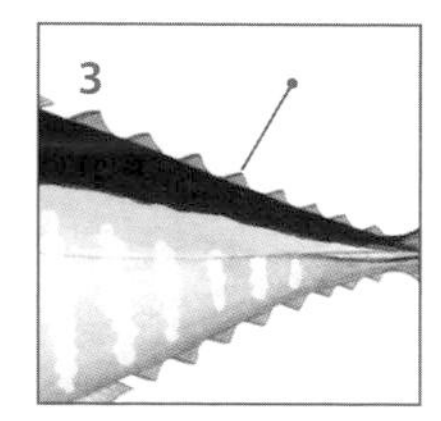

UNIQUE FEATURE: The Blackfin Tuna is the smallest species of tuna and has the fewest gill rakers of any of the tunas.

DIAGNOSTIC FEATURES: The Blackfin Tuna is the smallest species of *Thunnus*, with a maximum size of less than 110 cm fork length. The body is deepest near the middle of first dorsal-fin base. **Gill rakers:** It has fewer gill rakers than the other species of the genus, only 19–25 on the first gill arch. **Fins:** The pectoral fins are moderate in length, usually 22–31% of fork length. **Liver:** The ventral surface of the liver is not striated and the right lobe is longer than the central and left lobes. **Swim bladder:** A small swim bladder is present. **Vertebrae:** Vertebrae number 19 precaudal plus 20 caudal instead of 20 plus 19 as in the other species of *Thunnus*. The osteology was thoroughly described and illustrated by de Sylva (1955) and the osteology and soft anatomy were described by Gibbs and Collette in their 1967 revision of the genus. Illustrations of the sagittal otoliths were published by Baremore and Bethea (2010). **Color:** The back is a metallic dark blue; the lower sides are uniformly silvery gray or with pale streaks at least partly in vertical rows. The belly is milky white. The first dorsal fin is dusky; the second dorsal and anal fins are also dusky, with a silvery luster. The dorsal and anal finlets are dusky, with a trace of yellow.

GEOGRAPHIC RANGE: Blackfin Tuna are restricted to the western Atlantic Ocean from Martha's Vineyard in Massachusetts south to 26° S in Brazil, including Bermuda, the Caribbean Sea, the Gulf of Mexico, Fernando de Noronha, the Vitória-Trindade Seamount Chain, and the Saint Peter and Saint Paul Archipelago, 500 miles from the Brazilian coast.

SIZE: The IGFA all-tackle game fish record is of a 49 lb, 6 oz (22.4 kg) fish caught off Marathon, Florida, in April 2006. The maximum recorded length is 110 cm FL. Individuals sampled from the sport fishery in southeast Florida ranged in size from 34.5–87 cm FL and 0.75–11.79 kg. Analysis of dorsal spines produced estimates of ages up to 5 years, while ages up to 7.5 years (an 80 cm FL fish) were obtained from otolith analyses, and maximum ages ranging from 3 to 7 or 8 were determined, depending on the hard structure examined. For Blackfin Tuna caught in Rio Grande do Norte, northeast Brazil, the mean size of males (64 cm) was greater than females (61.1 cm), and individual weights ranged from 1.5 to 8.4 kg for males and 1.0 to 5.0 kg for females. Similarly, males reached larger sizes than females, 86.0 cm FL versus 72.5 cm FL, for fish from Baía Formosa, Rio Grande do Norte.

HABITAT AND ECOLOGY: Blackfin Tuna occur in coastal waters with temperatures above 20°C. They are epipelagic, often found over reefs, in bays, and offshore. They sometimes occur in large schools, often with Skipjack Tuna. Blackfin Tuna tagged with pop-up satellite archival tags in the Gulf of Mexico showed a preference for the surface mixed layer, spending 90% of their time in the upper 57 m of the water column. They undertook brief dives to depths in excess of 200 m, demonstrated a shallower nighttime distribution, and spent most of their time in waters warmer than 21.9°C. Limited tag return data (11 returns of 787 tagged) showed recaptures near release sites in the coastal waters of St. Vincent and the Grenadines after times of liberty of 5 to 1,230 days. Tag returns from individuals tagged near Bermuda provide support for an annual migration through the Sargasso Sea to the Bermuda Seamount. There is evidence of genetic differentiation between fish from the Gulf of Mexico and the western North Atlantic.

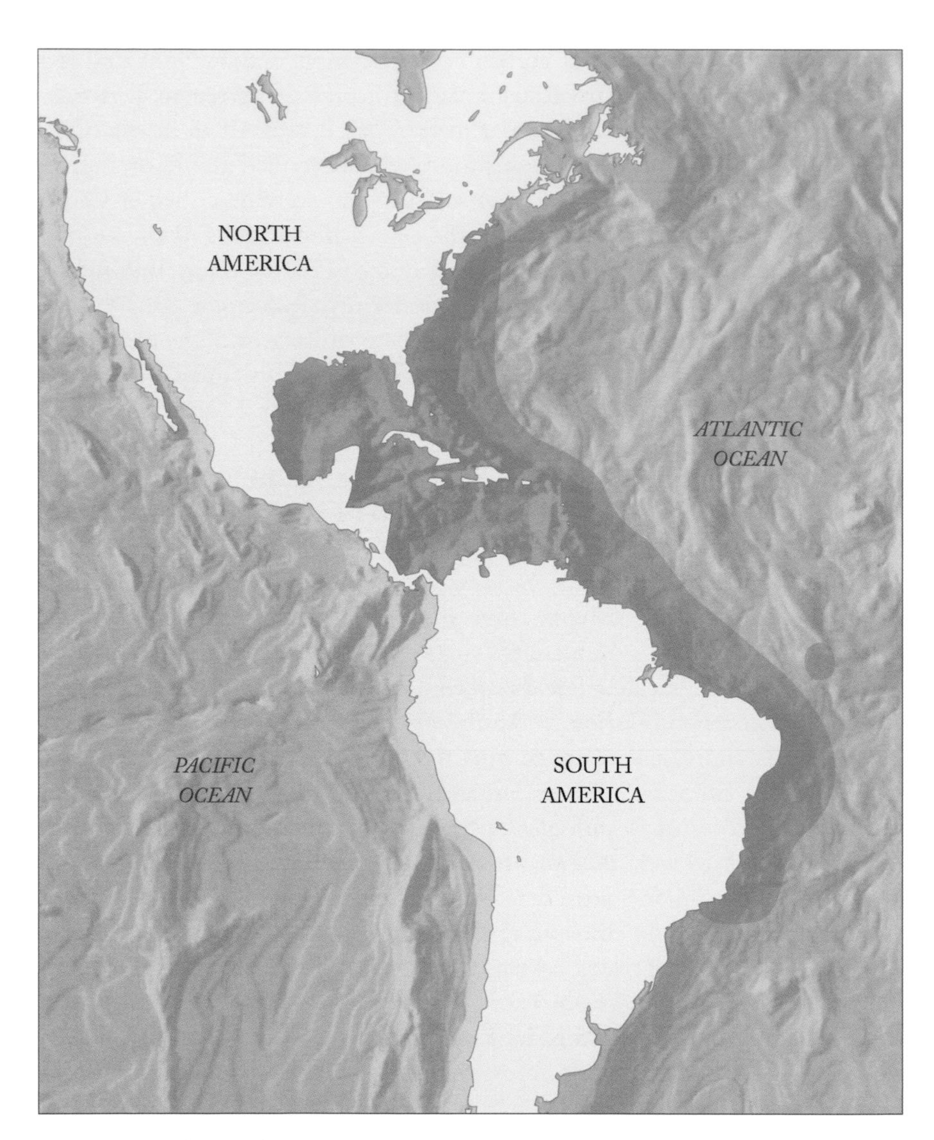

Right: Blackfin Tuna range map.

FOOD: Fishes, squids, and crustaceans were the primary diet items in Blackfin Tuna from the southeastern United States, Bermuda, Cuba, Tobago, and northeastern Brazil. Fishes were the most important prey item in all five areas, constituting 41–85% by volume, followed by squids at 8–34% and crustaceans at 4.9–36%. As Blackfin Tuna size increases, larger food items such as adult fishes occur more frequently and smaller food items such as crustaceans and juvenile fishes decrease. The most abundant fishes in stomachs from Cuba were triggerfishes and surgeonfishes. The most commonly eaten fish in Tobago was the Key Anchovy (*Anchoa cayorum*), which accounted for 50.5% by numerical abundance of all fishes eaten and 43.6% by weight and was found in 46.2% of stomachs. Squids were found in 32.8% of Blackfin Tuna stomachs sampled from the South Atlantic Bight.

REPRODUCTION: Around Florida, spawning season for Blackfin Tuna extends from April to November, with a peak in May, while in the Gulf of Mexico spawning apparently occurs from June to September. Males and females were ripe from June through September in Cuba. In Brazil, spawning peaks in November and December, and fish migrate to Formosa Bay to spawn from October to January. During the spawning season, there is an annual concentration of *T. atlanticus* along the southern coast of Rio Grande do Norte during the second half of the year. Around the Saint Peter and Saint Paul Archipelago, size at first maturity was estimated at 48 cm FL for females, 55 cm FL for males. In northeastern Brazil, length at 50% maturity was reported as 49.8 cm FL for females and 52.1 cm FL for males. The estimated average length for female gonadal maturation is 50–51 cm. Batch fecundity estimates for Blackfin Tuna from Cuba ranged from 199,000 to 957,000 eggs, 192,000–320,000 eggs for females 58–66 cm FL off Brazil, and 272,025 and 1,140,584 eggs for females 56 and 64 cm FL from the Saint Peter and Saint Paul Archipelago.

EARLY LIFE HISTORY: Three Blackfin Tuna larvae ranging from 5.1 mm notochord length to 8.5 mm standard length were described and illustrated in Richards' 2006 identification guide to larval western central Atlantic fishes. Pigmentation is present on the midbrain, gut, jaw tips, and first dorsal fin after reaching a length of about 5 mm. Development of the caudal complex and vertebral column was described from a series of 51 specimens 5.1–85.0 mm SL. Blackfin Tuna larvae were the most abundant of four species of tuna, billfish, and dolphinfish collected in summer ichthyoplankton surveys in the northern Gulf of Mexico from 2007–11, accounting for 87% of the larvae of the four selected taxa. They also dominated the tuna larvae taken in the Straits of Florida. Among the 84 species of fishes eaten by Sooty Terns and Brown Noddies in the Dry Tortugas of Florida were 48 juvenile Blackfin Tuna, 32–125 mm SL.

FISHERIES INTEREST: Blackfin Tuna are one of the most common tuna species in the western central Atlantic. They are a main source of food and are important to commercial and recreational fisheries in Bermuda, Cuba, the French West Indies, and Brazil. Blackfin Tuna are fished by trolling or drifting longline with live bait, and fish aggregating devices (FADs) are also used to increase the capture of this species. Blackfin Tuna are of only minor importance in the US commercial longline fisheries, and catches are generally not retained due to their low market value. The species is a common target for the recreational rod-and-reel fishery and represents an important component of sport fisheries in southeastern US waters extending up to North Carolina. Males are caught more often than females; in one report males constituted 80% of the catch. Reported landings fluctuated between 1,000 and 3,000 mt from the late 1950s until the mid-1980s, increased to between 2,400–5,200 mt from 1986–2004, and have varied between 1,000 and 2,000 mt for the past 10 years. Reported landings are an underestimate of the actual catch, as several countries with fisheries for the species do not report landings, artisanal landings are not fully reported, and it is likely that recreational catches that may comprise a large portion of the fishery in some areas are not reported. Blackfin Tuna was the species with the highest abundance in the pelagic longline fishery in northeast Brazil, with an average catch per unit effort (CPUE) of 0.32 individuals / 100 hooks, representing 56.2% of all caught tunas. Irradiation has been used to extend the shelf life of refrigerated Blackfin Tuna.

Landings of Blackfin Tuna have been reported all along the northeastern Brazilian coast by handline artisanal fisheries, with catches by state as follows: Bahia (57.6%), Rio Grande do Norte (23.7%), Alagoas and Pernambuco (17.7%), Ceará (0.8%), and Piauí (0.1 %). There is an annual concentration of this species along the southern coast of Rio Grande do Norte State during the second half of the year that is targeted by an artisanal fleet. It is fished along the Brazilian central coast (from southern Bahia to northern Rio de Janeiro State) by trolling and handline fisheries. It is also caught by sport fisheries off the coast of São Paulo State.

THREATS: Landings data show fluctuations without evidence of consistent decline, although some of the major fishing nations for this species (Cuba, Dominican Republic, and Martinique) have ceased reporting landings. The countries that are reporting landings do not show evidence of decline. Blackfin Tuna are listed as Least Concern on the IUCN Red List.

CONSERVATION: There are no formal stock assessments for Blackfin Tuna throughout its range. The landings data are incomplete, and ICCAT is evaluating the use of length composition data as indicators of stock status. There are no known species-specific conservation measures in place for Blackfin Tuna. Suggestions have been made for a capture-release study involving Caribbean and Brazilian northeastern and southeastern researchers to better establish fishing mortality rates and understand migratory patterns. Better data are needed from fisheries landings to specifically identify and record this species.

REFERENCES: Adams and Kerstetter 2014; Amorim and Silva 2005; Báez-Hidalgo and Bécquer 1994; Baremore and Bethea 2010; Beardsley and Richards 1970; Beebe 1936; Bezerra et al. 2011, 2013; Brusher and Palko 1987; Carles Martín 1971, 1974; Collette 2010; Collette and Nauen 1983; Collette et al. 2011; Cruz and Paiva 1964; de Sylva 1955; Doray et al. 2004; Fenton et al. 2015; Freire et al. 2005; Friedlander et al. 1994; García Coll 1988a, 1988b; Gibbs and Collette 1967; Guevara Carrió 1984; Guevara Carrió et al. 1988; Headley et al. 2009; Hensley and Hensley 1995; Lessa et al. 1998; Lesson 1831; Luckhurst 2014; Manooch and Mason 1983; Monte 1964; Monteiro et al. 2009; Mowbray 1920; Neilson et al. 1994; Pinheiro et al. 2015; Potthoff 1975; Potthoff and Richards 1970; Poveromo 2014; Rawlings 1953; Richards 2006; Richardson et al. 2006; Rooker et al. 2013; Saxton 2009; Singh-Renton and Renton 2007; Smith-Vaniz et al. 1999; Staudinger et al. 2012; Suárez Caabro and Duarte Bello 1961; Taquet et al. 2000; Vieira et al. 2005a, 2005b.

Southern Bluefin Tuna

Thunnus maccoyii (Castelnau, 1872)

COMMON NAMES: English – Southern Bluefin Tuna
French – Thon Rouge du Sud
Spanish – Aleta Azul del Sur, Atún Rojo del Sur

ETYMOLOGY: Named *maccoyii* after Frederick McCoy, "who did all in his power to assist me," by the career diplomat and naturalist Francis Louis Nompar de Caumont, Comte de Laporte de Castelnau, while he was living in Melbourne (Castelnau 1872).

SYNONYMS: *Thynnus maccoyii* Castelnau, 1872; *Thunnus phillipsi* Jordan and Evermann, 1926; *Thunnus thynnus maccoyii* (Castelnau, 1872); *Thunnus thynnus orientalis* (non Temminck and Schlegel, 1844); *Thunnus maccoyii* (Castelnau, 1872)

TAXONOMIC NOTE: This species was previously considered a subspecies of the Pacific Bluefin Tuna, *Thunnus orientalis*, or Atlantic Bluefin Tuna, *Thunnus thynnus*, but was validated morphologically as a distinct species by Godsil and Holmberg (1950) and Gibbs and Collette (1967) and confirmed genetically by Ward et al. (1995) and Tseng et al. (2011).

FIELD MARKS: 1 The Southern Bluefin Tuna differs from the other two species of bluefin tunas in having the caudal keels yellow instead of dark.

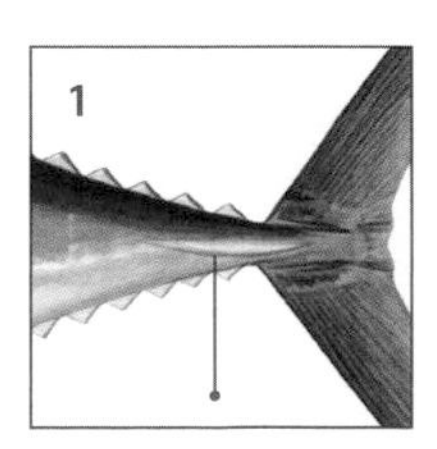

UNIQUE FEATURE: Unlike the other two species of bluefin tunas that are primarily inhabitants of northern waters, the Southern Bluefin Tuna is restricted to the cool waters of the Southern Hemisphere (except for its spawning migration into warm waters), like the Butterfly Mackerel and the Slender Tuna.

DIAGNOSTIC FEATURES: The Southern Bluefin Tuna is a very large tuna, deepest near the middle of the first dorsal-fin base **Gill rakers:** Gill rakers number 31–40 on the first arch, average 33.7. **Fins:** There are 12–14 spines in the first dorsal fin, 13–15 rays in the second dorsal fin, and 8–10 dorsal finlets. There are 22–24 rays in the anal fin, followed by 7–9 anal finlets. The pectoral fins have 31–36 rays, are very short (less than 80% of head length or 20.1–23% of fork length), and never reach the interspace between the dorsal fins. **Liver:** Striations cover the ventral surface of the liver, and the right lobe is not significantly longer than the central and left lobes. **Swim bladder:** A swim bladder is present. **Spleen:** The spleen is on the right side of the body cavity and the stomach is on the left, as in the other species of the genus, except for the Albacore. **Body cavity:** The bulge in the roof of the body cavity extends almost to the body wall, leaving a deep lateral pocket in the pectoral region on each side in medium and large specimens. **Vertebrae:** Vertebrae number 18 precaudal plus 21 caudal, for a total of 39. Descriptions and illustrations of the osteology and soft anatomy are presented by Godsil and Holmberg (1950) and Gibbs and Collette (1967). **Color:** Lower sides and belly silvery white, with colorless transverse lines alternating with rows of colorless dots (the latter dominate in older fish), visible only in fresh specimens; first dorsal fin yellow or bluish, anal fin and finlets dusky yellow wedged with black; median caudal keel yellow in adults. A rare waxy white Southern Bluefin Tuna was sold in Tokyo's Tsukiji fish market in November 2014.

GEOGRAPHIC RANGE: The Southern Bluefin Tuna is found in the southern parts of the Atlantic, Indian, and Pacific oceans. It primarily occurs in temperate and cold seas, mainly between 30° S and 50° S to nearly 60° S, but moves up into subtropical waters to spawn.

SIZE: The maximum size of the Southern Bluefin Tuna is 225 cm FL and 200 kg. The IGFA all-tackle game fish record is of a 369 lb, 4 oz (167.5 kg) fish caught off Tathra, Australia, in July 2009. Use of bomb radiocarbon dating suggests that a large percentage of fish greater than 180 cm FL are at least 20 years of age and that Southern Bluefin can live to ages in excess of 30 years. Otolith-based aging found ages of 0–41 years. Males grow faster than females, but females have a higher asymptotic growth, so they catch up in size by age 35. The generation length is conservatively estimated to be 12 years.

HABITAT AND ECOLOGY: As adults, most Southern Bluefin Tuna lead an oceanic, pelagic existence. Mature fish migrate to a single spawning ground between Java and Australia. After the larval stage, juveniles leave the spawning ground, moving south along the coast of Western Australia, reaching the south coast by age 1 and spending the austral summer there. Most juveniles continue to spend austral summers in the Great Australian Bight until about age 5. While in the Bight, telemetry indicates that fish spend much of the day in the upper 10 m and tend to move into deeper waters at night. During the winters, they migrate across the Great Australian Bight and around Tasmania to 45° S, along the southeastern Australian coastline to about 30° S off northern New South Wales, or as far east as New Zealand. Fish older than 5 years disperse widely across the southern oceans from the Tasman Sea across the Indian Ocean to the western Atlantic. Occurrence of juveniles 2–4 years of age on opposite sides of the Indian Ocean during the austral summer months confirms that not all juveniles are restricted to the southern coastal waters of Australia. However, a large archival tagging study indicated that the number of juveniles straying from Australian waters was quite small. Electronic tagging indicates that juveniles make sharp descents and ascents called spike dives around dawn and dusk. It has been suggested that these dives, along with anatomical development of the pineal organ, may aid in navigation. Southern Bluefin Tuna are remarkably well adapted to low dissolved oxygen, exhibiting a critical low-oxygen level between 1.57 and 2.49 mg/l, and are capable of aerobically supporting maintenance metabolism and routine swimming.

Below Southern Bluefin Tuna range map.

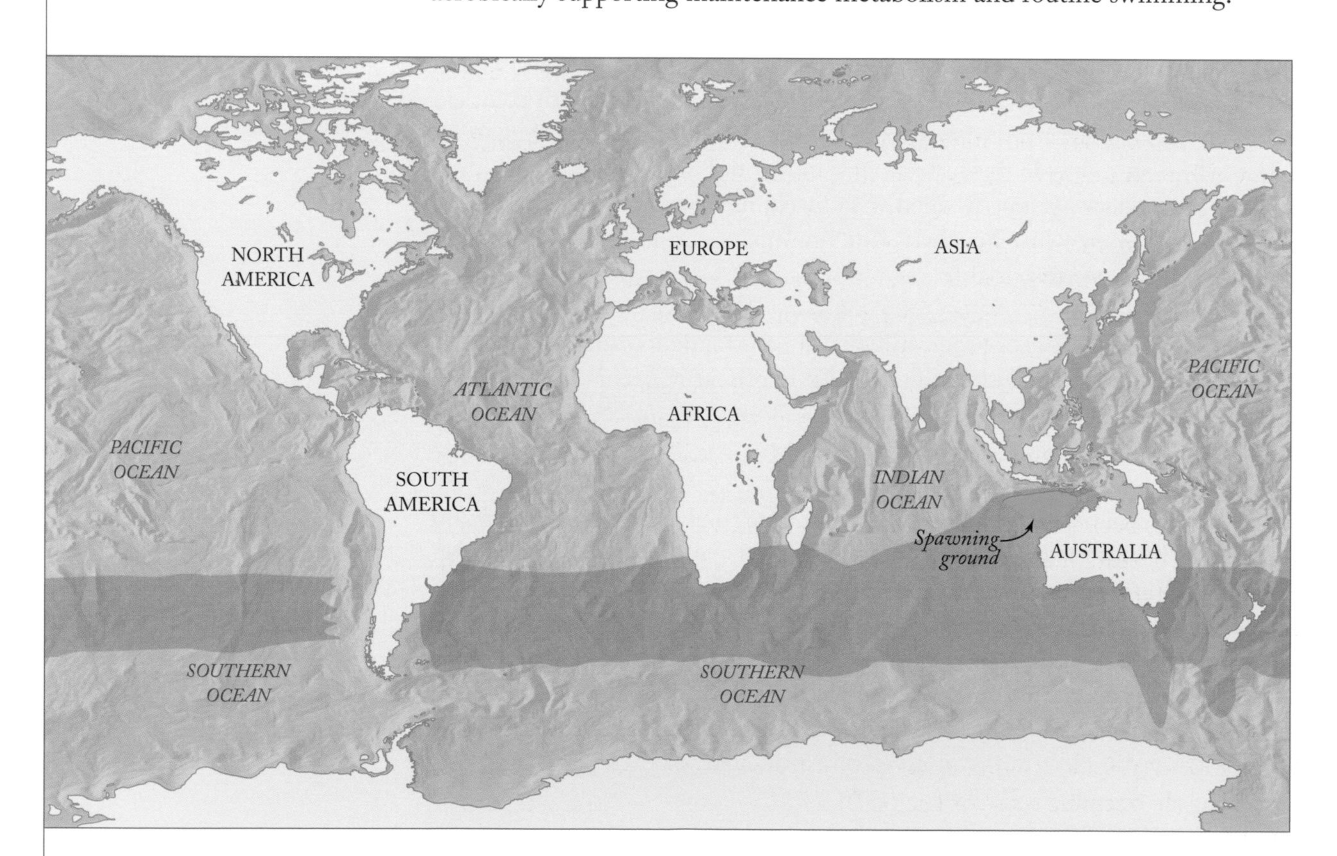

FOOD: Like other tunas, the Southern Bluefin Tuna is an opportunistic feeder, preying on a wide variety of fishes, crustaceans, cephalopods, salps, and other marine animals. It forages in ambient water temperatures ranging from 4.9–22.9°C and depths ranging from the surface to 672 m. The number of feeding events varies with the time of day, with the greatest number occurring around dawn. In southeastern Australian waters, it feeds on Jack Mackerel (*Trachurus declivis*), Australian Sardine (*Sardinops sagax*), squids, and Antarctic Krill (*Euphausia superba*). Juveniles up to 4 years old congregate in the eastern Great Australian Bight in the austral summer to feed on bony fishes, especially Australian Sardine, Australian Anchovy (*Engraulis australis*), Blue Mackerel (*Scomber australasicus*), and Jack Mackerel as well as squids and crustaceans. In the inshore waters off Tasmania, the main prey items were fishes and juvenile squid, but offshore the prey diversity was greater, reflecting the diversity of macronekton in those waters. Older fish, dispersed widely across the southern oceans, feed on a much larger proportion of squids, 51% of the prey weight compared to 46% by bony fishes.

REPRODUCTION: Large Southern Bluefin Tuna migrate north to spawn in a relatively small area off northwestern Australia in the eastern tropical Indian Ocean. Sexually mature fish have been taken on the spawning grounds every month except July, with peaks in abundance during October and February at water surface temperatures in excess of 24°C. Once females start spawning, they appear to spawn daily, similar to other tunas. On the spawning grounds, smaller fish (150–69 cm) are better represented in deep catches, while larger fish (190–209 cm) and spawning fish are better represented in shallow catches. Large fish tend to arrive on the spawning ground slightly earlier and depart later in the spawning season relative to small fish. Spent fish were rarely encountered on the spawning grounds, as they move south shortly after spawning. Southern Bluefin are asynchronous indeterminate spawners, with a batch fecundity of 57 oocytes per gram of body weight, similar to that found for Yellowfin Tuna. Fecundity of a 158 cm female, with each ovary weighing about 1.7 kg, was estimated at about 14–15 million eggs. It is not known whether all mature fish spawn each year, every few years, or at a lower frequency. Females outnumber males in catches of juveniles, but the sex ratio is reversed in adults, with males outnumbering females. Age at first maturity of females varies between 5 and 7 years at 110–25 cm FL, suggesting a decrease from the 6–7 years found earlier, but the onset of maturity more commonly occurs at 130 cm FL or about 8 years of age. Size at 50% maturity has been estimated to be 152 cm FL, and the mean age of maturity is 10–11 years and does not differ significantly between the sexes.

EARLY LIFE HISTORY: Southern Bluefin larvae have only been caught within latitudes 7°–20° S and longitudes 102°–124° E from September to March. The larvae closely resemble those of Pacific Bluefin Tuna, but the chromatophores on the dorsal surface of the body of Southern Bluefin Tuna are smaller. Lower jaw pigments are the earliest to develop among the bluefins, appearing in larvae as small as 3.00–3.49 mm SL. Larvae of Southern Bluefin Tuna, Albacore, and Skipjack are evenly dispersed in the mixed layer at night, but larvae of Southern Bluefin and Albacore move into surface layers during the day for feeding. Larvae feed only in the daytime, with feeding peaks in the early morning and late afternoon. Larval prey vary with the size of larvae and are strongly correlated with an increase in the size of the mouth. Copepod nauplii were the most abundant prey in smaller larvae (4.0 mm TL or smaller), with larger copepods becoming dominant in larger larvae (longer than 7 mm). Around the Nansei spawning grounds, copepods were the most abundant prey, 97% by number. Cannibalism was found in postflexion larvae. Growth rates of larvae from about 7 to 18 days old were relatively slow, like those of temperate-water species. Larval growth rates of Southern Bluefin Tuna are reported to be positively correlated with feeding but not with temperature, suggesting that larvae were food limited to a varying degree.

STOCK STRUCTURE: Genetic and otolith composition analyses of Southern Bluefin from South Africa, Western Australia, South Australia, and Tasmania showed no significant heterogeneity, consistent with the hypothesis of a single stock and spawning unit.

FISHERIES INTEREST: The Southern Bluefin Tuna has been intensively fished since the 1950s, primarily being taken on longlines, although purse seines now account for approximately 40% of the catch. In 1992, the first commercial farms began fattening purse seine-caught Southern Bluefin Tuna in large mesh enclosures moored offshore near Port Lincoln, South Australia. Most of the tuna transferred to tuna farms are immature fish (age 2–4 years) and are fattened in enclosures for about 150 days. Canning was the most important form of local utilization of this highly esteemed fish until the early 1980s. Currently, Japan is the primary market, with a very high percentage of the world catches consumed as sashimi within Japan. Worldwide reported landings show a gradual but variable increase from 13,716 mt in 1952 to a high of 81,416 mt in 1969, gradually decreasing to 10,961 mt in 1991. Catches from 1992 to 2012 were relatively stable, averaging around 16,000 mt per year, and steadily increased to 23,312 mt in 2016.

THREATS: The dramatic decline in the spawning biomass of Southern Bluefin Tuna to 8–12% of the unfished level has been well documented by the Commission for the Conservation of Southern Bluefin Tuna (CCSBT). The species was considered depleted by 2007 and seriously overfished by 2009. Very substantial and continuous unreported catches were taken since at least the early 1990s, and a significant proportion of these were taken by Japanese longliners. Cohorts from the 1980s grew substantially faster at young ages than cohorts from the 1960s, consistent with density-dependent responses to the history of exploitation. Although catches are now regulated with quotas, fishing capacity greatly exceeds the available quota. Ranching of Southern Bluefin Tuna in Australia has complicated estimation of catches that are transferred to pens, and the operation is associated with high mortality rates of fish during transport. In addition, parasites have been a problem with ranching operations. The scuticociliate *Uronema nigricans* is responsible for parasitic encephalitis that results in atypical swimming behavior followed by rapid death. The Southern Bluefin Tuna is listed as Critically Endangered on the IUCN Red List.

CONSERVATION: Southern Bluefin Tuna is a highly migratory species listed in Annex I of the 1982 Convention on the Law of the Sea. The species is managed by the CCSBT even though most catches occur in areas managed by three other tuna RFMOs: IATTC, ICCAT, and WCPFC. Based on the results of a 2014 stock assessment, Southern Bluefin Tuna is severely overfished, with a spawning stock biomass estimated to be less than 8–12% of the unfished level. However, estimates of adult abundance using novel close-kin mark-recapture methodology suggest an adult biomass two to three times higher than the assessment. The 2014 assessment indicates that Southern Bluefin Tuna is not experiencing overfishing, with current fishing mortality about two-thirds of that necessary to produce the estimated maximum sustainable yield of 33,000 mt. Several management measures are in place, including a harvest control rule. The CCSBT agreed that the status of the stock is at a critical stage and that a meaningful reduction in the total allowable catch (TAC) was necessary in order to rebuild the stock, and the CCSBT is working toward an interim rebuilding target reference point of 20% of the original spawning stock biomass. The CCSBT has adopted additional conservation measures, including requirements for fleets to monitor and submit data to show compliance with TACs, development of scientific observer programs, monitoring of farming operations, and port inspection of catches. The CCSBT has also implemented a Trade Information Scheme to collect more accurate and comprehensive data on Southern Bluefin Tuna fishing by monitoring trade and illegal, unregulated, and unreported fishing. Although the stock is heavily overfished, overfishing is no longer occurring due to measures taken in a rebuilding plan.

REFERENCES: Bestley et al. 2008; Bravington et al. 2016; Castelnau 1872; Caton 1994; CCSBT 2016; Chambers et al. 2014, 2017; Collette 2010; Collette and Nauen 1983; Collette et al. 2011; Davis and Farley 2001; Davis and Stanley 2002; Davis et al. 1990; Farley and Davis 1998; Farley et al. 2007, 2014, 2015; Fitzgibbon et al. 2010; Gibbs and Collette 1967; Godsil and Holmberg 1950; Grewe et al. 1997; Gunn et al. 2008; Hearn and Polacheck 2003; Hobday et al. 2016; IGFA 2018; ISSF 2018; Itoh and Sakai 2016; Itoh et al. 2011; Jenkins and Davis 1990; Jenkins et al. 1991; Jordan and Evermann 1926; Joseph 2009; Kalish et al. 1996; Lin and Tzeng 2010; Majkowski 2007; Munday et al. 1997; Nishikawa 1985; Nishikawa et al. 1985; Polacheck 2012; Polacheck et al. 2004; Proctor et al. 1995; Schaefer 2001; Serventy 1956; Shimose and Farley 2016; Thorogood 1986; Ueyanagi 1969a, 1969b; Uotani et al. 1981; Willis and Hobday 2007; Yabe et al. 1966; Young and Davis 1990; Young et al. 1997.

Bigeye Tuna

Thunnus obesus (Lowe, 1839)

COMMON NAMES: English - Bigeye Tuna; Spanish - Patudo; French - Thon Obèse; Japanese - Mebachi

ETYMOLOGY: The Bigeye Tuna was described as *Thynnus obesus* by Richard Thomas Lowe (1839:78), a British naturalist and clergyman who became a pastor in Madeira. He distinguished the "Atum Patudo" as being shorter and thicker, or more obese, than the Bluefin Tuna (as *Thynnus vulgaris*).

SYNONYMS: *Thynnus obesus* Lowe, 1839; *Thynnus sibi* Temminck and Schlegel, 1844; *Thunnus mebachi* Kishinouye, 1915; *Germo sibi* (Temminck and Schlegel, 1844); *Thunnus obesus* (Lowe, 1839); *Neothunnus obesus* (Lowe, 1839); *Thunnus obesus sibi* (Temminck and Schlegel, 1844); *Parathunnus obesus mebachi* (Kishinouye, 1915); *Thunnus obesus mebachi* Kishinouye, 1915; *Orcynus sibi* (Temminck and Schlegel, 1844); *Parathunnus mebachi* (Kishinouye, 1915); *Thunnus sibi* (Temminck and Schlegel, 1844); *Parathunnus obesus* (Lowe, 1839); *Germo obesus* (Lowe, 1839); *Parathunnus sibi* (Temminck and Schlegel, 1844)

TAXONOMIC NOTE: The genus *Thunnus* is divisible into two subgenera, *Thunnus* and *Neothunnus,* but the Bigeye Tuna does not neatly fit in either subgenus.

FIELD MARKS:

1 Pectoral fins moderately long, extending just beyond the second dorsal-fin origin.
2 The body is usually deepest in the middle of the body.
3 Live specimens lack vertical lines of light spots below the lateral line.

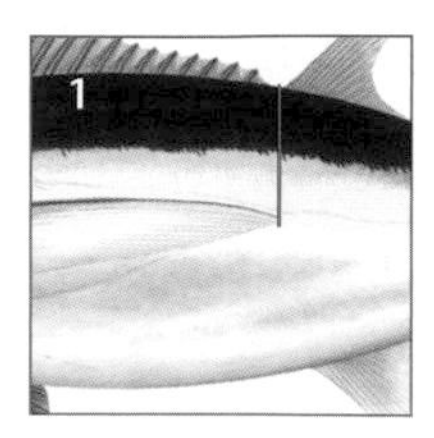

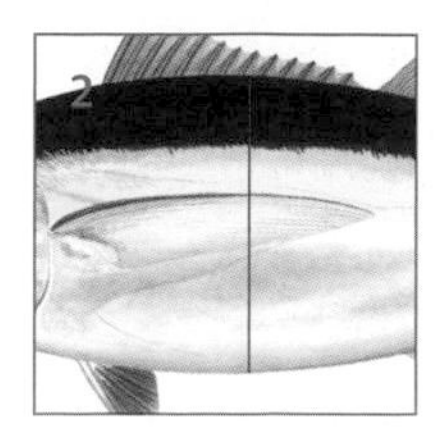

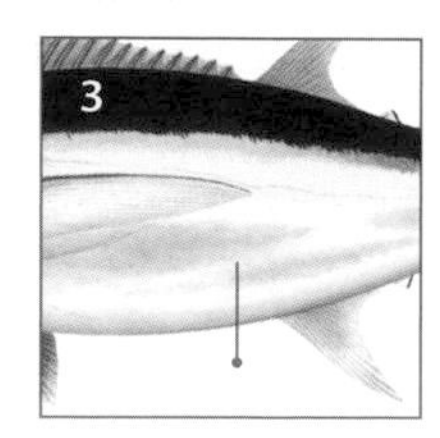

UNIQUE FEATURE: The Bigeye Tuna inhabits tropical waters, like three species of the subgenus *Neothunnus*, but spends much of its time in cooler waters, like four species of the subgenus *Thunnus*, by going into deeper waters rather than into northern or southern cold waters. It can do this because of adaptations to its blood and thermoregulatory system.

DIAGNOSTIC FEATURES: Bigeye Tuna are large, with the body deepest near the middle of the first dorsal-fin base. **Gill rakers:** Gill rakers number 23 to 31 on the first arch, average 27. **Fins:** There are 13 or 14 spines in the first dorsal fin, 14–16 rays in the second dorsal fin, and 8–10 dorsal finlets. There are 11–15 rays in the anal fin, followed by 7–10 anal finlets. The pectoral fins have 31–35 rays and are moderately long (22 to 31% of fork length) in larger individuals and relatively longer in smaller individuals. **Liver:** Striations cover the ventral surface of the liver, and the right lobe is not significantly longer than the central and left lobes. **Swim bladder:** A swim bladder is present. **Vertebrae:** Vertebrae number 18 precaudal plus 21 caudal, for a total of 39 vertebrae. Descriptions and illustrations of the osteology and soft anatomy are contained in the revision of the genus by Gibbs and Collette (1967). **Color:** The back is metallic dark blue, changing to whitish on lower sides and belly. A lateral iridescent band runs along the sides in live individuals. The first dorsal fin is deep yellow, the second dorsal and anal fins are light yellow, and the dorsal and anal finlets are bright yellow, edged with a narrow black border.

GEOGRAPHIC RANGE: Bigeye Tuna are circumglobal in tropical and temperate seas but are not known from the Mediterranean Sea. They occur as far north as 55–60° N in the northern summer and as far south as 45–50° S in the austral summer, but young fish and reproductively active fish primarily occur in tropical waters. There are three British records, all in relatively warm waters in the southwest, the most northerly in Wales at 51°40′ N, 4°15′ W from August 2006.

SIZE: The IGFA all-tackle game fish record for Bigeye Tuna is of a 435 lb (197.31 kg) fish caught in the Pacific Ocean off Cabo Blanco, Peru, in April 1957. The all-tackle record for the Atlantic is a 392 lb, 6 oz (178 kg) fish caught in the Canary Islands in July 1996. The maximum age is 16 years. The generation length in the Atlantic is 4.7 years, estimated from the age structure of the population.

HABITAT AND ECOLOGY: Bigeye Tuna are pelagic and oceanodromous and occur in surface waters with temperatures ranging from 13°–29°C, but the optimum is between 17°C and 22°C, coinciding with the temperature range of the permanent thermocline. Variation in occurrence is closely related to seasonal and climatic changes in surface temperature and thermocline. Juveniles and small adults school within the upper 50 m in monospecific groups or mixed with other tunas of similar size and may be associated with floating objects. Adults occupy warm surface waters to depths of 50–100 m during the night but typically make repeated dives to cooler waters at depths of up to 500 m or more during the day to exploit the deep scattering layer. Some dives exceeded 1,000 m where temperatures were less than 3°C. Such dives expose adults to a wide range of temperatures and in certain areas, waters of low oxygen concentrations. Bigeye Tuna can rapidly alter their body thermal conductivity. Their heat exchangers can be disengaged to allow rapid warming as they ascend from deep cold water into warmer surface waters and can be reactivated to conserve heat when they return to the depths. Bigeye Tuna are unique among the tunas in being able to withstand low oxygen concentrations, and they have a higher blood oxygen affinity.

Below: Bigeye Tuna range map.

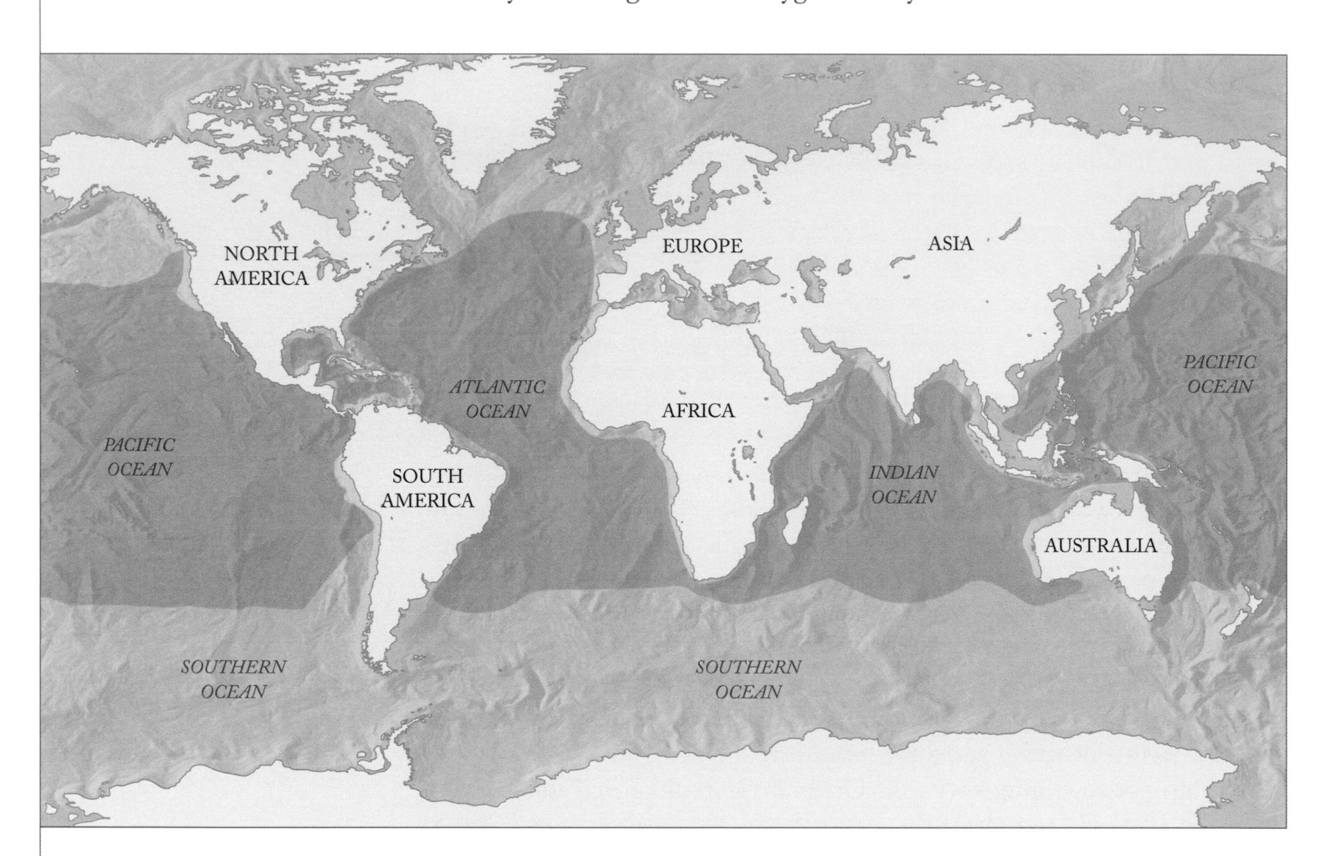

MOVEMENTS: Bigeye Tuna are a highly migratory species. Conventional tag data demonstrate that individuals can undertake movements in excess of 7,000 km; however, the vast majority of tag returns are for much smaller distances. In the eastern Pacific, greater than 95% of conventional tag recoveries have been within 1,894 km of the release location, and similar frequencies of long-distance movements have been observed in other ocean basins. Conventional and electronic tagging in the Pacific demonstrate greater longitudinal movements than latitudinal movements. While trans-Atlantic movements have been noted for Bigeye Tuna, they occur at a reduced frequency relative to Yellowfin Tuna.

FOOD: Bigeye Tuna are opportunistic predators that feed on a wide variety of fishes, cephalopods, and crustaceans during the day and at night. Food studies in the Atlantic were reviewed by Dragovich. Large juveniles and adults exploit the deep scattering layer, consuming mesopelagic fishes, cephalopods, and crustaceans at depths to 300–500 m during the day and in surface waters at night. Off northeastern Brazil, 83 prey items were found in stomachs of Bigeye Tuna, of which 46 were fishes, mostly epipelagic or mesopelagic species, 20 were cephalopods, and 13 were pelagic crustaceans. The Caribbean Pomfret (*Brama caribbea*) was the most important food item, followed by other mesopelagic fishes and a squid (*Ornithoteuthis antillarum*), the most preyed-on cephalopod.

Below Juvenile Bigeye Tuna, approx. 30 cm FL.

REPRODUCTION: Bigeye Tuna spawn between 15° N and 15° S. The Gulf of Guinea in the Atlantic Ocean is their best-known breeding and nursery area. Elsewhere, spawning has been recorded between 10° N and 10° S in the eastern Pacific. Sexual maturity for Bigeye Tuna is reached at 100–30 cm FL at an age of about 3 years. However, minimum length at sexual maturity for females can be 80–102 cm FL and predicted length at 50% maturity 102–35 cm FL. Spawning peaks from April through September in the Northern Hemisphere and between January and March in the Southern Hemisphere. Spawning occurs primarily at night between between 1900 and 0400 hours. Females are batch spawners, with a spawning frequency of 2.6 days. Mean fecundity is 24 oocytes/g body weight, with estimated batch sizes of 2.9–6.3 million eggs. Although spawning apparently occurs widely across the equatorial Pacific Ocean, the greatest reproductive potential appears to be in the eastern Pacific, based on maturation, size frequencies, and catch per unit effort.

EARLY LIFE HISTORY: Eggs and larvae are pelagic. Three larvae ranging from 5.1 mm notochord length to 8.5 mm SL were described and illustrated in Richards' 2006 identification guide to western central Atlantic fishes. Pigmentation is present on the midbrain, gut, jaw tips, ventral margin of the tail, and the first dorsal fin after reaching a length of about 5 mm. Flexion occurs in larvae at about 6 mm SL. Larval distribution in equatorial waters is transoceanic the year round, but there are seasonal changes in larval density in subtropical waters. Larvae probably occur only in warm waters, above the thermocline.

STOCK STRUCTURE: Samples of Bigeye Tuna from the Atlantic are genetically distinct from those found in the Indian and Pacific oceans. Two highly divergent mitochondrial haplotype clades coexist in the Atlantic Ocean. One is almost exclusively confined to the Atlantic Ocean, while the other is also found in the Indo-Pacific. Genetic studies have shown no population structuring within the Indian Ocean. For management purposes, four stocks of Bigeye Tuna are recognized: Atlantic Ocean, eastern Pacific Ocean, western and central Pacific Ocean, and Indian Ocean. Results from tagging studies demonstrate that a small fraction (~5%) of individuals undertake large (>1500 km) displacements (which could facilitate gene flow), but most individuals remain within the general area in which they were tagged. Based on tagging data, as many as nine ecological stocks have been proposed for the Pacific Ocean.

FISHERIES INTEREST: Global catches of Bigeye Tuna were less than 100,000 mt through the 1960s, increased to about 200,000 mt by 1980, and then quickly rose, peaking at over 500,000 mt in 1998. Current catches have been close to 400,000 mt. Bigeye Tuna are caught primarily by pelagic longlines and purse seines, with minor landings from bait boats and other gear types. Pelagic longlines accounted for the vast majority of the catches until the 1990s, at which time purse seine catches began to steadily increase. Pelagic longlines typically catch adult Bigeye Tuna, while the purse seine fishery takes juveniles that often form mixed schools with Skipjack and Yellowfin Tuna under FADs. With the increase in purse seine fishery, there has been a change in the overall selectivity of the fishery, resulting in lower MSYs. The smaller fish taken in the purse seine fishery are typically destined for canneries, while the larger fish from the pelagic longline fishery enter the sashimi market.

Atlantic Ocean: Bigeye Tuna catches in the Atlantic Ocean increased steadily from the 1960s, reaching a maximum of almost 135,000 mt in 1994. Catches over the past five years have been about 76,000 mt. The pelagic longline has been the dominant gear type, and even though longline catches decreased by more than half since they peaked at just over 80,000 mt in the mid 1990s, this gear type still accounts for 47% of the landings. Purse seine effort increased dramatically in the 1990s and currently accounts for 37% of the landings, while traditional bait boat fisheries produce 13% of the catch.

Eastern Pacific Ocean: Catches of Bigeye Tuna by pelagic longline vessels in the eastern Pacific increased in the 1950s, reaching more than 60,000 mt by the early 1960s, and peaked at close to 100,000 mt in the early 1990s. While there were some modest catches by purse seine vessels from the mid-1970s to the mid-1980s, catches by this gear type increased dramatically through the 1990s, peaking at about 90,000 mt in the early 2000s. Currently, purse seines account for about 61% of the landings, and pelagic longlines take the other 39%. The change in selectivity of the fishery to smaller fish has resulted in a decrease to the MSY of almost 50%. In 2016, the total catch was 94,100 mt, close to the five-year average of about 97,000 mt.

Western and Central Pacific Ocean: Catches of Bigeye Tuna in the western and central Pacific steadily increased from the 1950s through a peak of more than 200,000 mt in the early 2000s and have averaged about 149,000 mt over the past five years. Pelagic longlines accounted for the majority of the catch through the mid-2000s, but since that time longline catches have dropped by about 25,000 mt. The purse seine fishery increased dramatically through the 1980s and 1990s and now has catches comparable to those of the pelagic longline fleet. Each accounts for about 44% of the total catch. Due to a change in selectivity of the fishery resulting from increased catches of small fish by the purse seine fleet, the current MSY is less than half of that prior to the 1970s.

Indian Ocean: Catches of Bigeye Tuna in the Indian Ocean increased steadily from the 1950s, reaching a maximum of more than 150,000 mt in the late 1990s. Pelagic longline has been the dominant gear type throughout, although the overall share has decreased since the early 2000s. Pelagic longlines now account for about 54% of the catches, purse seines 28%, and other gear types 18%. Longline catches decreased significantly after the early 2000s, largely due to a shift in the location of the fishery to avoid piracy.

THREATS: Increased catches of small Bigeye Tuna by purse seines fishing on FADs has resulted in a large change in the selectivity of the fishery and decreased yields per recruit in all areas. In most oceans, the estimated MSY has dropped by 50% or more due to the change in selectivity of the fishery, and longline catches of larger fish have declined. Management of the catches of small fish is complicated by that fact that small Bigeye Tuna co-school with small Yellowfin Tuna and Skipjack under FADs. To date, the tuna RFMOs have had difficulty effectively managing the purse seine fishery and achieving an acceptable balance between the purse seine and longline fisheries. The Atlantic stock is currently overfished and experiencing overfishing, and the permissible catch could greatly exceed the TAC. Bigeye Tuna was classified as Vulnerable on the IUCN Red List in 2011.

CONSERVATION: The Bigeye Tuna is listed as a highly migratory species in Annex I of the 1982 Convention on the Law of the Sea. It is managed as four stocks by the relevant tuna RMFOs: in the Atlantic by ICCAT, in the eastern Pacific by IATTC, in the western and central Pacific by WCPFC, and in the Indian Ocean by IOTC.

Atlantic Ocean: ICCAT assessed Atlantic Bigeye Tuna in 2010, 2015, and 2018. In contrast to the relatively optimistic assessments prior to and including the 2010 assessment, the 2015 assessment indicated that the stock was overfished (SSB = 0.67 SSBMSY), and overfishing was occurring (F = 1.28 FMSY). MSY has decreased over time due to a changing selectivity of the fishery resulting from increased catches of small fish by purse seines fishing on FADs. ICCAT manages Atlantic Bigeye Tuna with a TAC of 65,000 mt, country-specific quotas for the major harvesting nations, and catch limits for minor harvesters.

However, the total permissible catch exceeds the TAC. Other conservation measures include a two-month annual time/area closure in the Gulf of Guinea for fishing in association with FADs, capacity limits on the number of purse seine and longline vessels for some countries, and an FAD management plan.

Eastern Pacific Ocean: The IATTC 2017 stock assessment update indicated that Bigeye Tuna in the eastern Pacific are not overfished (SSB = 1.23 SSBMSY) and overfishing is not occurring (F = 0.87 FMSY). Catches over the past five years have been below the estimated MSY of 106,200 mt. The IATTC has adopted a harvest control rule for tropical tunas. Current conservation measures for Bigeye Tuna include a 72-day closure for large purse seine vessels (>182 mt capacity), a time/area closure west of the Galapagos Islands, and limits on the number of FADs a vessel can deploy.

Western and Central Pacific Ocean: The results of an assessment of Bigeye Tuna in the western and central Pacific conducted by the South Pacific Commission in 2017 indicated that the stock is not overfished (SSB = 1.23 SSBMSY) and overfishing is not occurring (F = 0.83 FMSY). The estimated MSY of 156,000 mt is greater than recent catches that have averaged 149,200 mt over the past five years. Conservation measures include a three-month closure on FAD fishing in tropical waters, limits on the total number of vessel days, an FAD management plan, and a prohibition on discards of tropical tunas.

Indian Ocean: The 2016 assessment of Bigeye Tuna by the IOTC indicated that the stock is not overfished (SSB = 1.29 SSBMSY) and overfishing is not occurring (F = 0.76 FMSY). Catches over the past five years have averaged 101,000 mt, just below the estimated MSY of 104,000 mt. While the IOTC has no specific management measures for Bigeye Tuna, there are some in place for tropical tunas, including an FAD management plan and a ban on the discards of tropical tunas by purse seine vessels.

REFERENCES: Aires-da Silva et al. 2016; Alvarado Bremer et al. 1998; Appleyard et al. 2002; Calkins 1980; Chiang et al. 2008; Chow et al. 2000; Collette 1995, 2001, 2010; Collette and Nauen 1983; Collette et al. 2011; Cox et al. 2002; Dagorn et al. 2000; Dragovich 1969; Durand et al. 2005; Evans et al. 2008; Farley et al. 2006; Fonteneau and Hallier 2015; Fuller et al. 2015; Gibbs and Collette 1967; Harley et al. 2010; Hillary and Mosqueira 2006; Holland and Sibert 1994; Holland et al. 1992; Kailola et al. 1993; Kikawa 1966; Kishinouye 1915; Lee et al. 2005; Lowe 1839; Lowe et al. 2000; Lu et al. 2001; Maigret and Ly 1986; Majkowski 2007; Martínez et al. 2006; McKechnie et al. 2017; Miyabe 1994; Nikaido et al. 1991; Powell et al. 2009; Richards 2006; Schaefer 2001; Schaefer and Fuller 2002, 2006, 2009, 2010; Schaefer et al. 2005, 2014; Stéquert and Conand 2004; Sun 2003; Vaske et al. 2012; Zhu et al. 2009, 2010.

Pacific Bluefin Tuna

Thunnus orientalis (Temminck & Schlegel, 1844)

COMMON NAMES: English - Pacific Bluefin Tuna
French - Thon Bleu du Pacifique
Spanish - Atún Aleta Azul del Pacifico

ETYMOLOGY: Coenraad Jacob Temminck, a Dutch ornithologist, illustrator, and collector, and Hermann Schlegel, a German-born zoologist who spent much of his life in the Netherlands, named this tuna *orientalis* due to its occurrence in Japan (Temminck and Schlegel 1844:94).

SYNONYMS: *Thynnus orientalis* Temminck & Schlegel, 1844; *Orcynus schlegelii* Steindachner, 1884; *Thunnus saliens* Jordan & Evermann, 1926; *Thunnus thynnus orientalis* Temminck & Schlegel, 1844

TAXONOMIC NOTE: The Pacific Bluefin Tuna had been recognized as a subspecies (*Thunnus thynnus orientalis*) of the Atlantic Bluefin Tuna (*Thunnus thynnus thynnus*) but is now recognized as a separate species based on morphology (Collette 1999) and genetics (Tseng et al. 2011).

FIELD MARKS: 1 The median caudal keel of the Pacific Bluefin Tuna is black in adults, not yellow as in the Southern Bluefin Tuna.

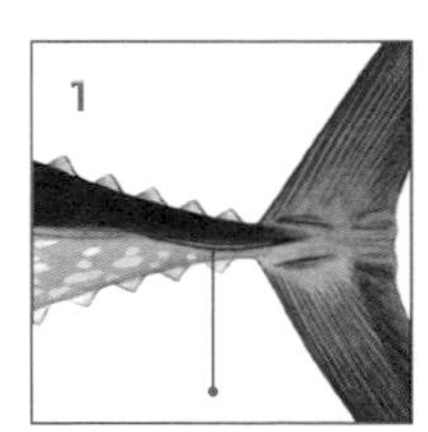

UNIQUE FEATURE: This is the most expensive fish in the world. The price of the first big Pacific Bluefin Tuna brought into the Tokyo Tsukiji market each year increased for three years: $396,000 for a 341 kg fish in 2011; $735,000 for a 269 kg fish in 2011; and an astonishing $1,800,000 for a 222 kg fish in 2013. This auction is an annual event intended to generate publicity and the price paid greatly exceeds vessel prices, which are approximately $25/kg.

DIAGNOSTIC FEATURES: The Pacific Bluefin Tuna is a very large species whose body is deepest near the middle of the first dorsal-fin base. **Gill rakers:** Gill rakers number 32–40 on the first gill arch, mean 36. **Fins:** The second dorsal fin is higher than the first dorsal. The pectoral fins are very short, less than 80% of head length (16.8–21% of fork length), never reaching posteriorly to the interspace between the dorsal fins. **Liver:** The ventral surface of the liver is striated. **Swim bladder:** A swim bladder is present. **Body cavity:** The bulge in the roof of the body cavity is relatively narrow in large specimens. **Vertebrae:** Vertebrae number 18 precaudal plus 21 caudal, as in all but one of the species of *Thunnus*. Descriptions and illustrations of the osteology and soft anatomy are presented by Godsil and Holmberg (1950) and Gibbs and Collette (1967). **Color:** The back is metallic blue-black, the lower sides and belly silvery white with colorless transverse lines alternating with rows of colorless dots visible only in fresh specimens. The first dorsal fin is yellow or bluish; the second dorsal is reddish brown. The anal fin and anal finlets are dusky yellow edged with black. The median caudal keel is black in adults, not yellow as in the Southern Bluefin Tuna.

GEOGRAPHIC RANGE: The Pacific Bluefin Tuna is present throughout the Pacific and enters the eastern Indian Ocean. In the eastern Pacific, it is found from the Gulf of Alaska to southern California and Baja California, Mexico. In the western Pacific, it is known from Sakhalin Island in the southern Sea of Okhotsk south to the northern Philippines. Although basically a temperate species, it also ventures into tropical waters south to French Polynesia and New Zealand, but there is no evidence of spawning in these areas.

SIZE: The largest Pacific Bluefin Tuna recorded by the Far Seas Fisheries Research Laboratory, Shimizu, Japan, weighed approximately 555 kg and was about 3 m long. It was caught about 300 miles south of Kyushu Island, Japan, in April 1986. The IGFA all-tackle game fish record is of a 907 lb, 6 oz (411.6 kg) fish caught off Three Kings, New Zealand, in February 2014. While most Pacific Bluefin Tuna caught off California are in the 5–25 kg weight range, fish as large as 457.7 kg and 271.2 cm FL have been caught off the Channel Islands during the winter months. Growth is most rapid in the first few years of life and asymptotic length is reached after approximately 20 years or older. Longevity is at least 26 years, based on a 260 cm FL fish. As with the other two species of bluefin tunas, males tend to grow larger than females. Based on maturity and longevity studies, the generation length is estimated to be 7–9 years.

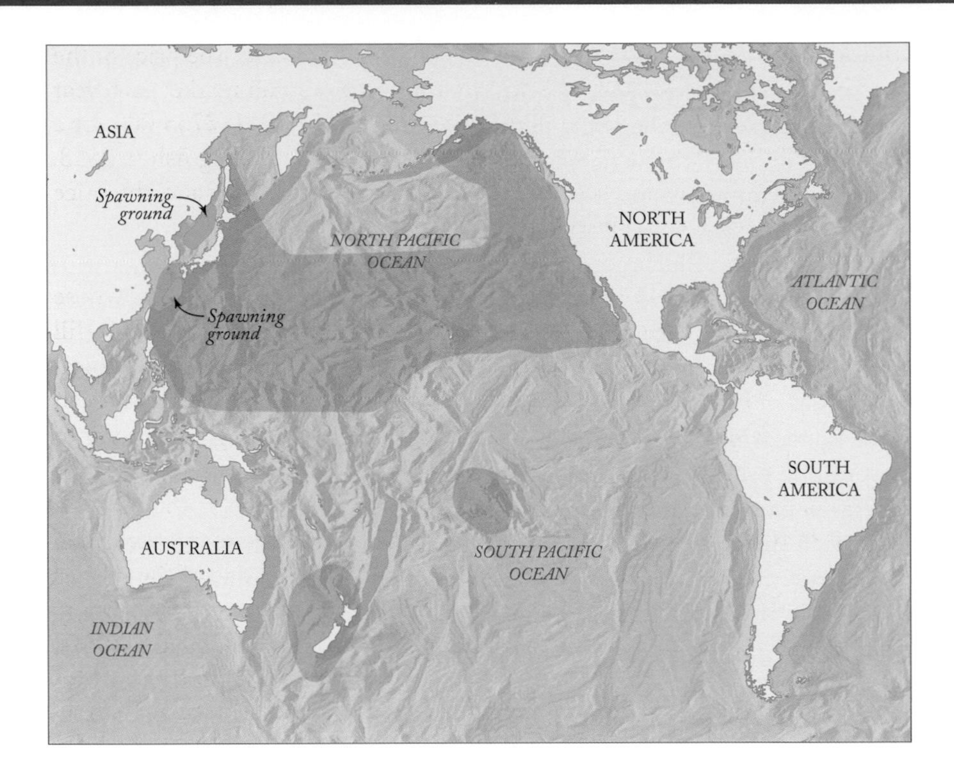

Left: Pacific Bluefin Tuna range map.

HABITAT AND ECOLOGY: The Pacific Bluefin Tuna is epipelagic and usually oceanic, but seasonally comes close to the shore. It is found from the surface to depths of at least 550 m and tolerates wide temperature ranges. It spends the majority of its time in the upper mixed layer, but makes frequent dives below the thermocline. Pacific Bluefin Tuna schools by size, sometimes in mixed-species schools with other scombrids. The Pacific Bluefin Tuna has undergone evolutionary changes in three visual pigment genes that may have contributed to enhanced detection of blue-green contrast and measurement of the distance to prey in the blue pelagic waters. Hearing in Pacific Bluefin Tuna is most sensitive from 400 to 500 Hz, similar to Yellowfin and Kawakawa, the other two tunas that have been tested.

MOVEMENTS: Some juvenile Pacific Bluefin Tuna undertake trans-Pacific migrations from nursery areas in the western Pacific into the eastern Pacific, spending up to four years before returning to the western Pacific to spawn. It was originally thought that the proportion of Pacific Bluefin Tuna that underwent this migration was relatively small, but recent research using intrinsic tracers in tissues suggests that in some years it is likely a much larger proportion of the population. Electronic tagging studies indicate that juvenile fish take a direct path across the North Pacific, primarily utilizing waters of the Subarctic Frontal Zone, and crossings have been completed in as few as 66 days. In the eastern Pacific, juveniles move northward in the spring from off the coast of Baja California, Mexico, to waters off central and northern California by the early fall. Their trans-Pacific migrations are facilitated by long body muscle tendons that reduce the metabolic energy needed during rapid and continuous swimming. A musculo-vascular complex, consisting of fin muscles, bones, and lymphatic vessels, is involved in the hydraulic control of the median fins and may have evolved in response to the demand for swimming and maneuvering control in this high-performance species.

FOOD: The Pacific Bluefin Tuna is a voracious predator that feeds on a wide variety of small schooling fishes, squids, crabs, and other macronekton. Larvae (2.3–14.6 mm TL) feed during the day on small zooplankton including copepods, cladocerans, and appendicularians. In the Tsushima Current region (Sea of Japan), small Pacific Bluefin Tuna (20–25 cm FL) preyed upon small squid, while larger fish (25–35 cm FL) gradually shifted their diet to mesopelagic fishes. In the Kuroshio region of the western Pacific, small Pacific Bluefin Tuna preyed mostly on small plankton (crustacean larvae), and larger individuals shifted to epipelagic clupeoid fishes such as sardines and anchovies. In the eastern Pacific, juvenile Pacific Bluefin Tuna feed primarily on fishes, which constitute 93% of gut contents by volume, but they also eat small quantities of crustaceans and squids. In some years, the Northern Anchovy (*Engraulis mordax*) was the primary food item, with the Red Swimming Crab (*Pleuroncodes planipes*) and Saury (*Cololabis saira*) the second and third most important food items, respectively. A 458 kg fish captured off California had fed on Chub Mackerel (*Scomber japonicus*) and Market Squid (*Loligo opalescens*).

REPRODUCTION: Pacific Bluefin Tuna are known to spawn in only two areas of the western North Pacific: from the Philippine Islands north to the Ryuku Islands from April to June and further north in the Sea of Japan from June to August. The distribution of sizes of fish on the two spawning grounds differs, with larger fish (modal size 220 cm FL) on the southern grounds and with smaller fish (modal size 150 cm FL) spawning in Sea of Japan. The difference in size distributions results in a difference in estimated age at 50% maturity: age 11 for the southern grounds and age 4 for the northern grounds. The fraction of females in spawning condition increases steadily on the spawning grounds through the spawning season, as females leave the area immediately after spawning ceases. The sex ratio is about 1:1. Females are batch spawners and can spawn almost every day, with an average batch fecundity of 6.4 million oocytes. Batch fecundity increases with length, from about 5 million eggs at 190 cm FL to about 25 million eggs at 240 cm FL. The life cycle of the Pacific Bluefin Tuna has been completed under aquaculture conditions, the first for a large species of tuna.

EARLY LIFE HISTORY: Eggs of Pacific Bluefin Tuna are spherical, with a mean diameter of 1.02 mm, and contain a single oil globule 0.26 mm in diameter. Larvae are similar in appearance to larval Southern Bluefin and have 1–4 melanophores on the dorsal margin of the trunk and 1–5 on the ventral margin. Illustrations and descriptions of the development of 7 laboratory-reared Pacific Bluefin Tuna larvae, 2.91 mm to 13.55 mm BL, from newly hatched larva to

25-day old juvenile, were published by Miyashita et al. Growth rates of cultured larvae averaged 0.33 mm per day and increased with age. Average body length of newly hatched larvae was 2.83 mm, and larvae grew on average to 5.80 mm by 10 days, 10.62 mm by 20 days, and 35.74 mm by 30 days after hatching. At 3 days after hatching and 3.3 mm SL, the mouth developed and the larvae began feeding. The shift from preflexion to flexion in another laboratory-reared series was first noted at 3.5–4.0 mm SL; flexion occurred at 6–8 mm BL, and postflexion began at 6.0–6.5 mm SL. Larvae in the postflexion stage grew to approximately 3.8 mm SL by day 5, 6.1 mm SL by day 10, and 8.0 mm by day 15. Larvae are found in somewhat warmer waters than other species of tunas, 24–28°C surface temperature. Patches of larvae are transported via the Kuroshio Current to nursery areas off coastal Japan. Larvae exhaust their food supply in the yolk and oil globule by the third to fifth day after hatching and must feed by then or starve. Differentiation of gastric glands occurred in 18-day-old 7.6 mm SL larvae. Numerous pyloric caeca developed in 30-day-old 12.6 mm SL larvae.

STOCK STRUCTURE: Pacific Bluefin Tuna comprise a single, pan-Pacific population, with no significant genetic differentiation between samples from Japan and Mexico or from Taiwan and New Zealand. Multiple lines of evidence demonstrate significant movement of juveniles and adults from the western North Pacific to the eastern North Pacific, with adults returning to the western North Pacific for spawning.

FISHERIES INTEREST: The Pacific Bluefin Tuna is a high-value species in the global fresh-fish markets, particularly in the sashimi and sushi markets of Japan, and aquaculture production is being intensively studied in Japan. Most of the catch is from the western Pacific Ocean, with 20–30% occurring in the eastern Pacific. Purse seines are the dominant gear type, accounting for about two-thirds of the catch, while coastal set nets, troll lines, and pelagic longlines account for most of the reminder. Juveniles have represented the majority of the landings since the early 1950s, with catches of age 0 fish increasing since the 1990s. FAO worldwide reported landings show a sharp increase from 1,452 mt in 1950 to 40,094 mt in 1956, as the purse seine fishery developed. Landings remained high (25,000–40,000 mt) for the next 10 years. Since that time, landings have shown considerable variation, likely resulting from fluctuations in recruitment, with highs approaching the levels of the 1960s and a low of 8,797 mt in 1990. Landings for the past several years have averaged around 14,000 mt. Japan is responsible for the majority of the landings in the western Pacific, with some catches by Taiwan and Korea. The United States was responsible for the largest catches in the eastern Pacific through the late 1970s, fishing in waters off Baja California. However, US effort and landings decreased when Mexico established its exclusive economic zone in 1976. Mexico has accounted for the majority of eastern Pacific landings since the mid-1980s, and a significant fraction of the purse seine catch of juvenile fish is transported to coastal pens for fattening.

THREATS: Pacific Bluefin Tuna are the most overfished of any of the tunas, with an estimated spawning stock biomass at 2.6% of the unfished condition. The capture of large numbers of juvenile fish results in significant overfishing. The Pacific Bluefin Tuna is now listed as Vulnerable on the IUCN Red List, downgraded from Least Concern because old stock assessments had indicated that the population was relatively stable.

CONSERVATION: Management of the Pacific Bluefin is divided geographically between two tuna regional fisheries management organizations (RFMOs), the Western and Central Pacific Fisheries Commission (WCPFC) and the Inter-American Tropical Tuna Commission (IATTC). The 2016 assessment indicates that Pacific Bluefin Tuna SSB has varied over the past 65 years, but there was a steady decline from 1996–2010. It appears that the decline has stopped in recent years, but the current SSB is estimated to be at or near a historic low, 2.6% of the unfished stock. The estimated fishing mortality, which has decreased in recent years, was still higher than most biological reference points, indicating that either the stock was at the threshold of overfishing or that overfishing is still occurring. The WCPFC implemented a multi-year rebuilding plan in 2015 to rebuild SSB to the median level (1952–2014) by 2024, with a 60% probability. The rebuilding plan reduces total fishing effort north of 20° N and reduces catches of both juveniles (<30 kg) and adults (>30 kg) from 2002–4 levels. The IATTC implemented a measure in 2016 that reduces commercial catches and encourages member nations to limit their catch of juvenile bluefin to no more than 50% of the total catch.

REFERENCES: Bayliff 1994; Bell 1986; Boustany et al. 2010; Chen et al. 2006; Collette 1999; Collette and Nauen 1983; Collette et al. 2014; Dale et al. 2015; Foreman and Ishizuka 1990; Furukawa et al. 2014; Gibbs and Collette 1967; Godsil and Holmberg 1950; Hsu et al. 2000; Huff 2017; IGFA 2018; ISSF 2013, 2017; Itoh et al. 2003; Jordan and Evermann 1926; Kaji et al. 1996, 2001; Kitigawa et al. 2009, 2010; Kodama et al. 2017; Madigan 2015; Madigan et al. 2017; Miyashita et al. 2001; Muhling et al. 2017; Murua et al. 2017; Nakae et al. 2013; Nakamura et al. 2013; Nishikawa 1985; Nomura et al. 2014; Ohshimo et al. 2017; Okochi et al. 2016; Pavlov et al. 2017; Pinkas 1971; Sawada et al. 2005; Schaefer 2001; Shimose et al. 2009, 2013, 2018; Steindachner and Döderlein 1884; Tanaka et al. 2006, 2007; Temminck and Schlegel 1844; Tseng and Smith 2012; Tseng et al. 2011; Ueyanagi 1969; Uotani et al. 1990; Yabe et al. 1966.

Atlantic Bluefin Tuna

Thunnus thynnus (Linnaeus, 1758)

COMMON NAMES: English - Atlantic Bluefin Tuna
French - Thon Rouge de l'Atlantique
Spanish - Atún Aleta Azul
Japanese - Kuromaguro

ETYMOLOGY: The Swedish naturalist Carl Linnaeus (1758:297), who devised the binomial nomenclature system we use, selected the name *thynnus* for the Atlantic Bluefin Tuna, based on the Greek name for tuna, *thynnos*.

SYNONYMS: *Scomber thynnus* Linnaeus, 1758; *Thynnus mediterraneus* Risso, 1827; *Thynnus vulgaris* Cuvier in Cuvier and Valenciennes, 1832; *Thynnus secundodorsalis* Storer, 1855; *Thynnus linnei* Malm, 1877; *Orcynus thynnus* (Linnaeus, 1758); *Thunnus thynnus* (Linnaeus, 1758); *Thunnus thynnus thynnus* (Linnaeus, 1758)

TAXONOMIC NOTE: The Atlantic Bluefin Tuna had been recognized as a subspecies (*Thunnus thynnus thynnus*) closely related to the Pacific Bluefin Tuna (*Thunnus thynnus orientalis*), but both are now considered to be separate species based on morphology (Collette 1999) and genetics (Tseng et al. 2011).

FIELD MARKS:

1 Atlantic Bluefin Tuna has a few more gill rakers on the first arch, 34–43 (mean 38.9) than the Pacific and Southern Bluefins, 31–40 (means 33.7 and 35.9, repectively).
2 Pectoral fins are very short, less than 8% of head length.

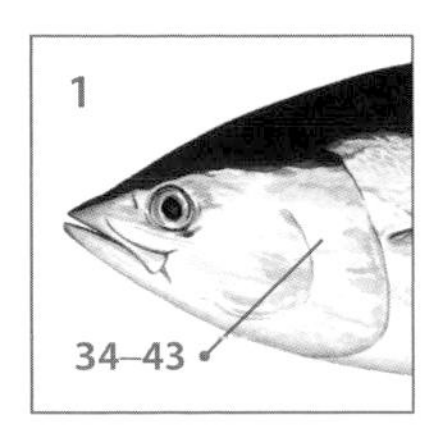

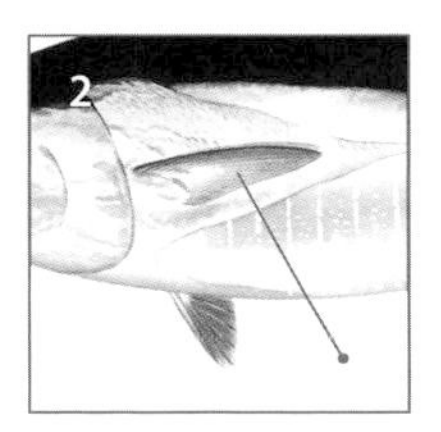

UNIQUE FEATURE: Atlantic Bluefin Tuna have the broadest thermal niche of all the species in the family Scombridae and are capable of tolerating temperatures of 1–29°C.

DIAGNOSTIC FEATURES: The Atlantic Bluefin Tuna is a very large tuna whose body is deepest near the middle of the first dorsal-fin base. **Gill rakers:** Gill rakers number 34–43 on the first gill arch, mean 38.9. **Fins:** The second dorsal fin is higher than the first dorsal. The pectoral fins are very short, less than 80% of head length (17.0–21.7% of fork length), never reaching posteriorly to the interspace between the dorsal fins. **Liver:** The ventral surface of the liver is striated, its three lobes subequal in length. **Swim bladder:** A large swim bladder is present. **Body cavity:** The bulge in the roof of the body cavity is wide, with no lateral concavity, instead of narrow with a lateral concavity as in Pacific Bluefin. **Vertebrae:** Vertebrae number 18 precaudal plus 21 caudal as in all but one of the species of *Thunnus*. Descriptions and illustrations of the osteology and soft anatomy are presented by Godsil and Holmberg (1950) and Gibbs and Collette (1967). **Color:** The back is metallic blue-black, the lower sides and belly silvery white with colorless transverse lines alternating with rows of colorless dots that are visible only in fresh specimens. The first dorsal fin is yellow or bluish; the second dorsal is reddish brown. The anal fin and anal finlets are dusky yellow edged with black. The median caudal keel is black in adults as in Pacific Bluefin, not yellow as in the Southern Bluefin Tuna.

GEOGRAPHIC RANGE: Atlantic Bluefin Tuna occurs throughout much of the North Atlantic and have occasionally been reported from the South Atlantic. Historically, it was present in the western Atlantic from Canada to Brazil, including the Gulf of Mexico and the Caribbean Sea, although the bulk of the population off Brazil has disappeared, with very few catches in the area over the last 40 years. In the eastern Atlantic, it is present from Norway to the Canary Islands. Rising water temperatures have brought Atlantic Bluefin Tuna into Greenland waters, and it has begun to migrate to the Icelandic Basin in the autumn. It has been reported from Mauritania and off South Africa. It is present in the Mediterranean Sea and the southern Black Sea. Atlantic Bluefin Tuna was well documented from the Black Sea in ancient times, and there was an annual migration from those waters to eastern Mediterranean spawning grounds. However, after World War II, environmental conditions in the Black Sea deteriorated, and sightings of Atlantic Bluefin Tuna have become rare. An 2011 analysis of historical and present ranges concluded that the Atlantic Bluefin showed the largest range contractions at that time (46% since 1960) of any pelagic species.

SIZE: The IGFA all-tackle game fish record is of a 1,496 lb (678.58 kg) fish caught in Aulds Cove, Nova Scotia, in October 1979. This species has a maximum size over 300 cm FL, but is more common to 200 cm. Maximum age is at least 35 years and possibly as much as 50 years.

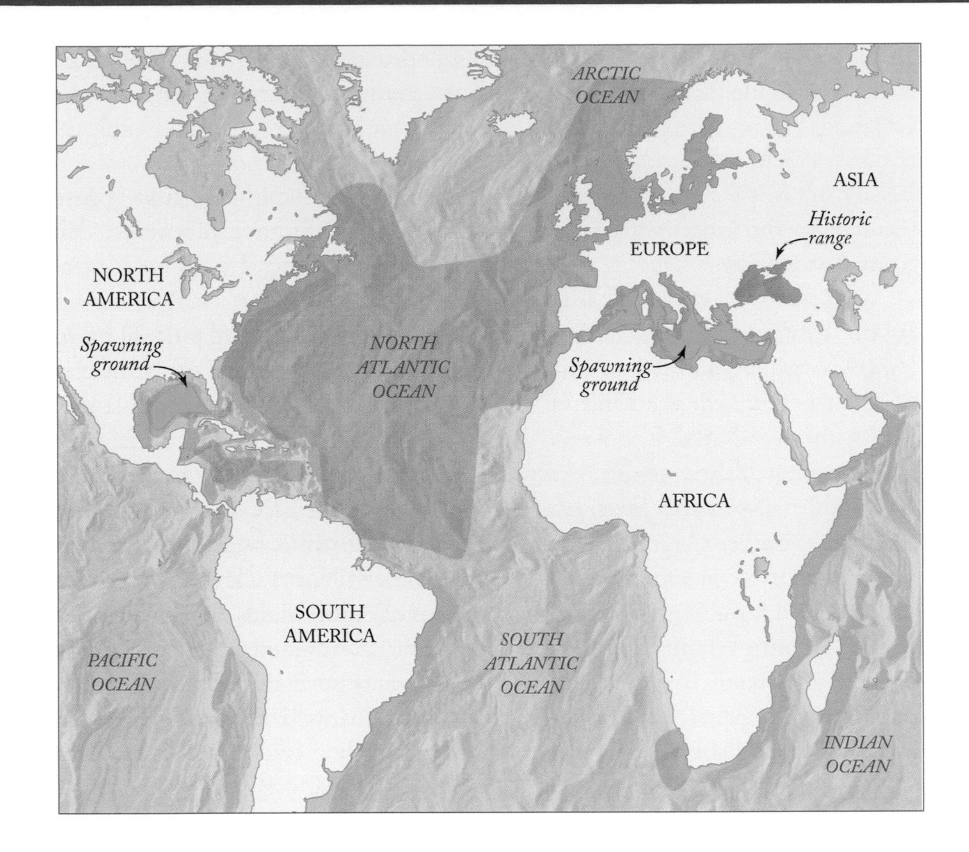

Right: Atlantic Bluefin Tuna range map.

HABITAT AND ECOLOGY: Atlantic Bluefin Tuna are pelagic and oceanodromous. Seasonally, they can be found close to shore. Adults can tolerate the widest range of temperatures of any of the tunas, 1–29°C, and show a marked preference and tolerance for high-salinity waters. The wider temperature tolerance in larger adults allows them to feed in the northern latitudes, up to the Norwegian Sea in the eastern Atlantic and to the Gulf of Saint Lawrence in the western Atlantic. They school by size, sometimes together with Albacore, Yellowfin, Bigeye, or Skipjack. Electronic tagging studies indicate that Atlantic Bluefin Tuna spend considerable time in near-surface waters, but are capable of diving to depths in excess of 800 m. In the western North Atlantic, juveniles (age 2–5) occupied shallower depths during the summer (individual mean depths of 5–12 m) than in the winter (mean depths of 41–58 m).

FOOD: Atlantic Bluefin Tuna are opportunistic, non-specialized visual predators that prey on small schooling fishes (anchovies, sauries, and hakes), squids, and red crabs. They often occur at the vicinity of thermal and chlorophyll fronts. In the Mediterranean, juveniles prey mainly on zooplankton and small pelagic and coastal fishes, sub-adults prey on medium-sized pelagic fishes, shrimps, and cephalopods, and adults prey mainly on cephalopods and larger fishes. A total of 47 prey taxa were identified from stomachs of Atlantic Bluefin Tuna from the eastern Mediterranean, including 34 species of fishes, 11 of squids, and 2 of crustaceans. Most of the important fish families were deep-sea fishes and jacks.

In the eastern Atlantic, krill, anchovies, and horse mackerel are important prey items of juvenile tuna in the Bay of Biscay. During autumn migrations to the Icelandic Basin, Atlantic Bluefin forage at pelagic and mesopelagic depths. Of the 36 prey species found in their stomachs, the most important were European Flying Squid (*Todarodes sagittatus*), barracudinas (Paralepididae), and gonate squid. Hyperiid amphipods were more abundant but of minor importance due to their small size.

In the western Atlantic, pelagic schooling fishes such as Atlantic Herring (*Clupea harengus*) and Atlantic mackerel (*Scomber scombrus*) dominated the diet of Atlantic Tuna in the Gulf of St. Lawrence. Along the US mid-Atlantic Bight, they feed mainly on sand lance. Larger fish in the Gulf of Mexico feed on bony fishes, squids, crustaceans, and a pelagic tunicate. Longnose Lancetfish (*Alepisaurus ferox*) was the most important fish, and cephalopods from the families Argonautoidae and Onychoteuthidae were important in terms of occurrence and number. Along the coast of North Carolina, stomachs contained 14 families of bony fishes, 5 species of crabs, squids, and a shark. By weight, Atlantic Menhaden (*Brevoortia tyrannus*) dominated the diet. Offshore in the Gulf Stream, they feed under sargassum mats for fishes like triggerfishes, seahorses, pipefishes, and small jacks, while individuals feeding deeper eat pomfrets, lancetfishes, and squids.

REPRODUCTION: Maturity schedules appear to differ between eastern and western Atlantic Bluefin Tuna. Reproductive analyses of fish sampled in the Mediterranean Sea (near or on the spawning grounds) are in general agreement, with first maturity occurring at age 3 (100 cm FL), 50% maturity at age 4 (115 cm FL), and 100% maturity by age 5 (135 cm). Maturity schedules for western Atlantic Bluefin Tuna suggest later ages of maturation, and there is considerable range among studies, with age at 50% maturity ranging from 5 to 12 or more years.

Eastern and western Atlantic Bluefin Tuna exhibit fidelity to respective spawning grounds in the Mediterranean Sea and the Gulf of Mexico. Within the Mediterranean Sea, spawning is known to occur near the Balearic Islands, in the South Tyrrhenian Sea, and in the waters around Malta and eastern Sicily plus the northern Levantine Sea. Fish enter the Mediterranean through the Straits of Gibraltar in May–June and leave in July–August. The onset of spawning near the Balearic Islands is between the end of May and the beginning of June.

Western Atlantic Bluefin Tuna spawn in the Gulf of Mexico and the Straits of Florida from mid-April to early June at temperatures of 22.6–27.5°C. Results of electronic tagging studies suggest distinct behaviors during the spawning time, most noticeably with changes in diving times and depths. As they entered and exited the Gulf of Mexico, they dove to daily maximum depths of 568–580 m but during the putative breeding phase, they made significantly shallower daily maximum dives, 203 ± 76 m. Some larvae have been collected northeast of Campeche Bank, suggesting spawning outside of the Gulf of Mexico, and recently, collections of larvae in the Slope Sea off the US Atlantic coast during July demonstrate some spawning activity in that region.

Atlantic Bluefin Tuna are multiple batch spawners, with a spawning frequency of 1–2 days in the Mediterranean. Estimated relative batch fecundity is greater than 90 oocytes/g of body weight, slightly higher than that estimated for other tunas in the genus *Thunnus*. Females weighing between 270 and 300 kg produce as many as 10 million eggs per spawning season.

Above: Juvenile Atlantic Bluefin Tuna, approx. 25 cm FL.

EARLY LIFE HISTORY: Illustrations of an egg and five larval Atlantic Bluefin Tuna, 3.0 mm notochord length to 8.5 mm SL, were provided by Richards. The larvae differ from other species of the genus in pigment pattern and location of the first haemal arch on the vertebrae. The eggs are about 1.00–1.22 mm in diameter and have one oil globule. At 24°C, embryo development lasts about 32 hours. Larval length at hatching is 2.8–3.0 mm. Larval stages last about 30 days. In reared systems, they grow very fast, reaching 4–5 cm at the end of the first month. Transformation of the gut from an undifferentiated canal at hatching to a complex juvenile-like digestive tract occurred at three weeks post hatch.

Larvae feed mostly during daylight hours. In the Mediterranean, copepod nauplii dominated the diet of small larvae, but cladocerans became increasingly more abundant at sizes greater than 5 mm SL. Unlike Mediterranean larvae, larvae from the central Gulf of Mexico (3.0–6.7 mm BL) feed on different invertebrate zooplankton, copepods, cirripeds, and cladocerans. Ontogenetic dietary shifts between larvae from the Gulf of Mexico and the Mediterranean were observed with nitrogen stable isotope ratios that may have important implications in growth during their early life stages. Unlike other scombrid larvae, larval Atlantic Bluefin Tuna are not piscivorous at these small sizes in either region. Larvae from the Balearic Sea and the Gulf of Mexico show similar growth trajectories at younger stages, but Gulf of Mexico larvae grow much faster after a size of about 6 mm SL. This may be because of the warmer temperatures in the Gulf of Mexico compared to the Balearic Sea or the differences in diet between larvae in the two regions.

STOCK STRUCTURE: Atlantic Bluefin Tuna are thought to comprise distinct western and eastern Atlantic stocks, which primarily spawn in the Gulf of Mexico and Mediterranean Sea, respectively. The eastern Atlantic stock is about an order of magnitude larger than the western Atlantic stock. Analyses of early life history stages and adults taken from the spawning grounds indicate that the two stocks are genetically distinct, and genetic heterogeneity has been demonstrated within the Mediterranean Sea. Individuals appear to demonstrate fidelity to either the Gulf of Mexico or Mediterranean Sea spawning grounds, but outside of the spawning grounds there can be considerable mixing of the two stocks throughout much of the North Atlantic. Electronic tagging and analyses of otolith stable isotopes and organochlorine pollutants suggest that there is substantial variation in movements and stock composition both spatially and temporally. ICCAT recognizes the western and eastern stocks of Atlantic bluefin tuna, and while it is understood that there is mixing of the stocks, for management purposes they are separated at 45° W meridian.

FISHERIES INTEREST: Fisheries for Atlantic Bluefin Tuna date back literally thousands of years, as fish were caught on hand lines and with beach seines in the Mediterranean Sea during the Phoenician and Roman eras. The species is currently taken with several gear types, including purse seines, longlines, traps, and a variety of hand gears. Atlantic Bluefin Tuna is also an important game fish throughout its range. It is a highly valued species, with much of the catch entering the sashimi market. Reported landings of Atlantic Bluefin Tuna were relatively stable from 1950 to 1993, fluctuating between 15,000 to 39,000 mt per year. Reported and unreported landings increased with a buildup of fisheries within the Mediterranean Sea in the 1990s and early 2000s to a peak of 62,638 mt in 2007. Implementation of management measures reduced catches to less than 15,000 mt from 2010 to 2014. Since that time there has been a gradual increase in landings, with a total of 21,997 mt reported for 2016.

Western Atlantic: Landings of Atlantic Bluefin Tuna in the western Atlantic rapidly increased from about 1,000 mt in 1960 to a peak of 18,608 mt in 1964, with the expansion of the Japanese longline fleet that was targeting large fish off Brazil and the development of the US purse seine fishery targeting juvenile fish off the US Atlantic coast. Catches quickly dropped off to less than 3,000 mt by 1969 with the loss of the concentration of large fish off Brazil. Landings gradually increased through the 1970s and early 1980s to about 6,000 mt due to the expansion of longline effort in the Gulf of Mexico and central North Atlantic and an increase in purse seine catches of adult fish for the growing sashimi market. Management measures implemented in 1981 reduced landings by about 50%, and since that time, they have mostly fluctuated between 1,800 and 3,000 mt, averaging about 1,900 mt over the past five years. Currently, the majority of the catch is taken by hand gears (rod and reel, handlines), with longlines accounting for a little more than a third of the landings.

Eastern Atlantic and Mediterranean: Fisheries for Atlantic Bluefin Tuna have existed in the Mediterranean for thousands of years. Originally taken in beach seines or with hand gear, trap fisheries developed in the western Mediterranean by the 1600s, with annual catches ranging from 7,000–30,000 mt, but falling off in the late 1970s. A handline fishery for juveniles developed in the Bay of Biscay during the mid-1800s and continues today with bait boats. Norway and other countries developed a purse seine fishery for adults in northern waters during the 1950s. Purse seine catches reached 16,000 mt in the mid-1950s, but the fishery rapidly collapsed in 1963. The expansion of sashimi markets in the 1980s and 1990s increased demand and fishing effort for Atlantic Bluefin Tuna. The development of tuna caging operations, which allowed for holding and fattening of purse seine–caught fish in the Mediterranean Sea, increased the profitability of the fishery, and catches by this gear type alone rose quickly to 25,000 mt in the late 1990s. Overall catches in the eastern Atlantic and Mediterranean increased from about 20,000 mt in the late 1980s to more than 50,000 mt by the mid-1990s. There was significant underreporting of catches during this time. It is estimated that annual landings exceeded 50,000 mt from the mid-1990s through 2007 and likely exceeded 60,000 mt in 2007. Management measures implemented in 2008 decreased fishing effort, resulting in catches of about 10,000 mt from 2010–12. Since that time catches have been increasing, with reported landings of almost 22,000 mt in 2016. Currently, purse seines take 59% of the catch, traps 20%, pelagic longlines 15%, and assorted gear types the remaining 6%. Japan is a major importer of bluefin tuna species, and it is estimated that approximately 80% of the Japanese market is supplied by the countries around the Mediterranean Sea.

THREATS: The eastern stock of Atlantic Bluefin Tuna is fished by many nations, and achieving consensus on management measures, especially allocation issues, is extremely difficult, which greatly increases management response time. Data deficiencies remain, and the lack of information for CPUE time series compromises assessments. More information is needed from the tuna caging operations, including better data on size at capture as well as mortalities associated with transport and caging. The lack of understanding of stock-recruit relationships and the effects of regime shifts on recruitment levels has resulted in the loss of biomass-based reference points in both the east and west.

The warm ambient temperatures on their breeding ground in the Gulf of Mexico potentially present a distinct threat to these large, endothermic fish, and this potential will increase with increasing water temperatures.

Substantial habitat loss for both adult and larval Atlantic Bluefin Tuna is predicted for the main spawning ground in the northern Gulf of Mexico as temperatures warm. Oil spills also pose a threat to spawning grounds in the Gulf of Mexico. Oil slicks have been shown to have toxic effects on the buoyant embryos.

It became rare, relative to historical levels, because of massive overfishing. The Center for Biological Diversity petitioned the US Government to list the Atlantic Bluefin Tuna under the US Endangered Species Act, and the US government agreed to conduct a status review for this species but decided not to list it as an endangered species. It was listed as globally Endangered on the IUCN Red List in 2011, as was the Mediterranean population, but due to the large reductions in catches starting in 2008, the status of the eastern Atlantic stock was upgraded to Near Threatened in the European Red List.

CONSERVATION: The Atlantic Bluefin Tuna is listed as a highly migratory species in Annex I of the 1982 Convention on the Law of the Sea.

Western Atlantic: Assessments of western Atlantic Bluefin Tuna have been problematic due to mixing, uncertainty about stock-recruit relationships, and the potential for regime shifts to have altered environmental conditions and recruitment levels. In previous assessments, ICCAT evaluated both a "high recruitment" and a "low recruitment" hypothesis, the results of which offered very different perspectives on stock status. In 2014, the estimated SSB in the high-recruitment model was 0.48 SSBMSY, while it was 2.2 SSBMSY in the low-recruitment model, with corresponding fishing mortalities of 0.88 and 0.36 FMSY, respectively. Realizing that such disparate results were of limited use to fishery managers, in 2017, ICCAT assessed western Atlantic Bluefin Tuna without estimating biomass-based reference points, and simply estimated fishing mortality relative to F0.1, a proxy for FMSY. The estimated fishing mortality was 0.59 F0.1, indicating that overfishing is not occurring.

ICCAT significantly reduced catches of Atlantic Bluefin Tuna in the western Atlantic in 1982 with the adoption of a TAC and a sharing arrangement specifying country-specific quotas. The TAC has fluctuated between 1,800 and 2,700 mt based on scientific advice following assessments every two or three years, and a formal rebuilding plan was adopted in 1998 that includes a 30 kg minimum size (with an allowance of small fish for the US recreational fishery), a ban on directed fishing in the Gulf of Mexico, and a host of other measures. There are numerous domestic management measures within the United States for the commercial and recreational fisheries.

Eastern Atlantic and Mediterranean: There has been considerable uncertainty associated with assessments of the eastern Atlantic and Mediterranean Sea stock of Atlantic Bluefin Tuna due to issues with the catch per unit effort data, misreporting of catch during the late 1990s and 2000s, a lack of understanding of stock-recruit relationships and potential recruitment levels. Overfishing occurred during the late 1990s and 2000s, resulting in a decrease in the SSB. Meaningful management measures implemented in 2008 reduced catches by more than half and have resulted in an increase in SSB. Prior to 2017, ICCAT assessments of the stock produced biomass-based and fishing mortality–based reference points. To account for uncertainties regarding recruitment, the 2014 assessment considered high-recruitment, medium-recruitment, and low-recruitment scenarios, resulting in SSB estimates of 0.67, 1.1, and 1.6 SSBMSY, respectively. In all three scenarios, overfishing was not occurring. Similar to the assessment for the western Atlantic stock, in 2017 the assessment of the eastern Atlantic and Mediterranean stock did not consider biomass-based reference points and estimated fishing mortality relative to F0.1 as a proxy for FMSY. The estimated fishing mortality was 0.34 F0.1, indicating that overfishing is not occurring.

Management of the eastern Atlantic and Mediterranean stock of Atlantic Bluefin Tuna has been especially challenging because of the large number of countries involved. ICCAT adopted measures in the late 1990s that greatly curtailed the catch of small fish in traditional fisheries within the Mediterranean Sea and banned the use of drift nets in 2004. A multifaceted rebuilding program was adopted in 2006, which included the establishment of a TAC, country-specific quotas, significant time / area closures for various fleets, and a host of monitoring, control, and surveillance measures. The rebuilding plan was amended in 2008 to include a significant reduction in the TAC and corresponding capacity limitations that eventually reduced catches from more than 60,000 mt to less than 15,000 mt in two years. The development of an electronic catch documentation system for Atlantic Bluefin Tuna has reduced illegal, unreported, and unregulated (IUU) fishing, as it provides a means for countries to exclude IUU fish from their markets.

REFERENCES: Abdul Malak et al. 2011; Alvarado Bremer et al. 2005; Arrizabalaga et al. 2015; Baglin 1982; Bakun 2012; Block et al. 2001, 2005; Boustany et al. 2008; Brette et al. 2014; Brill et al. 2002; Butler et al. 2010, 2015; Carlsson et al. 2004, 2007; Catalán et al. 2011; Collette 1999; Collette and Nauen 1983; Corriero et al. 2003; Cuvier and Valenciennes 1832; Diaz 2011; Dickhut et al. 2009; Doumenge 1998; Dragovich 1969, 1970; Druon et al. 2011, 2016; Fromentin and Powers 2005; Galuardi and Lutcavage 2012; Gibbs and Collette 1967; Godsil and Holmberg 1950; Gordoa and Carreras 2014; Heffernan 2014; IGFA 2018; Karakulak et al. 2009; Laiz-Carrión et al. 2015; Linnaeus 1758; Logan et al. 2011; MacKenzie et al. 2009, 2014; MacKenzie and Mariani 2012; Malca et al. 2017; Malm 1877; Matthews et al. 1977; Muhling et al. 2012, 2015, 2017; Neilson and Campana 2008; Nieto et al. 2015; Olafsdottir et al. 2016; Oray and Karakulak 2005; Pleizier et al. 2012; Porch 2005; Ravier and Fromentin 2002; Riccioni et al. 2010; Richards 2006; Richardson et al. 2016; Risso 1827; Rooker et al. 2007, 2008; Safina and Klinger 2008; Santamaria et al. 2009; Sarà and Sarà 2007; Schaefer 2001; Sinopoli et al. 2004; Siskey et al. 2016; Sissenwine et al. 1998; Storer 1853; Takeuchi et al. 2009; Teo et al. 2007a, 2007b; Tilley et al. 2016; Tseng et al. 2011; Yúfera et al. 2014; Varela et al. 2014; Wilson et al. 2016; Worm and Tittensor 2011.

Longtail Tuna

Thunnus tonggol (Bleeker, 1851)

COMMON NAMES: English - Longtail Tuna
French - Thon Mignon
Spanish - Atún Tongol

ETYMOLOGY: Named *tonggol* by the prolific Dutch ichthyologist and army surgeon Pieter Bleeker (1851:356–357), based on the Malay common name Ikan Tonggol.

SYNONYMS: *Thynnus tonggol* Bleeker, 1851; *Thunnus rarus* Kishinouye, 1915; *Kishinoella rara* (Kishinouye, 1915); *Neothunnus tonggol* (Bleeker, 1851); *Kishinoella tonggol* (Bleeker, 1851); *Thunnus nicolsoni* Whitley, 1936; *Neothunnus rarus* (Kishinouye, 1915); *Thunnus tonggol* (Bleeker, 1851)

TAXONOMIC NOTE: The Longtail Tuna was confused with the Southern Bluefin Tuna in Australia for many years until the Australian ichthyologist D. L. Serventy clearly distinguished the two species.

FIELD MARKS:

1 Longtail Tuna have fewer gill rakers than any other species of *Thunnus* in the Indo-West Pacific, only 19–27 on the first gill arch.

2 Juveniles 17–30 cm can be distinguished from juvenile Pacific Bluefin Tuna by the lower gill raker count (20–25 versus 35–37) and a shorter pectoral fin, 48–56% of head length versus 67–95%. The low gill raker count plus the longer right lobe of the liver and lack of liver striations are useful in distinguishing Longtail Tuna from Pacific Bluefin even at relatively small sizes, 20–30 cm FL.

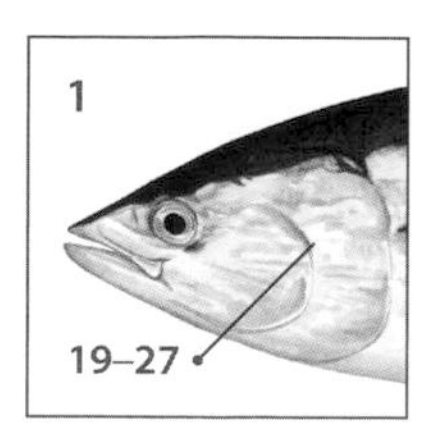

UNIQUE FEATURE: The Longtail Tuna is the only member of its genus that inhabits turbid inshore waters.

DIAGNOSTIC FEATURES: Longtail Tuna is the second smallest species of tuna. The body is deepest near the middle of the first dorsal-fin base. **Gill rakers:** It has fewer gill rakers than any other species of *Thunnus* in the Indo-West Pacific, 19–27. **Fins:** The second dorsal fin is higher than the first dorsal fin. The pectoral fin is short to moderately long, 22–31% of fork length in smaller individuals (under 60 cm FL) and 16–22% in larger individuals. **Liver:** The ventral surface of the liver is not striated and the right lobe is significantly longer than the central and left lobes. **Swim bladder:** A swim bladder is either rudimentary or absent. **Vertebrae:** Vertebrae number 18 precaudal plus 21 caudal, for a total of 39. Descriptions of the osteology and soft anatomy are contained in the revision of the genus by Gibbs and Collette (1967). **Color:** The back is metallic blue-black, and the lower sides and belly are silvery white with colorless elongate oval spots arranged in horizontally oriented rows. The dorsal, pectoral, and pelvic fins are blackish. The tips of the second dorsal and anal fins are washed in yellow. The anal fin is silvery. The dorsal and anal finlets are yellowish with grayish margins. The caudal fin is blackish with yellow-green streaks.

GEOGRAPHIC RANGE: Longtail Tuna inhabit coastal waters in the Indo-West Pacific region between 47° N and 33° S from the Red Sea and East Africa west to Papua New Guinea, north to the Philippine Islands and Japan, and south to Australia. Described as *Thunnus rarus* by Kishinouye in 1915, Longtail Tuna were not reported from Japan again until two large catches were made in sardine set nets in Wakasa Bay in 1968; now there is a regular fishery for it.

SIZE: The maximum size of Longtail Tuna is about 130 cm FL. The IGFA all-tackle game fish weight record is of a 79 lb, 2 oz (35.9 kg) fish caught off Montague Island, New South Wales, Australia, in April 1982, but the longest game fish is a 110 cm FL fish caught off Moreton Island, Queensland, in June 2012. In Australia, longevity is estimated to be about 10 years, but they may live as long as 18 years in the central Indo-Pacific.

HABITAT AND ECOLOGY: Unlike other species of the genus, Longtail Tuna occupy neritic areas close to shore. While other species of tunas avoid turbid waters, diet analysis indicates that Longtail Tuna may be more tolerant of turbid waters, allowing them to remain within the coastal area during the wet season in Australia to exploit locally abundant prey. In the Sea of Japan catch rates are highest at sea surface temperatures of 24°C.

FOOD: Like other tunas, Longtail Tuna feed on a variety of fishes, cephalopods, and crustaceans. In waters off the coast of India, their food is 82% fishes, 13.4% mollusks, and 4.6% crustaceans. Important fishes in their diet include sardines, anchovies, scads (*Decapterus* and *Selar*), cutlassfishes (*Trichiurus*), flyingfishes, halfbeaks, and frigate tunas (*Auxis*). In the Sea of Japan, important diet items included Jack Mackerel, Japanese Anchovy, and Round Herring. Small schooling fishes like sardines and anchovies constituted the major part of the diet in terms of both biomass and frequency of occurrence, 90–93% by weight, in two regions of Australian waters. Demersal fishes, squids, and crustaceans were also present in smaller quantities, for a total of 101 different prey animals. Comparison of stomach contents from day-collected and night-collected Longtail Tuna indicates that most feeding takes place in the daytime.

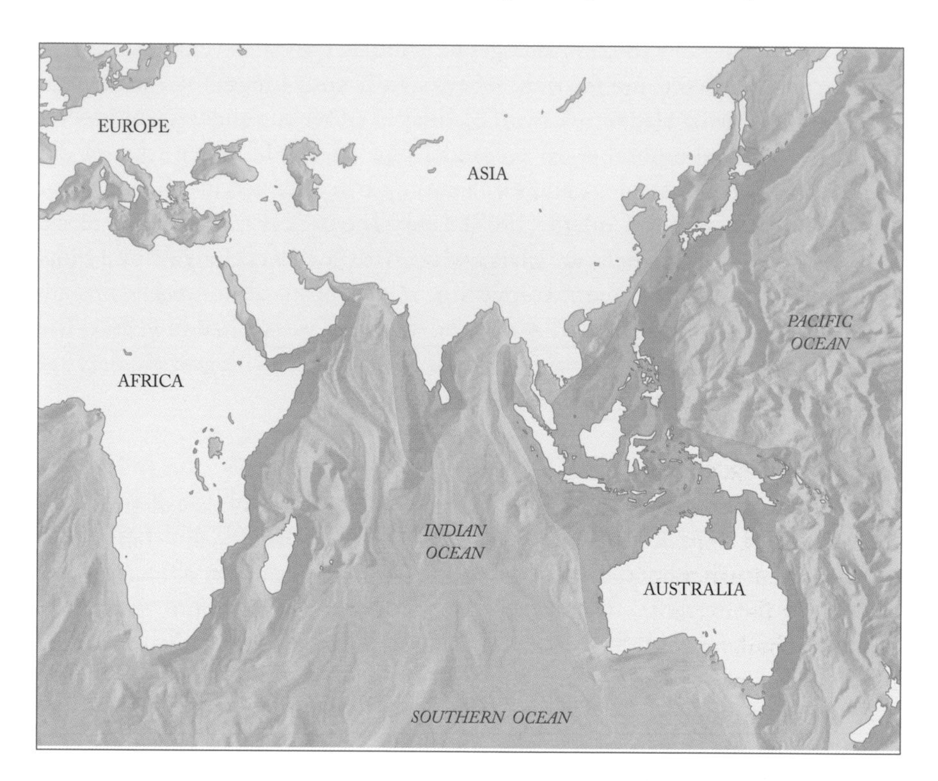

Right: Longtail Tuna range map.

REPRODUCTION: In the Gulf of Thailand, 50% of female Longtail Tuna were mature at 39.6 cm FL, and age at first maturity was estimated to be 2 years. Fecundity of fish ranging in size from 43.8–49.1 cm varies from 1.2–1.9 million eggs. Spawning is reported to be confined to coastal waters, based on the occurrence of their larvae collected at surface water temperatures of 28°C. Longtail Tuna spawn year-round in waters off the coast of India, with a major peak in October–November. It appears that there are two distinct spawning seasons for this species off the west coast of Thailand: a major spawning period during the northeast monsoon from January–April and a minor spawning period during the southwest monsoon in August–September. Spawning is also apparently seasonal for this species off Papua New Guinea and off New South Wales, occurring during the austral summer. It is possible that there is an additional spawning ground between Japan and South Korea based on captures of a near-ripe female and juveniles that appear to be too small to have drifted in from known spawning grounds. Mature females in a ripening stage were found off Taiwan but none were found in a spawning stage.

EARLY LIFE HISTORY: Five larvae 3.75 to 5.76 mm SL from Western Australia and Thailand were described and illustrated by Nishikawa and Ueyanagi. Pigmentation on the tip of the lower jaw first appeared in larvae larger than 8.75 mm. Juveniles 125–166 mm FL were taken in the Sea of Japan at 35°30′ N, 131°00′ E at a surface temperature of 24°C in September. The smallest of these juveniles was considered to be within a month of hatching. The number of daily growth increments of otoliths of juveniles 13–49 cm FL from the South China Sea increased from 33 to 434 with increasing fork length.

STOCK STRUCTURE: Gill raker counts indicate differences between populations in the western Indian Ocean, with a modal number of 26, and those in the eastern Indian Ocean and western Pacific, with a mode of 23. No genetic differences were found between two collections of Longtail Tuna from northwest India, and there was no differentiation from analysis of the mitochondrial DNA D-loop region, nor among collections from Indonesia, Vietnam, and the Philippines around the South China Sea. However, genetic differences were reported between collections from the Indian Ocean and South China Sea, but not within basins of the Indo-Pacific. It is highly likely that there are at least two discrete populations that should be managed separately.

FISHERIES INTEREST: Longtail Tuna are caught mainly by gill net and to a lesser extent by artisanal purse seines and bait boats. As a result of their coastal distribution, Longtail Tuna are heavily exploited by small-scale commercial and artisanal fisheries in at least 17 countries in the Indo-Pacific. There are two major fishing grounds for Longtail Tuna, one off the South China Sea coast of Thailand and Malaysia and the other off countries bordering the North Arabian Sea. They are also taken off Japan in the Sea of Japan, primarily during the summer.

Global catch of Longtail Tuna increased dramatically from about 30,000 mt in 1980 to more than 150,000 mt in 1990, peaking at 291,000 mt in 2007. Since that time landings have fluctuated but remain in excess of 200,000 mt. Within the Indian Ocean, landings increased steadily from the mid-1950s, reaching 20,000 mt in the mid-1970s, 50,000 mt in the 1980s, 90,000 mt in the late 1990s, peaking at 175,000 mt in 2012. Major fishing nations for Longtail Tuna in the Indian Ocean are Iran (43% of landings) and Indonesia (17%). Also, the catch of Longtail Tuna is probably underestimated because some is landed as Yellowfin Tuna; there is a least one verified report of four samples of fish labelled as "Bluefin Tuna fillet" that were identified as Longtail Tuna using DNA barcodes.

Although of little commercial importance in Australia, Longtail Tuna are an important sport fish there, highly regarded because of their relatively large size and fighting ability and because they can be targeted from small vessels near shore.

THREATS: Longtail Tuna grow more slowly and live longer than other tuna species of similar size. Coupled with their restricted neritic distribution, Longtail Tuna may be vulnerable to overexploitation by artisanal fisheries. There are issues with data reporting, and landings of Longtail Tuna may be misreported as Yellowfin Tuna or combined with seerfishes (*Scomberomorus*). Caution needs to be exercised in managing the species until more reliable biological and catch data are collected to assess the status of the populations. Longtail Tuna are listed as Data Deficient on the IUCN Red List, but this rating should be reevaluated using recent stock assessments from the Indian Ocean and Australia. More information is needed on the status of this species' population, including better catch data and effort information and a comprehensive stock assessment.

CONSERVATION: The dramatic increase in the fisheries effort for Longtail Tuna in the Indian Ocean has reduced spawning stock biomass. Standardized CPUE (catch per unit effort) in the northwest Indian Ocean suggested that the Longtail Tuna stock was entering an overfishing status. Surplus production models indicate that the Indian Ocean stock is being exploited at a rate that has exceeded FMSY in recent years. An assessment in 2013 indicated that the biomass was close to BMSY and that increases in fishing mortality in recent years had exceeded FMSY. A subsequent assessment in 2016 indicated that the stock is currently overfished and that overfishing is occurring. In order to recover the Indian Ocean stock to levels above the MSY reference points, the Scientific Committee of the Indian Ocean Tuna Commission has recommended that catches be reduced by 30% of current levels, which corresponds to catches slightly below MSY, in order to recover the status of the stock.

An assessment of Longtail Tuna in Australian waters indicated that overfishing was not occurring and that there was some scope for increased fishing mortality. As a reflection of their importance to recreational fisheries in Australia, Longtail Tuna were declared a "recreational only" species by the Commonwealth government in December 2006. However, an annual catch limit of 70 mt is permitted for Australian commercial fisheries.

REFERENCES: Abdussamad et al. 2012; Al-Kiyumi et al. 2014; Bleeker 1851; Boonragsa 1987; Collette 2010; Collette and Nauen 1983; Collette et al. 2011; Fukusho and Fujita 1972; Gibbs and Collette1967; Griffiths 2007; Griffiths et al. 2007, 2009, 2010; IGFA 2018; IOTC 2015a, 2015b; Itoh et al. 1999; Kishinouye 1915; Kobayashi 2004, 2005; Kumar and Kocour 2015; Kunal et al. 2014; Martin and Sharma 2015; Mohri et al. 2009; Nakamura 1969; Nishida and Iwasaki 2015; Nishikawa and Ueyanagi 1991, 1992; Phuoc 2009; Serventy 1941, 1942, 1956; Willette et al. 2016; Wilson 1981; Yesaki 1994.

Family XIPHIIDAE
Swordfish

Family ISTIOPHORIDAE
Billfishes

Swordfish

Xiphias gladius Linnaeus, 1758

COMMON NAMES: English – Swordfish
French – Espadon
Spanish – Pez Espada
Japanese – Meka, Mekajiki

ETYMOLOGY: The Swedish naturalist Carl Linnaeus (1758:148), who devised the binomial nomenclature system still in use, selected the generic name *Xiphias*, meaning "swordfish" in Greek, and the species name *gladius*, taken directly from the Latin gladius, meaning "sword," so as Richard Ellis (2013) has pointed out, the translation of the scientific name is the redundant "swordfish with a sword."

SYNONYMS: *Xiphias gladius* Linnaeus, 1758; *Xiphias imperator* Bloch and Schneider, 1801; *Tetrapterus imperator* (Bloch and Schneider, 1801); *Xiphias rondeletti* Leach, 1818; *Phaethonichthys tuberculatus* Nichols, 1923; *Xiphias estara* Phillips, 1932; *Xiphias thermaicus* Serbetis, 1951; *Xiphias gladius estara* (Phillips, 1932)

TAXONOMIC NOTE: The generic name of the Swordfish, *Xiphias*, should not be confused with the similar-sounding name *Ziphius*, for *Ziphius cavirostris*, the Cuvier's or Goose-beaked Whale, a member of the family Ziphiidae.

FIELD MARKS:
1 The bill is extremely long but flat and sword-like in cross section, unlike the rounded bill of istiophorid billfishes.
2 There are no teeth in the jaws.
3 Pelvic fins are absent.
4 There is a single large keel on the caudal peduncle.

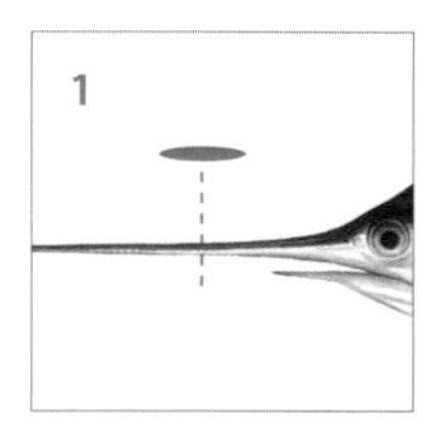

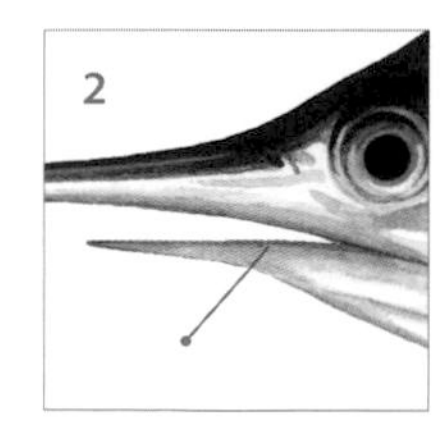

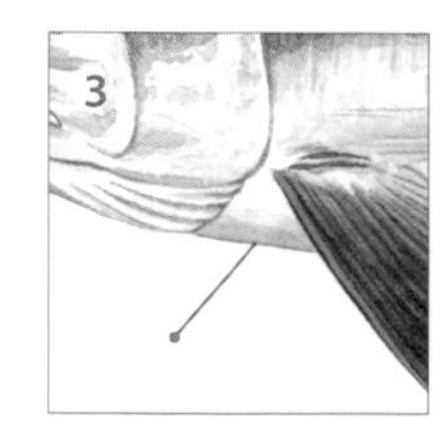

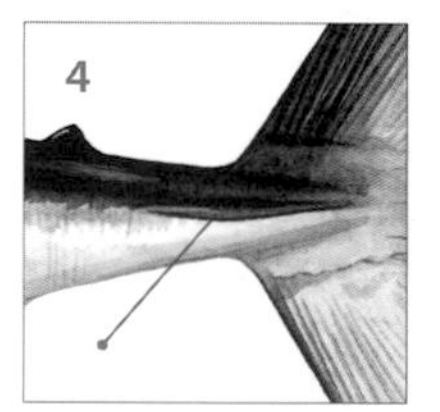

UNIQUE FEATURE: The Swordfish is the only fish with a flattened bill ("sword") in front of its head, unlike the rounded bill of istiophorid billfishes.

DIAGNOSTIC FEATURES: The body is elongate and cylindrical, not compressed laterally. The upper jaw is prolonged into a long bill, flat-oval in cross section. The eyes are large. The nape is conspicuously elevated. The left and right branchiostegal membranes are united basally but free from the isthmus. Gill rakers are absent. Fine file-like teeth are present in specimens up to about 1 m in length but disappear with growth. **Fins:** There are two widely separated dorsal fins, the first much larger than the second, with 34–49 spines. The anterior lobe of the first dorsal fin is higher than the body depth. The second dorsal fin has 4–6 rays, and its position is slightly posterior to that of the second anal fin. There are two anal fins, the first much larger than the second and with 13 or 14 rays, the second with 3 or 4 rays and similar in size and shape to the second dorsal fin. The pectoral fins are falcate, situated low on the body, and with 16–18 rays. Pelvic fins and pelvic girdle are absent. The caudal fin is large and lunate. **Caudal peduncle:** The caudal peduncle has a single strong keel on each side and a deep notch on both the dorsal and ventral surfaces. **Anus:** The anus is situated near the origin of the first anal fin. **Lateral line:** The lateral line is absent in adults but recognizable in specimens up to about a meter in body length. **Scales:** Adults lack scales, but juveniles have small scales with small spines until about a meter in body length, when they become overgrown by skin. **Vertebrae:** Vertebrae number 26 (15 or 16 precaudal and 10 or 11 caudal). **Swim bladder:** The swim bladder is large and has a single chamber, unlike the many small, bubble-shaped chambers present in the swim bladders of istiophorid billfishes. **Color:** Back and sides of the body iridescent bluish black to blackish brown, gradually fading to light brown to silvery on the ventral surface. The membranes of the first dorsal fin are dark blackish brown. The other fins are brownish or blackish brown.

Below: Juvenile Swordfish, approx. 3 ft FL.

GEOGRAPHIC RANGE: Swordfish occur worldwide in tropical, temperate, and sometimes cold waters of the Atlantic, Indian, and Pacific oceans, including the Mediterranean Sea, the Sea of Marmara, the Black Sea, and the Sea of Azov.

SIZE: The IGFA all-tackle game fish record is of a 1,182 lb (536.15 kg) fish caught off Iquique, Chile, in May 1953.

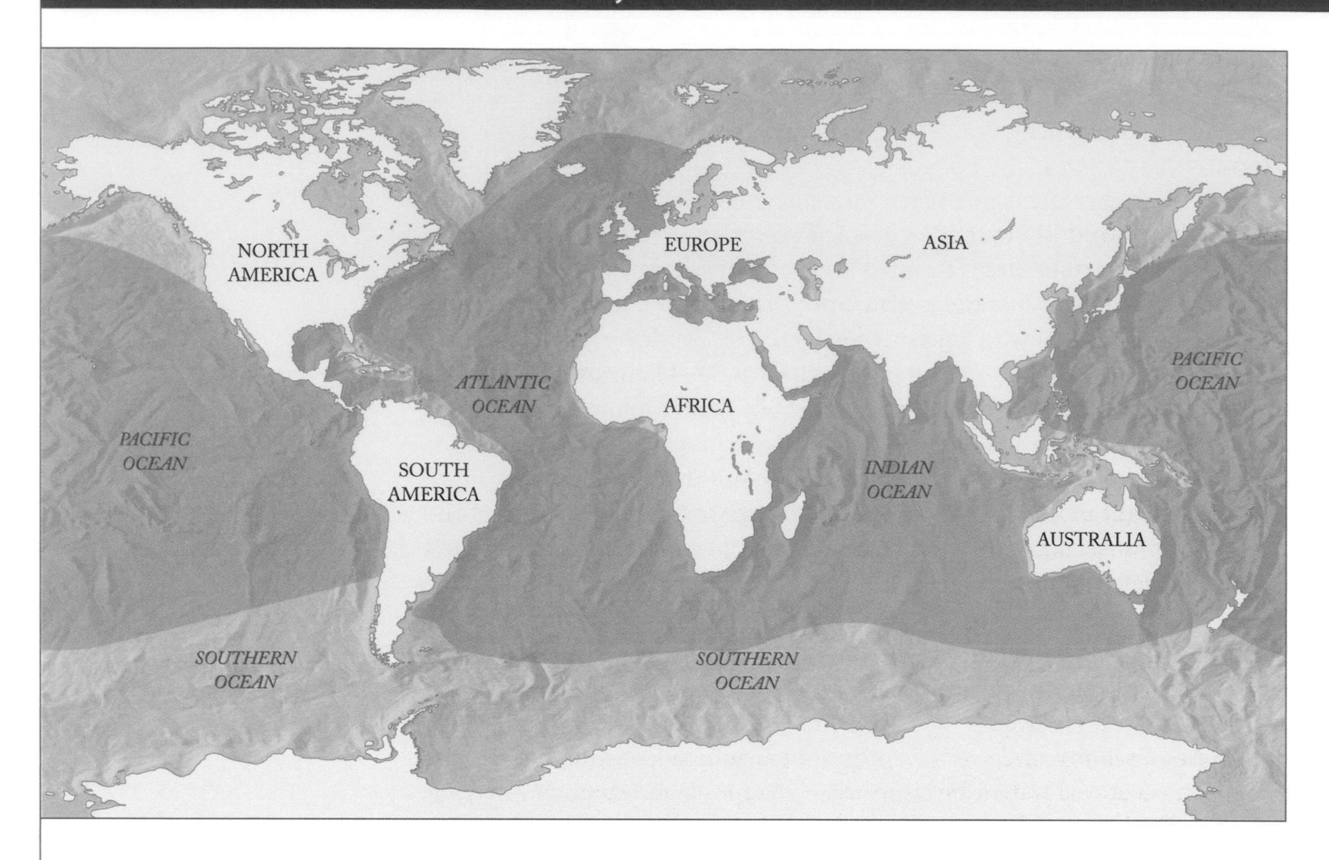

Above: Swordfish range map.

HABITAT AND ECOLOGY: Swordfish are oceanic, but are sometimes found in coastal waters, generally above the thermocline, preferring temperatures of 18–22°C. They migrate toward temperate or cold waters for feeding in the summer and back to warm waters in the fall for spawning and overwintering. Swordfish typically forage in deep waters during the day and stay in the mixed layer to feed at night, and there is a tendency to remain closer to the surface with lower lunar light levels. Based on records of forage organisms taken by Swordfish, depth distribution in the western North Pacific normally ranges from the surface to a depth of about 550 m, but there are depth records down to 2,878 m. In the western Atlantic, tagged individuals occupied surface waters less than 100 m during the night and depths greater than 400 m during daylight hours, with a maximum depth record of 1,448 m. The maximum depth of Swordfish tagged in the eastern South Pacific was 1,136 m, and five of six Swordfish dove deeper than 900 m. Temperature and depth ranges for tagged fish in the central North Pacific were 3.2–28.8°C and 0–1,227 m; five fish dove as deep as 1,200 m.

MOVEMENTS: Conventional tagging studies of Swordfish showed that individuals can undertake long-distance movements, but recent pop-up satellite tagging (PSAT) studies have demonstrated that some movements are cyclical in nature. For example, Swordfish tagged in the western North Atlantic off Canada move southward to the Caribbean during the fall and return to

Canada in the late spring, often with very high site fidelity. To date, there has been little evidence for major east-west movements of these fish. Evidence for cyclical movements has also been reported for Swordfish in other ocean basins.

FOOD: Adult Swordfish are opportunistic feeders, known to forage from the surface to the bottom. They typically forage in deep water during the day and stay in the mixed layer at night. Swordfish can use their sword to kill prey. Off the Bay of Biscay, fishes were found in 95% of the stomachs of 86 Swordfish, representing 40.5% by mass of the diet, while cephalopods were found in 76% of the stomachs, representing 59% by reconstituted mass. Two species of lanternfishes accounted for fairly high proportions of the diet, as did three species of squids. There is little dietary overlap with Mediterranean Spearfish caught in the same area. Unidentified fishes formed nearly half the food of Swordfish taken off Brazil. As top-level predators, large individuals may accumulate large percentages of mercury in the flesh.

REPRODUCTION: Swordfish are reported to spawn in the upper layers of the water column, from the surface to a depth of 75 m. The distribution of larval swordfish indicates that spawning occurs in waters with a temperature of 24°C or more and salinity of 33.8 to 37.4 ppt. Spawning appears to occur in all seasons in equatorial waters, but is restricted to spring and summer at higher latitudes. Pairing of solitary males and females is thought to occur when spawning. Mature fish showed a well-defined pattern of spawning behavior in the Mediterranean, with pairs exhibiting coordinated circular swimming patterns at sea surface temperatures of 24°C or above. Swordfish first spawn at 5–6 years of age in the Pacific. In the Atlantic the age at 50% maturity for females is 5 years, 3.5 in the Mediterranean. Swordfish reach sexual maturity at about 179 cm LJFL in the north Atlantic and at about 156 cm LJFL in the south Atlantic. Spawning in southern Brazil occurs from November to February between 20° S and 28° S and 40° W to 47° W. Estimates of egg numbers vary considerably, from 1 million to 16 million in a 168 kg female and 29 million in a 272 kg female. Pelagic eggs measure 1.6–1.8 mm, and the newly hatched larvae is 4 mm long.

AGE AND GROWTH: Determination of swordfish age is difficult since the otoliths are very small and scales are buried under the epidermis in adults. Age estimates based on external features of the sagittal otolith suggest a maximum lifespan of at least 14 years for males and 32 years for females. Age estimates from sectioned sagittae ranged from 50 days to 15 years, with a maximum of 9 years for males and 15 years for females. Females grow faster than males after an estimated age of 2 and reach a much larger size. Rings on the second anal-fin element suggest slower growth than the estimates from otoliths. The generation length is estimated to be 5.6 years.

EARLY LIFE HISTORY: Swordfish larvae have two growth phases. Larvae complete yolk and oil globule absorption 5 to 6 days after hatching. In the Mediterranean, larvae shorter than 8 mm feed exclusively on copepods, primarily of the genus *Corycaeus*. Larvae 9–11 mm eat copepods and chaetognaths. Growth accelerates at 13 days, following an abrupt change in diet and in jaw and gut structure. The sword is well developed by a length of 10 mm. Larvae longer than 11 mm feed on fish larvae. Scales first appear on the abdomen of larvae at about 6 mm SL and spread along the body until about 192 mm, when they start to be overgrown by a thickened dermis. Nine larval Swordfish 6.1–192.1 mm from the western Atlantic were illustrated by Arata.

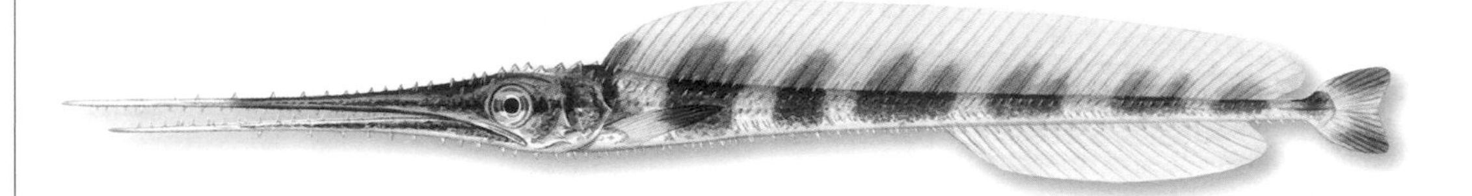

Above: Larval Swordfish, approx. 69 mm SL.

STOCK STRUCTURE: Although the Swordfish is considered to be highly migratory, genetic evidence indicates at least four major stocks: Mediterranean Sea, North Atlantic, and South Atlantic, with additional structure within the Indo-Pacific. More research is required to delineate the boundaries between the three Atlantic stocks, but the level of genetic differentiation between the Mediterranean Sea and the North Atlantic stocks clearly indicates that there is little genetic exchange occurring between the two. In addition, slight genetic differentiation has been reported between Swordfish from the eastern and western Mediterranean. Genetic heterogeneity has been reported among Swordfish from various locations in the Pacific, and evidence is consistent with four separate Pacific substocks: northwest, northeast, southwest, and southeast. Samples drawn from Madagascar and the Bay of Bengal differed genetically from samples from other parts of the Indian Ocean and the western Pacific.

FISHERIES INTEREST: Swordfish are a highly important food and game fish species. They are caught by pelagic longline, harpoon, drift gill net, set net, and other fishing gears in commercial fisheries. The largest catches of Swordfish occur on pelagic longline gear in directed fisheries or as incidental catch in fisheries targeting tunas. They are marketed fresh or frozen. Swordfish are also taken on rod-and-reel gear in sports fisheries in all oceans. Global Swordfish production exceeded 125,000 mt in 2014, with approximately 61,600 mt from the Pacific, 34,700 mt from the Indian Ocean, and 30,400 mt from the Atlantic Ocean.

Pacific Ocean: Swordfish are managed as four distinct stocks in the Pacific Ocean: eastern Pacific, western and central North Pacific, western South Pacific, and eastern South Pacific.

Eastern Pacific: In the eastern Pacific Swordfish landings were relatively low through the 1990s, varying from 2,000 to 7,000 mt. The fishery increased during the 2000s, peaking at 9,900 mt in 2012. The major Swordfish fishing nations in the eastern Pacific are Japan, China, Taiwan, and Spain.

Western and Central North Pacific: Catches of Swordfish in the western and central North Pacific Ocean were dominated by Japan in the early years and approached 22,000 mt in 1960. Landings quickly fell to half that level and were fairly stable through the 1970s. Fishing effort increased in the 1980s and early 1990s, peaking at 19,000 mt in 1993. Landings subsequently decreased and have fluctuated between 10,000 and 12,000 mt since. Japan, Taiwan, Korea, and the United States are the major fishing nations on this stock.

Western South Pacific: Incidental catches of Swordfish from pelagic longline fisheries targeting tunas comprised the majority of landings from this stock through the mid-1980s, with landings levels below 5,000 mt. As Australia and New Zealand developed directed Swordfish fisheries and effort in the pelagic longline fisheries for tunas increased, landings quickly rose to over 15,000 mt by the late 1990s. Current catches fluctuate between 12,000 and 15,000 mt, and the major fishing nations include Australia, New Zealand, Japan, Taiwan, Korea, and Spain.

Eastern South Pacific: This was a minor fishery until the mid-1980s, with landings less than 2,500 mt. The fishery quickly increased in the late 1980s, reaching 12,500 mt in 1991. Landings decreased over the next five years but then increased, peaking at 15,000 mt in 2003. Current catches remain near that level. The major fishing nations in the eastern South Pacific Swordfish fishery are Chile, Spain, and Japan.

Indian Ocean: The Swordfish fishery in the Indian Ocean was relatively small through the mid-1980s, with landings of less than 5,000 mt. The fishery increased to 10,000 mt by 1990 and exceeded 25,000 mt in 1995. The fishery peaked at 39,000 mt in 2005 and is currently about 30,000 mt.

Atlantic Ocean: Swordfish are managed as three stocks in the Atlantic Ocean: North Atlantic Swordfish, South Atlantic Swordfish, and Mediterranean Swordfish.

North Atlantic: The Swordfish fishery in the North Atlantic was originally a harpoon fishery, but pelagic longline gear became the dominant gear type by the 1960s. Landings increased to 9,000 mt in 1970, but dropped shortly thereafter due to concerns about mercury levels. There was a rapid increase in the fishery in the early 1980s, with landings increasing to over 20,000 mt in 1987, with Spain, Canada, Japan, and the United States being the major harvesters. The stock was overfished in the 1990s, but rebuilding efforts through the late 1990s and early 2000s were successful, and the current total allowable catch is 13,700 mt.

South Atlantic: Prior to 1980, landings of Swordfish from the South Atlantic were below 5,000 mt. The fishery steadily increased from 1980 to 1985, when landings peaked at more than 21,000 mt. Management measures reduced catches, allowing the stock to rebuild, and the current total allowable catch is 15,000 mt, although landings are considerably below this value. Spain, Brazil, Japan, Namibia, and São Tome and Principe are the major harvesters.

Mediterranean: Swordfish catches were below 3,000 mt until the mid-1960s and slowly increased to 7,000 mt by 1983. There was a very rapid buildup over the next few years and landings peaked at just over 20,000 mt in 1988. Landings fluctuated between 12,000 and 15,000 mt until 2011, when management actions such as enforcement of the drift net ban reduced effort, resulting in catches below 10,000 mt. While many countries around the Mediterranean report Swordfish landings, Italy and Spain are the major harvesters.

THREATS: Worldwide, swordfish abundance is estimated to be at 56% of 1950 levels. The species' life history characteristic of relatively late maturity makes it vulnerable to overexploitation. The majority of Swordfish are taken on pelagic longline gear that effectively takes both juvenile and adult Swordfish. The high hooking mortality of Swordfish diminishes the effectiveness of minimum size measures intended to reduce juvenile mortality. Although drift nets have been banned in many areas, they continue to be used and result in high juvenile mortality. Exploitation levels differ from region to region, but the species is globally listed as Least Concern on the IUCN Red List.

CONSERVATION: Swordfish are highly migratory and are listed in Annex I of the 1982 Convention on the Law of the Sea.

Eastern Pacific: This stock is not overfished, but in the late 1990s/early 2000s and in 2012, overfishing may have occurred. The current biomass remains well above the level to support maximum sustainable yield.

Western and Central North Pacific: Swordfish in the Western and Central Pacific are not overfished, and it is likely that overfishing is not occurring. This stock is considered as Least Concern on the IUCN Red List.

Western South Pacific: This stock showed a decline of 30% spawning biomass over 18 years (3 generation lengths), but spawning biomass is estimated to be well above that necessary for maximum sustainable yield. Catch levels have increased in recent years and based on some models, overfishing may be occurring. This stock is currently considered as Least Concern on the IUCN Red List. This stock should be carefully monitored, especially as increased catch levels may not be sustainable with increased effort.

Eastern South Pacific: This stock is not overfished, and overfishing is not occurring. In Chile there is a size limit for Swordfish, and total effort has declined as fishing moved from drift nets to longlines.

Indian Ocean: As of the last assessment in 2013, Swordfish in the Indian Ocean are not overfished, and overfishing is not occurring. The standardized CPUE in the Indian Ocean declined by more than 50% over the 20 years between 1986 and 2006, but management measures focused on controlling effort in the fishery and fleet redistribution have resulted in a sustainable fishery.

North Atlantic: This stock was overfished in the 1990s, with a biomass of 0.58 BMSY at the time of the 1996 assessment. Successive years of good recruitment, plus a combination of international management efforts, including individual country quotas, effort reductions, and minimum sizes, as well as domestic measures implementing time / area closures of the US east coast to protect juvenile swordfish, resulted in a relatively rapid rebuilding of the fishery. Current landings are below the total allowable catch. The fishery is currently managed with country-specific quotas and minimum sizes. The North Atlantic stock is considered Least Concern on the IUCN Red List.

South Atlantic: This stock was overfished, with overfishing occurring in the late 1990s / early 2000s, but at present, the stock in not overfished, and overfishing is not occurring. Current catches are well below the total allowable catch. The fishery is currently managed with country-specific quotas and minimum sizes. The South Atlantic Swordfish stock is considered Least Concern on the IUCN Red List.

Mediterranean: In the Mediterranean Sea, Swordfish are currently overfished, and overfishing is occurring. The high exploitation rate (taking into account the very large reported catch of nearly 10,000 mt taken in a small area) and the extremely large and uncertain catch of very small fish are causes for concern. However, landings statistics and population parameters indicate a certain stability over the past 20 years. Because of overfishing and abundance of juveniles in the catch, the population is considered Near Threatened on the IUCN Red List. Management measures include a two-month closed season for directed fisheries and a minimum size, but a greater reduction in fishing effort will be required to allow rebuilding of the stock.

FURTHER READING: Ellis, R. 2013. Swordfish: A biography of the ocean gladiator. University of Chicago Press, Chicago, 279 pp.

REFERENCES: Abascal et al. 2010; Abdul Marak et al. 2011; Abecassis et al. 2012; Alvarado Bremer et al. 1996, 2006; Amorin et al. 2011; Arata 1954; Berkeley and Houde 1983; Chancollon et al. 2006; Chow et al. 1997, 2007; Chow and Takeyama 2000; Collette et al. 2011; Cox et al. 2002; Ellis 2013; Govoni et al. 2003, 2004; Hinton and Alvarado Bremer 2007; Hinton and Maunder 2006; IGFA 2017; IOTC 2006; Kasapidis et al. 2008; Kotoulas et al. 1995; Lerner et al. 2013; Linnaeus 1758; Lu et al. 2006, 2016; Nakamura 1985; Neilson et al. 2009, 2013; Palko et al. 1981; Potthoff and Kelly 1982; Radtke and Hurley 1983; Reeb et al. 2000; Romeo et al. 2009, 2011, 2014; Rossel and Block 1996; Smith et al. 2015; Viñas et al. 2010; Wilson 1984; Wilson and Dean 1983.

Black Marlin

Istiompax indica (Cuvier, 1832)

COMMON NAMES: English – Black Marlin
French – Makaire noir
Spanish – Aguja negra
Japanese – Shirokajiki

ETYMOLOGY: Although no etymology was provided by French anatomist and ichthyologist Georges Cuvier (Cuvier and Valenciennes 1832), the species name *indica* comes from the Latin *indicus*, meaning Indian, in reference to its original description from Sumatra in the Indian Ocean.

SYNONYMS: *Tetrapturus indicus* Cuvier, 1832; *Tetrapturus australis* Macleay, 1854; *Histiophorus brevirostris* Playfair, 1867; *Makaira marlina* Jordan and Hill, 1926; *Istiompax australis* Whitley, 1931; *Makaira nigricans tahitiensis* Nichols and LaMonte, 1935; *Istiompax dombraini* Whitley, 1954; *Makaira xantholineatus* Deraniyagala, 1956

TAXONOMIC NOTE: This species was previously included in the genus *Makaira* but does not show close morphological or genetic affinity to the Blue Marlin, *Makaira nigricans*.

FIELD MARKS:
1 The pectoral fins of adults are rigid and cannot be adpressed against the sides of the body.
2 The body is not very compressed laterally.
3 The nape is highly elevated in adults.
4 The height of the anterior lobe of the first dorsal fin is less than the body depth in adults.
5 The second dorsal fin is slightly anterior to the second anal fin.

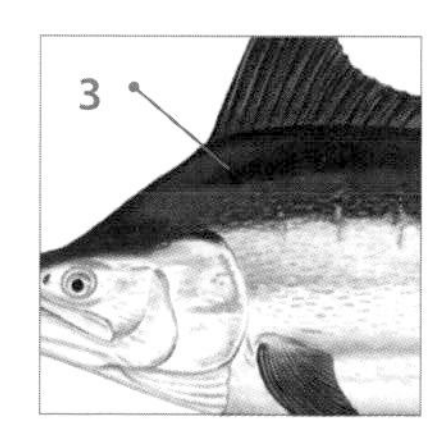

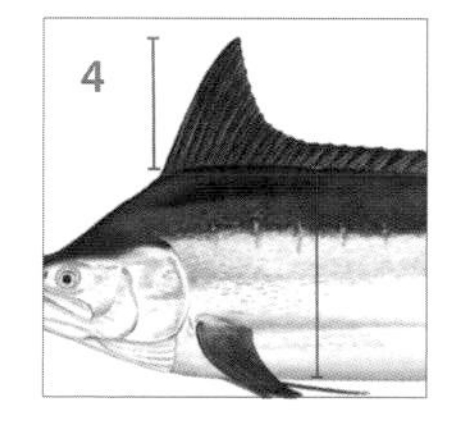

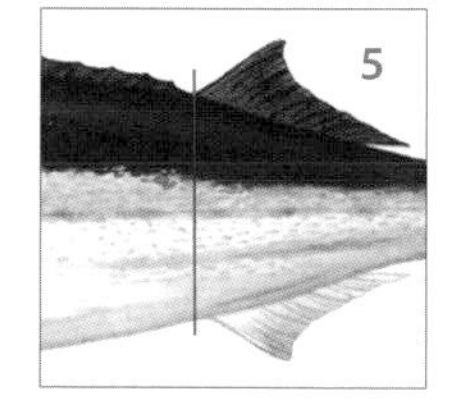

UNIQUE FEATURE: The only billfish with non-adpressible pectoral fins.

DIAGNOSTIC FEATURES: The body is not strongly compressed laterally. The bill is long, extremely stout, and round in cross section. The nape is conspicuously elevated. The left and right branchiostegal membranes are completely united but free from the isthmus. Gill rakers are absent. Both upper and lower jaws and the palatine bones have small file-like teeth. **Fins:** There are two dorsal fins, the first with 34–43 spines. The anterior lobe of the first dorsal fin is lower than the body depth. The first dorsal-fin base is long, ending close to the origin of the second dorsal fin. The second dorsal fin has 5–7 rays, and its position is slightly anterior to that of the second anal fin. There are two anal fins, the first with 10–14 rays, the second with 6 or 7 rays; the anal fins are similar in size and shape to the second dorsal fin. The pectoral fins have 12–20 rays, are rigid, and cannot be adpressed to the sides of the body. The pelvic fins are shorter than the pectoral fins, have a poorly developed membrane, and are depressible into deep ventral grooves. **Caudal peduncle:** The caudal peduncle is compressed laterally and slightly depressed dorsoventrally, with strong double keels on each side and a poorly developed notch on both the dorsal and ventral surfaces. **Anus:** The anus is situated near the origin of the first anal fin. **Lateral line:** The lateral line is single but obscured, especially in larger fish. **Scales:** The body is densely covered with thick, elongate bony scales, each with usually 1 or 2 (mostly 1) sharp posterior points. **Vertebrae:** Vertebrae number 24 (11 precaudal and 13 caudal). **Color:** The body is blackish to blue dorsally and silvery white ventrally, usually without blotches or dark stripes, although light-blue vertical stripes may occur. The first dorsal-fin membrane is blackish to dark blue. The other fins are usually dark brown, sometimes tinged with dark blue.

GEOGRAPHIC RANGE: Black Marlin are distributed throughout the tropical Indo-Pacific between latitudes 40° N and 40° S, but occasionally enter temperate waters. In the eastern Pacific, they are found from southern California to the southwestern and northeastern Gulf of California to Chile, including all of the oceanic islands. Strays migrate into the Atlantic Ocean by way of the Cape of Good Hope, but the existence of an Atlantic breeding stock is unlikely.

SIZE: The largest Black Marlin on record was caught off Tuticorin, India, in 2013 and was 422 cm total length (344 cm LJFL). The IGFA all-tackle game fish record is of a 1,560 lb (707.61 kg) fish caught off Cabo Blanco, Peru, in August 1953. Males and females are indistinguishable externally, but females attain a much larger size, with males rarely reaching 400 lb in weight or lower jaw fork lengths of 270 cm. Sex ratio varies with area and season.

HABITAT AND ECOLOGY: Black Marlin are pelagic and oceanodromous and are usually found in surface waters above the thermocline at temperatures from 15–30°C, often close to land masses, islands, and coral reefs. In the vicinity of the Great Barrier Reef, Black Marlin primarily occur in the upper 20 m and waters of 26–27°C.

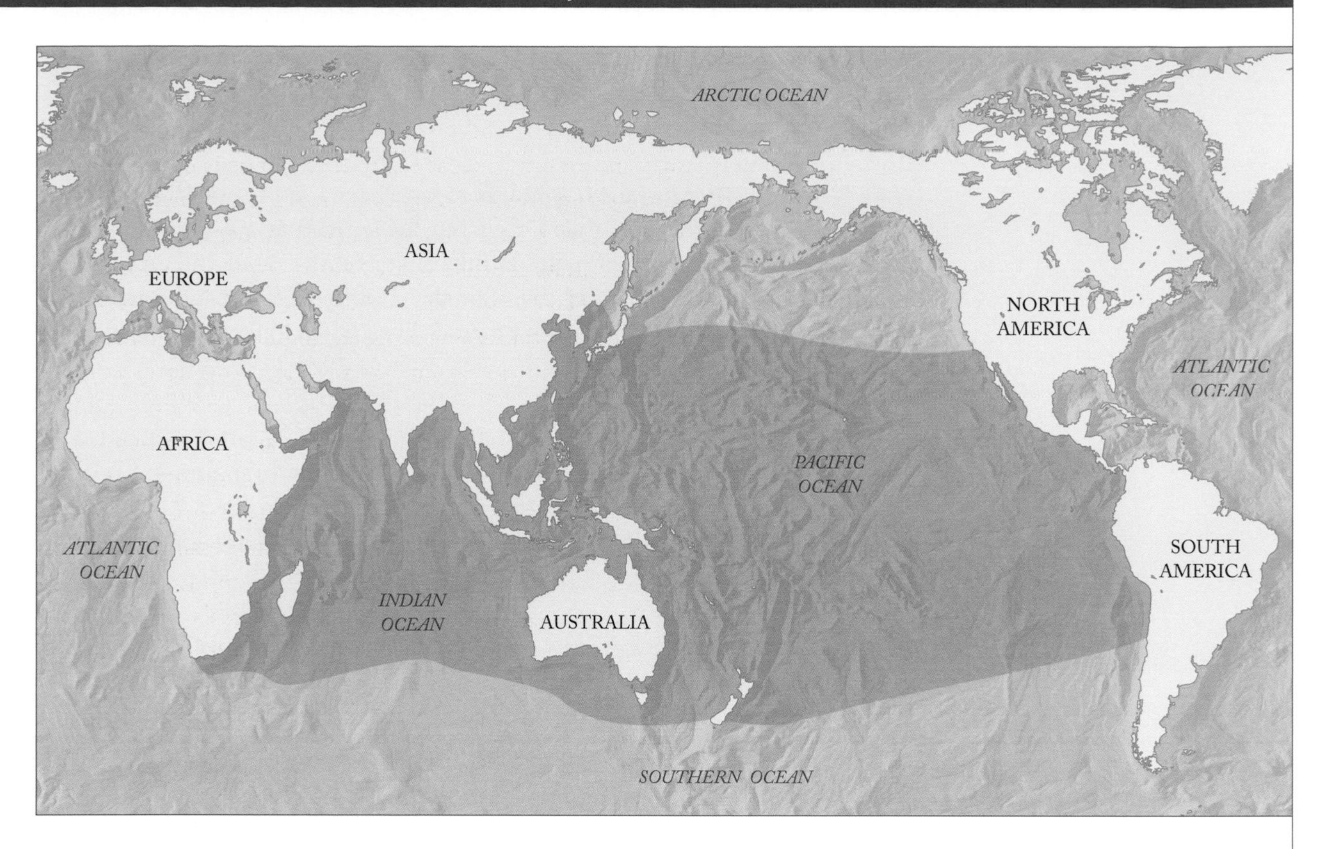

Above: Black Marlin range map.

MOVEMENTS: Conventional tagging studies have demonstrated that Black Marlin are capable of long-distance movements. Tag returns have confirmed trans-equatorial, trans-Pacific, and inter-ocean (Pacific Ocean to Indian Ocean) displacements. Analysis of net tag displacement by time at large suggests seasonal site fidelity or cyclical annual movements. Satellite tagging of animals caught during the spawning season off the Great Barrier Reef indicated a dispersal area of up to 5,000 km.

FOOD: Black Marlin are apex predators, feeding on fishes, squids, cuttlefishes, octopods, large decapod crustaceans, and especially on small tunas such as Skipjack, Yellowfin, Bigeye, and frigate tunas. The most important of 12 food items around Yonaguni Island, southwestern Japan, was Skipjack Tuna. Black Marlin sometimes use their bills for feeding. Analyses of stomach contents show that the prey is usually swallowed head first and often bears slashes inflicted by the predator's bill. Talbot and Penrith reported on a 480 lb Black Marlin that had speared and swallowed a 75 lb Yellowfin Tuna.

REPRODUCTION: Spawning is known to occur in the South China Sea near Taiwan and in waters adjacent to the Great Barrier Reef of Australia. Off Australia, intensive spawning occurs during October and November at water temperatures of 27–28°C. Length of males at first maturity was about 140 cm LJFL and was 230 cm LJFL for females. Egg counts of ripe females total about

40 million. The morphology of Black Marlin sperm is typical of the simple type I teleost sperm morphology, not like the advanced type II found in tunas, mackerels, and their relatives.

AGE AND GROWTH: Dorsal-fin spine rings have been used to age Black Marlin, with maximum ages of 12 years observed for females and 6 years for males. In Taiwan waters, length of males at first maturity was about 140 cm, 230 cm for females. Males were larger than females during the first four years of life, while females attained larger sizes by age 5, lived much longer, and grew much larger than males.

EARLY LIFE HISTORY: Ueyanagi and Yabe reported on 400 larval specimens of Black Marlin from the western Pacific and Indian oceans ranging in size from 2.7 to 19.5 mm SL. A developmental series of eight specimens 2.9 to 23.2 mm SL was described and illustrated by Ueyanagi and Yabe. Larval Black Marlin have a short snout, large eye diameter, a high dorsal fin, and long pelvic fins.

STOCK STRUCTURE: Black Marlin population structure is not well understood. Preliminary analyses of mitochondrial DNA control-region restriction fragments and three microsatellite loci did not reveal heterogeneity between samples from Australia and the eastern Pacific, while sequence analysis of two mitochondrial DNA gene regions demonstrated significant heterogeneity among samples from six Indo-Pacific sampling locations.

FISHERIES INTEREST: Black Marlin is taken as an incidental catch in pelagic longline operations targeting tunas and Swordfish, although a small directed harpoon fishery exists off Taiwan. As an incidental catch, there is generally poor catch reporting for Black Marlin, and it is often not identified to species (included as "marlins"). The world catch for this species is variable but shows a gradual increase from around 5,000 mt in the 1950s to 10,826 mt in 2006.

Black Marlin are taken in sport fisheries throughout the species' range, and there has been a directed fishery on the spawning aggregation of the Great Barrier Reef for more than 50 years. There is also a fishery for young-of-the-year Black Marlin in waters inshore of the Great Barrier Reef.

THREATS: Black Marlin are caught mainly as bycatch by pelagic longliners, trolling, and harpooning and sometimes by set and gill nets. This species may potentially be threatened by artisanal and commercial longline fisheries, as it is commonly taken as bycatch in purse seiners and is sometimes discarded. It is an important species in sport fisheries and is targeted by artisanal fisheries.

CONSERVATION: Black Marlin are highly migratory and are listed in Annex I of the 1982 Convention on the Law of the Sea. International management authority for the Black Marlin in the Pacific is shared by the Inter-American Tropical Tuna Commission (IATTC) and the Western and Central Pacific

Fisheries Commission (WCPFC), and in the Indian Ocean the Indian Ocean Tuna Commission (IOTC) manages the stock. There has never been an assessment of the status of Black Marlin throughout the Pacific, but landings trends over the past 30 years have been generally declining. Within the Indian Ocean, CPUE exhibited dramatic declines since the beginning of the fishery in the 1950s, with catch rates in the initial fishing grounds decreasing substantially. Nominal CPUE in the northwestern Australian area has declined approximately 87% since 1977, while nominal CPUE in the Seychelles area has been very low and has declined from about 30% since the 1970s. A meta-analysis of four Black Marlin studies showed high post-release survival rates regardless of method of capture.

In the eastern Pacific, landings data from 1995–2005 has shown a rapid increase, from 300 mt to 1,400 mt. Landings data for Black Marlin taken by purse seine and longline in the eastern Pacific varied from 100 to 417 mt per year from 1978 to 2007, with a uniformly low catch.

Billfishes, including Black Marlin, cannot be offered for sale in the United States except in Hawaii and the Pacific Insular areas. In the eastern Pacific, billfishes cannot be taken commercially in Peru, Panama, Nicaragua, and Guatemala, and all recreational fishing is catch and release in Panama, Nicaragua, and Guatemala. Black Marlin are protected in Mexico by the 50-mile coastal zone area where this species cannot be taken commercially, and there is a larger conservation zone off of Baja California where it cannot be taken commercially. Commercial harvest in Costa Rica is limited to 15% of all landings.

FURTHER READING: McGlashan, A. 2015. Marlin on the flats: Chasing marlin on the flats of Queensland's Fraser Island. International Angler 78(2):22–29.

REFERENCES: Collette 2010; Collette et al. 2006, 2011; Cuvier and Valenciennes 1832; Domeier and Speare 2012; Falterman 1999; González-Armas et al. 2996; Matsumoto and Bayliff 2008; Musyl et al. 2014; Nakamura 1975, 1985, 2001; Pepperell 2010; Serafy et al. 2009; Shimose et al. 2008; Shomura and Williams 1975; Sivadas et al. 2013; Speare 2003; Sun et al. 2014; Talbot and Penrith 1962; Ueyanagi and Yabe 1959; Uozumi 1999; van der Straten et al. 2006; Wang et al. 2006; Wapenaar and Talbot 1964; Williams et al. 2014.

Family ISTIOPHORIDAE

Sailfish

Istiophorus platypterus (Shaw, 1792)

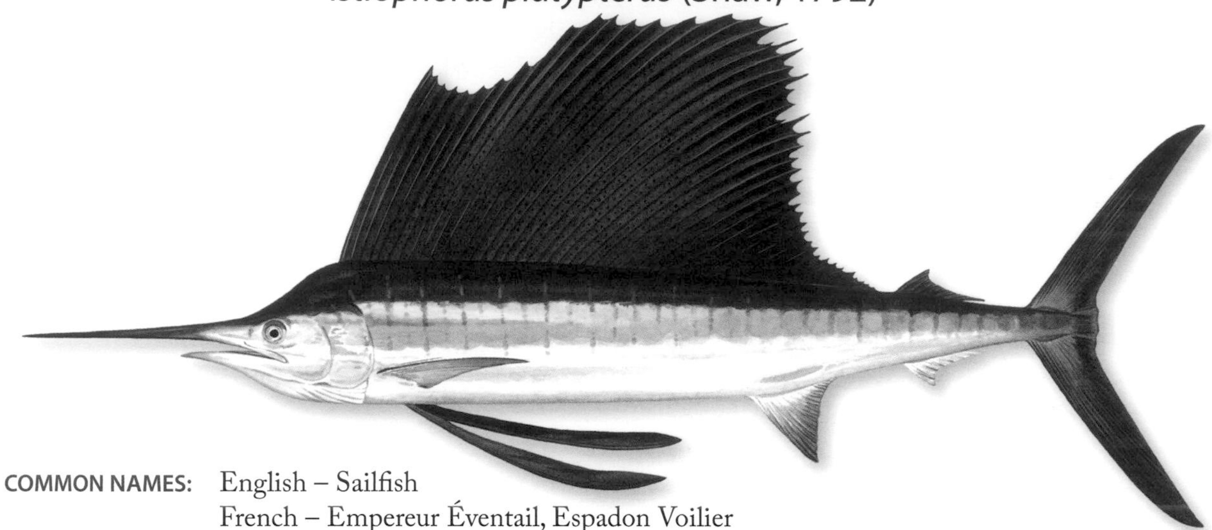

COMMON NAMES: English – Sailfish
French – Empereur Éventail, Espadon Voilier
Spanish – Aguja de Abanico, Banderón, Pez Velo

ETYMOLOGY: The original description by the British naturalist George Shaw (Shaw and Nodder 1792) was based on their plate 88. The species name *platypterus* is a combination of the Greek *plex*, "flat," and the Greek *ptero*, "fin," meaning "flatfin," an allusion to the sail-like dorsal fin.

SYNONYMS: *Xiphias platypterus* Shaw, 1792; *Scomber gladius* Bloch, 1793; *Istiophorus gladifer* Lacepède, 1801; *Xiphias velifer* Bloch and Schneider, 1801; *Makaira albicans* Latreille, 1804; *Histiophorus immaculatus* Rüppell, 1830; *Histiophorus americanus* Cuvier, 1832; *Histiophorus ancipitirostris* Cuvier, 1832; *Histiophorus gracilirostris* Cuvier, 1832; *Histiophorus indicus* Cuvier, 1832; *Histiophorus pulchellus* Cuvier, 1832; *Makaira velifera* Cuvier, 1832; *Skeponopodus guebucu* Nardo, 1833; *Histiophorus orientalis* Temminck and Schlegel, 1844; *Histiophorus granulifer* Castelnau, 1861; *Istiophorus triactis* Klunzinger, 1871; *Histiophorus dubius* Bleeker, 1872; *Istiophorus ludibundus* Whitley, 1933; *Istiophorus brookei* Fowler, 1933; *Istiophorus japonicus* Jordan & Thompson, 1914; *Istiophorus eriquius* Jordan, 1926; *Istiophorus greyi* Jordan & Evermann, 1926; *Istiophorus maguirei* Jordan and Evermann, 1926; *Istiophorus volador* Jordan and Evermann, 1926; *Istiophorus wrighti* Jordan and Evermann, 1926; *Histiophorus magnioci* Jordan, 1927; *Istiophorus amarui* Curtiss, 1944

TAXONOMIC NOTE: Although some authors maintained that the Indo-Pacific and Atlantic sailfish were distinct species, there is no genetic evidence to support this (Graves and McDowell 1995, McDowell 2002, Collette et al. 2006).

FIELD MARKS:
1 The first dorsal fin is sail-like and much taller than the body depth.
2 The pelvic fins are very long, with well-developed membranes, and reach nearly to the anus.

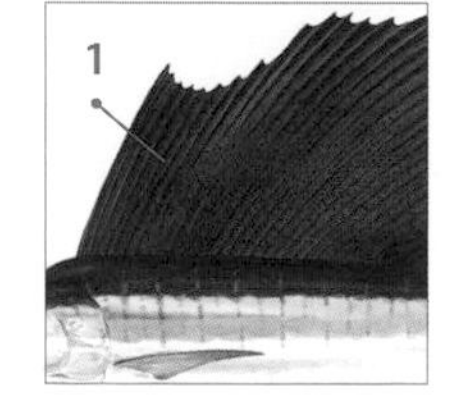

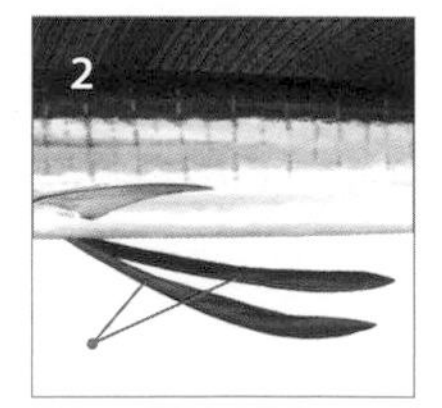

UNIQUE FEATURE: Sailfish use their expandable dorsal fin to herd prey fishes.

DIAGNOSTIC FEATURES: The body is fairly compressed. The bill is long, slender, and round in cross section. Both upper and lower jaws and the palatine bones (in the roof of the mouth) have small, file-like teeth in adults. Gill rakers are absent. The right and left branchiostegal membranes are united and are free from the isthmus. **Fins:** There are two dorsal fins: the first is large, with 42–46 rays; the second dorsal is small, with 6 or 7 rays. The first dorsal fin is sail-like, much higher than the body depth at mid-body level. Its base is long and close to that of the second dorsal fin. There are two anal fins, the first with 11–14 rays, the second with 6 or 7 rays. The position of the second anal fin is slightly more anterior than that of the second dorsal fin. The pectoral fins have 17–20 rays. The pelvic fins are extremely long, almost reaching to the anus, and depressible into a groove. They have one spine and several soft rays fused tightly together and with a well-developed membrane. **Caudal peduncle:** The caudal peduncle has double keels on each side and a shallow caudal notch on both the dorsal and the ventral surfaces. **Anus:** The anus is situated near the origin of the first anal fin. **Lateral line:** The lateral line is single and clearly visible. **Scales:** Scales vary in shape with growth. In adults they are somewhat sparse and imbedded in the skin, and each has a single, rather blunt point or two posterior points. **Vertebrae:** Vertebrae number 24 (12 precaudal and 12 caudal). **Color:** The body is dark blue to bluish green dorsally, light blue splattered with brown laterally, and silvery white ventrally. There are about 20 rows of vertical bars on the sides, each stripe composed of many light-blue round dots. The bases of the first and second anal fins are often tinged with silvery white. The membrane of the first dorsal fin is dark blue or blackish blue, with scattered small, round, black dots. The remaining fins are blackish blue, sometimes tinged with dark brown. The body and fins are capable of rapid color change and may display as dark brown to nearly black, particularly when hunting.

Below: Dark hunting coloration.

GEOGRAPHIC RANGE: In the Indian and Pacific oceans, Sailfish occur between approximately 45° N and 40° S in the western Pacific, 35° N and 35° S in the eastern Pacific, and 45° S in the western Indian Ocean and 35° S in the eastern Indian Ocean. In the eastern Pacific, Sailfish are found from southern California and the lower three-fourths of the Gulf of California to Peru, including around all of the oceanic islands. Sailfish are widely distributed in the tropical and temperate waters of the Atlantic Ocean from approximately 40° N in the northwest Atlantic and 50° N in the northeast Atlantic to 40° S in the southwest Atlantic and 32° S in the eastern South Atlantic, including a record from Ascension Island in the mid-Atlantic. Sailfish are reported to have entered the Mediterranean Sea from the Red Sea via the Suez Canal, although the records are few and mostly based on juvenile specimens. Furthermore, in at least one case a specimen recorded from the Mediterranean as a sailfish was subsequently identified as a Mediterranean Spearfish, *Tetrapturus belone*.

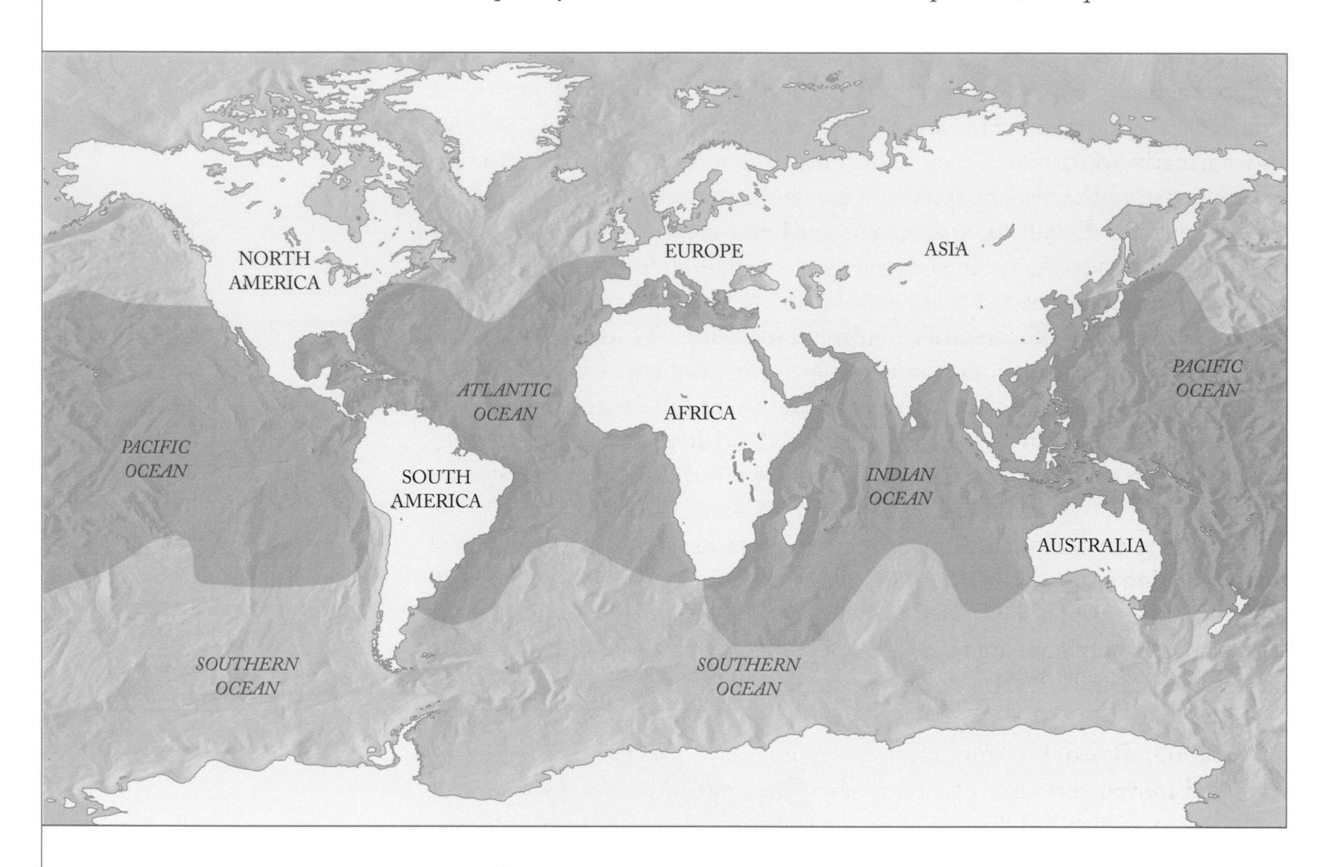

Above: Sailfish range map.

SIZE: Sailfish attain a larger maximum size in the Indo-Pacific than in the Atlantic. The IGFA all-tackle game fish record in the Indo-Pacific is a 221 lb (100.24 kg) fish caught off Santa Cruz Island, Ecuador, in 1947, and in the Atlantic it is a 142 lb, 6 oz (64.6 kg) fish caught off Lobito, Angola, in 2014. The majority of sport fishing catches in southern Florida range from about 102 to 140 cm body length (or 173–229 cm total length), with a considerable range in weights (6.0–49.4 kg). The length of individuals caught by pelagic longliners in the Atlantic ranges from about 125 to 210 cm LJFL (with most between 150 and 195 cm). Sailfish taken by drift gill nets off eastern Taiwan ranged 80–289 cm LJFL for females, 78–227 cm LJFL for males.

HABITAT AND ECOLOGY: Sailfish are oceanic and epipelagic, spending most of their time in the upper 10 m in a temperature range between 21°C and 28°C

However, on occasion they descend into waters deeper than 100 m, with one reported dive exceeding 500 m.

Sailfish are the least oceanic of the billfishes, with the greatest concentrations found in coastal waters. In the Atlantic, pelagic longline data indicate that Sailfish are continuously distributed across the ocean; however, catch rates in oceanic waters are much lower than those in waters closer to land. In the western Atlantic, Sailfish are concentrated in the Caribbean Sea, in the Gulf of Mexico, and around the West Indies and Florida. Distribution along the east coast of the United States appears to be influenced by water temperatures and meteorological conditions (wind). During summer, Sailfish move northward along with the extension of warm water, and with the beginning of cold weather and northerly winds they move southward, congregating in schools off the Florida coast. In the eastern Atlantic, shifts of the frontal zone of the Canaries Current and the Equatorial Countercurrent influence aggregation of Sailfish populations off West Africa. In spring the major concentration of Sailfish moves northward along the coast; it returns southward in autumn, following the 28°C isotherm. The period of increased abundance off the Ivory Coast coincides with the period of maximum surface water temperature, around 28°C. In both the eastern tropical Pacific and the eastern tropical Atlantic, Sailfish concentrate in shallower waters than in the western part of both oceans, due to hypoxia-based habitat compression over oxygen-minimum zones in the eastern tropical seas.

Observations on Sailfish diel activity off Florida have shown scattering of the population in the early morning, but by 0900 hours schools of up to 30 individuals begin to form and feed on concentrations of small forage fish. Conventional tagging, particularly in the eastern tropical Pacific, has been highly unsuccessful—31,103 releases with only 20 recoveries.

MOVEMENTS: Based on conventional tagging data, Sailfish exhibit more restricted movements than other billfishes. The maximum net displacement for a tagged sailfish is 3,861 km, and unlike many other species of billfishes, there are no records of trans-oceanic movements. Analyses of conventional and electronic tag data suggest that in some areas Sailfish undertake cyclical seasonal movements or possibly exhibit site fidelity. The majority of conventional tagging has been performed by recreational anglers in the western North Atlantic, and tag returns indicate connectivity from the east coast of the United States, the Gulf of Mexico, the Caribbean, and the northern coast of South America.

FOOD: Sailfish are a generalist pelagic predator and have a diet that overlaps with many other large pelagic fishes, such as other billfishes, tunas, sharks, and dolphinfishes. In the Atlantic, Sailfish overlap in geographical range and hence compete for food during certain seasons of the year, particularly with White Marlin and Blue Marlin. Around Florida, adults have been shown to feed mainly on pelagic fishes such as Little Tunny, halfbeaks, cutlassfishes, needlefishes, sardines, and jacks as well as paper nautilus (*Argonauta nodosa*) and squids. Off the coast of Mazatlán, Mexico, the most important food items by both frequency and weight were cephalopods, particularly the jumbo squid (*Dosidicus gigas*), which occurs in 67% of the stomachs. Paper nautilus occurs in 58%, followed by a triggerfish (40%) and frigate tunas (14%). Off northeastern Brazil, Atlantic Pomfret (*Brama brama*) and a squid (*Ornithoteuthis antillarum*) comprised 47% of the diet of a sample of more than 100 Sailfish. Thirty-one species of fishes were found in stomachs of Sailfish from Brazil, with sardines the most important (23.1%), followed by paper nautilus (12.7%). Off eastern Taiwan, stomach contents indicated that Sailfish were generalist predators feeding mainly on epipelagic fishes, especially Bullet Tuna. Major prey items in the Indian Ocean are fishes (81% occurrence), crabs (31%), and squids (25%). In the eastern Arabian Sea, diet was dominated by finfishes (flyingfishes, frigate tunas, snake mackerels, and jacks) in terms of weight and frequency of occurrence but by squids in terms of numbers.

Billfish have been observed to feed cooperatively on schooling fishes and to use their bill to strike prey. When feeding on schooling sardines in open water as a group, some Sailfish will exhibit dark coloration, swimming with their dorsal and pelvic fins erect, effectively "herding" the school of bait fish. Individually, a fish will depress its fins, acquire a bright blue coloration, and use its bill to either tap an individual prey target or to slash through the school. Sailfish also feed on bottom-dwelling organisms, including sea robins (Triglidae), cephalopods, and gastropods.

REPRODUCTION: Sailfish are dioecious, but there are no external morphological characters to distinguish males and females; large fish are usually females. Size at sexual maturity is 150 cm eye-FL off the southeast Pacific coast of Mexico, 166 cm LJFL off eastern Taiwan in the western Pacific, and 155 cm LJFL in the western South Atlantic. Sailfish appear to spawn throughout the year in tropical and subtropical waters, with peak spawning occurring in the respective local summer seasons. Spawning occurs with males and females swimming in pairs or with two or three males following a single female. Around Florida, United States, Sailfish often move inshore into shallow waters where females, swimming sluggishly with their dorsal fins extended and accompanied by one or more males, may spawn near the surface in the warm season. Spawning also occurs in offshore waters beyond the 100-fathom isobath from south of Cuba to the Carolinas, United States, from April to September. In the eastern Atlantic, spawning has been observed in West African shelf waters throughout the year, with peak intensity during the summer months in the Conakry–St. Luis region and from February to April in the Conakry-Freetown region.

Sailfish oocyte development is asynchronous, resulting in batch spawning and indeterminate fecundity. Fecundity increases sharply with size of the female. Off southeast Florida, a 33.4 kg female may shed up to 4.8 million eggs in three batches during one spawning season. Batch fecundity of Sailfish off the

Pacific coast of Mexico estimated for 21 females was 1,710,000 ± 600,000 eggs per spawning, and the average interval between spawnings for 93 mature females was 3.6 days.

AGE AND GROWTH: Cross sections of the fourth dorsal-fin spine were used to age Sailfish from southeast Florida, resulting in a maximum age of 7 years, although spines from very large Sailfish were not available. Sailfish grow rapidly, reaching 100 cm in their second year, which is more than 50% of the mean asymptotic eye–fork length. The greatest time at liberty for a tagged Sailfish is 17 years. Assuming a longevity of 13 years and an age of maturity of 2.5 years, the estimated generation length is 4.3 years.

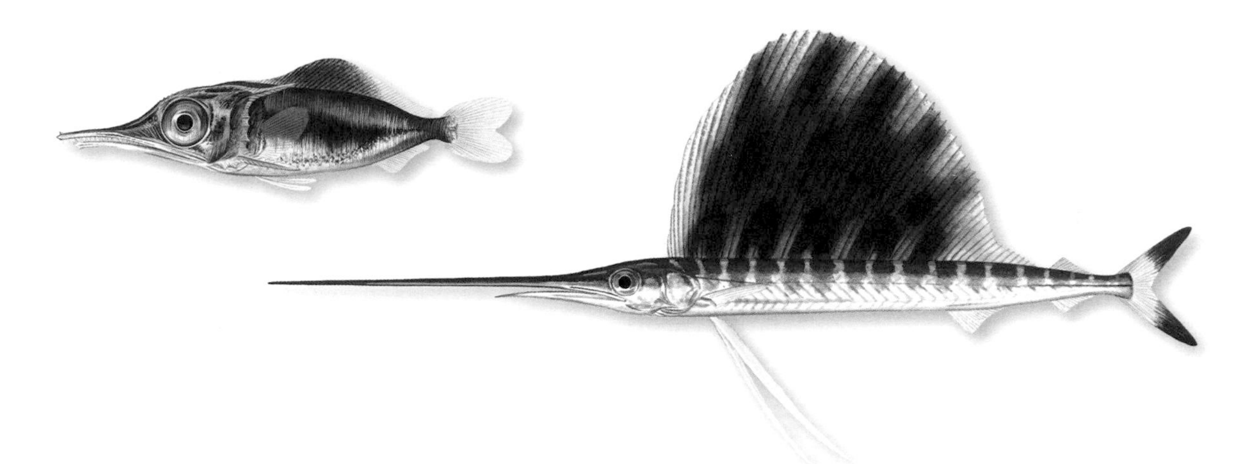

Above left: Larval Sailfish, approx. 5 mm LJFL.
Above right: Juvenile Sailfish, approx. 20 cm LJFL.

EARLY LIFE HISTORY: Molecular genetic techniques were used to identify larvae of three western Atlantic billfishes, including the Sailfish, by Luthy et al. Their paper also included a key to identification of postlarvae and illustrations of the heads of the three species. A summary of characteristics of the early stages of three of the western Atlantic billfishes, with illustrations, is presented by Richards and Luthy. Ten specimens from the Florida Current ranging in standard length from 3.9 to 70 mm were described and illustrated by Voss. At 15 mm, the dorsal fin has enlarged, the upper jaw has elongated, and the lower jaw is shorter than the upper jaw. Food consists primarily of copepods until about 20 mm but changes rapidly to include predominantly fishes at larger sizes. Larval Sailfish were found from August to November in the southern Gulf of California at sea surface temperatures of 27.5°C–31.5°C. Sailfish larvae ranging in size from 3.2–13.4 mm notochord or SL from the Straits of Florida were collected in waters between 26.1°C and 30.6°C. Growth in length was comparable to that of other pelagic fishes as larvae but slightly higher than that of Blue Marlin. A large collection of Sailfish larvae from the northern Gulf of Mexico (2,428 specimens 2.0–24.3 mm SL) that were spawned from May to September were collected at similar temperatures (26.4° C to 30.4°C) and demonstrated the importance of the northern Gulf of Mexico as an important spawning and nursery area.

STOCK STRUCTURE: Sailfish stock structure is not well understood. Population-level genetic divergence has been demonstrated between samples of Sailfish from the Atlantic and Indo-Pacific oceans based on analyses of mitochondrial DNA and nuclear microsatellite loci, but the hypothesis of a single stock of sailfish in the Atlantic Ocean could not be rejected. Based on analyses of mitochondrial DNA and a few nuclear microsatellite loci, Sailfish in the Arabian Gulf differ significantly from those in the western Indian Ocean, consistent with the observation that all tagged individuals have stayed within the Gulf. Recently, genetic heterogeneity of mitochondrial DNA haplotypes and microsatellite allele frequencies has been demonstrated across the North Pacific Ocean, suggesting significant within-ocean structuring.

FISHERIES INTEREST: Sailfish are taken in directed artisanal and recreational fisheries and as an incidental bycatch by pelagic longline and purse seine gear targeting tunas and Swordfish. There is poor reporting of catch data from artisanal fisheries, but considering the coastal distribution of Sailfish, it is thought that artisanal catches could represent a large share of total landings. The majority of reported catches are from the pelagic longline fleet. Reported global landings increased from less than 300 mt in 1950 to more than 8,000 mt in 1960 and exceeded 20,000 mt in 2000. Current landings are just under 40,000 mt. The flesh is dark red and usually not as good as that of marlins. It is marketed mostly frozen and often fresh in local markets.

The importance of Sailfish to local economies through recreational fishing and tourism is difficult to overstate. There are nearly year-round US recreational fisheries for Sailfish in the Florida Straits and southern Gulf of Mexico, and similar fisheries are increasing throughout the Caribbean Sea and western Central America as more countries develop tourist-based infrastructures. Ditton and Stoll estimated that over 230,000 anglers in the United States alone target billfish, for a combined 2,137,000 days annually. Billfish anglers are also among the highest-spending saltwater anglers, with an estimated $180 million in expenditures by billfish tournament anglers in 1989 alone.

Currently, Sailfish are divided into five stocks for management purposes: western Atlantic, eastern Atlantic, eastern Pacific, western and central Pacific, and Indian oceans.

THREATS: Sailfish management is challenged by a lack of data. There is poor reporting of catches from artisanal fisheries, which may represent a significant fraction of the overall landings. Many pelagic longline fleets do not report billfish catches to species, and Sailfish are included with other istiophorid billfishes. In addition, for those commercial fisheries that are required to release billfishes, there are typically poor records of releases. Sailfish are fully fished in the Atlantic, and several model runs suggested the stocks may be overfished, with overfishing occurring. Localized depletions have been noted in Central America and in Iran and India, but there are no stock assessments for these regions. Landings and effort data are not reliable for these regions as Sailfish catches are often aggregated with other billfishes. There is no current indication of widespread decline so the Sailfish is listed as Least Concern on the IUCN Red List.

CONSERVATION: Sailfish are a highly migratory species listed in Annex I of the 1982 Convention on the Law of the Sea.

Atlantic Ocean: Sailfish are managed as two stocks in the Atlantic: one in the western Atlantic, and one in the eastern Atlantic. There is considerable uncertainty regarding the status of Atlantic Sailfish stocks, but most models indicate that the eastern Atlantic stock is overfished and that overfishing is occurring, while the western Atlantic stock is possibly overfished, and overfishing may be occurring. The eastern stock is more productive than the western, probably providing greater maximum sustainable yield. Both eastern and western stocks suffered the greatest declines prior to 1990. Since 1990, there has been little concordance among trends in relative abundance from different fisheries. Over the past 10 years, reported catches of the eastern and western stocks have decreased from about 4,300 mt to 2,000 mt. Examination of length frequencies does not show changes in the average length or length distribution.

The United States requires release of all Sailfish from commercial gear. Regulations in the United States, the Bahamas, and Bermuda prohibit commercial sale of Sailfish, as do regulations in the eastern Pacific in Panama, Nicaragua, and Guatemala, although enforcement of billfish management policies in Central America is usually lacking. The Mexican government allows Sailfish to be taken only with sport fishing gear and controls the number of licenses and limits the catch to one billfish (marlin, sailfish, spearfish, or swordfish) per boat per day. The United States has a minimum size of 60 inches (152 cm) for Atlantic Sailfish taken in the recreational fishery, and anglers are required to use circle hooks when fishing with natural baits in billfish tournaments. Throughout the recreational fishery an increasing conservation awareness has resulted in very high release rates. Short-term survival of Sailfish captured by pelagic longline gear and released has been demonstrated, and a meta-analysis of seven Sailfish studies showed high post-release survival rates regardless of method of capture.

In the Atlantic, the International Commission for the Conservation of Atlantic Tunas Standing Committee on Research and Statistics recommended at the time of the last assessment that catches for the eastern stock should be reduced from current levels and that catches of the western stock of sailfish should not exceed current levels. Any reduction in catch in the West Atlantic is likely to help stock rebuild and reduce the likelihood that the stock is overfished.

Eastern Pacific: There has been no effort to assess the status of Sailfish or spearfish species in a comprehensive manner in the eastern Pacific. Reported catches have been fairly stable over the past 10–25 years at around 2,000 mt; however, catches are likely higher than reported, given that many Sailfish landings are reported as unidentified billfishes. There has been a reduction in directed fishing effort for this species recently. It is a very important sport fish in the eastern tropical Pacific. There are some indications of localized declines. Overall Sailfish abundance is 80% below the 1964 levels in Costa Rica, Guatemala, and Panama, and trophy fish sizes are 35% smaller than their unexploited sizes. Recent CPUE data from the recreational fishery off Central America has generated cause for concern.

Western Central Pacific: Data for Sailfish are not routinely recorded; however, it is inferred that no significant declines are occurring. In view of recent rapid increases in fishing effort off Taiwan, stock status should be closely monitored.

Indian Ocean: Catch reports often refer to total catches of all billfish species combined. Catches have greatly increased, from around 5,000 mt in the early 1990s to almost 29,000 mt in 2011, due largely to development of a gill net/longline fishery in Sri Lanka and extension of the area of operation of Iranian gill net vessels.

FURTHER READING: Tinsley, J. B. 1964. The Sailfish: Swashbuckler of the open sea. University of Florida Press, Gainesville, FL, 216 pp.

REFERENCES: Amorin et al. 2011; Arizmendi-Rodríguez et al. 2006; Beardsley et al. 1975; Ben-Tuvia 1966; Cerdenares-Ladrón de Guevara et al. 2011, 2013; Chiang et al. 2006, 2009, 2013; Collette 2010; Collette et al. 2006, 2011a, 2011b; De Croz 1994; de Sylva and Breder 1997; Ditton and Stoll 1998; Domenici et al. 2014; Ehrhardt and Fitchett 2006; Fisher and Ditton 1992; Ganga et al. 2011; González-Armas et al. 2006; Graves and McDowell 1995; Hedgepeth and Jolley 1983; Hoolihan 2006; Hoolihan and Luo 2007; Hoolihan et al. 2004; Jolley 1977; Kerstetter and Graves 2007; Kitchell et al. 2006; Luthy et al. 2005a, 2005b; McDowell 2002; Morrow and Harbo 1969; Mourato et al. 2014; Musyl et al. 2014; Nakamura 1985; Ortiz et al. 2003; Ovchinnikov 1970; Pepperell 2010; Prince et al. 1986, 2006, 2010; Richards and Luthy 2005; Shaw and Nodder 1792; Shomura and Williams 1975; Simms et al. 2010; Speare 1995; Tsai et al. 2014; Uozumi 1999; Varghese et al. 2013; Voss 1953; Wang et al. 2006; Wirtz et al. 2014.

White Marlin

Kajikia albida (Poey, 1860)

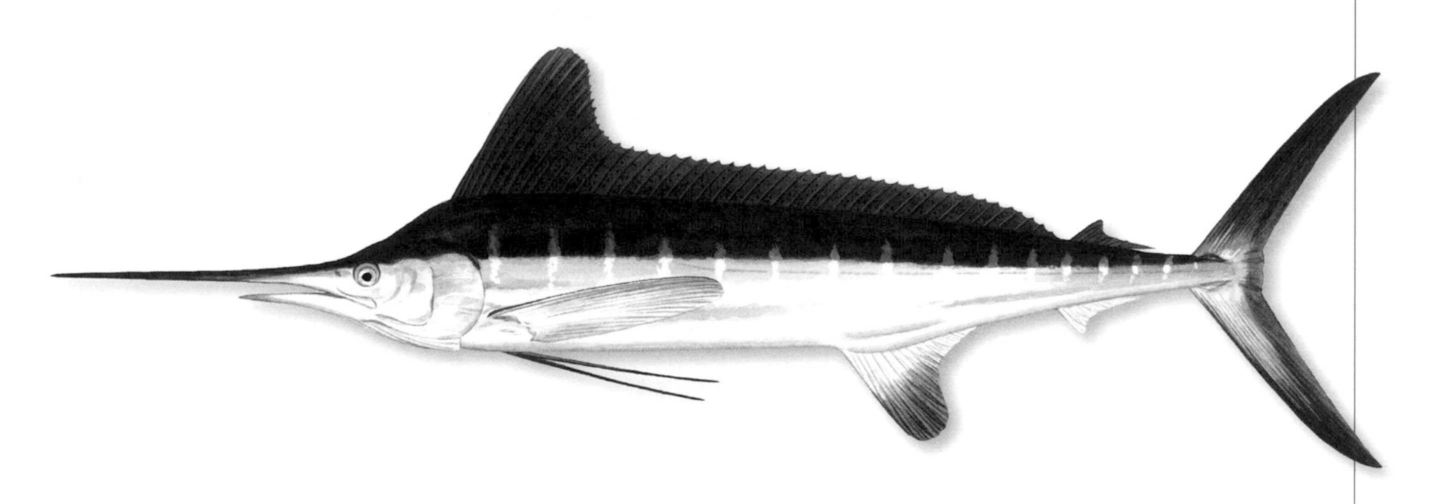

COMMON NAMES: English – White Marlin
French – Makaire Blanc
Spanish – Aguja Blanca

ETYMOLOGY: The White Marlin was described by the Cuban ichthyologist Felipe Poey (1860), who used an appropriate species name, *albida*, which comes from the Latin *albus*, meaning "white."

SYNONYMS: *Tetrapturus albidus* Poey, 1860; *Tetrapturus lessonae* Canestrini, 1861

TAXONOMIC NOTE: The White Marlin was previously known as *Tetrapturus albidus*. This species is genetically very similar to the Striped Marlin, *Kajikia audax*, but represents a distinct evolutionary unit.

FIELD MARKS:

1. Anterior lobe of first dorsal fin typically rounded and higher than remainder of fin, the height decreasing gradually backward.
2. The anal opening in the White Marlin is located close to the anal fin, separated by a distance of less than half the anal-fin height.
3. The branchiostegal membranes do not extend to the rear edge of the operculum.
4. The dorsal fin is spotted and scales on the mid-body are overlapping.

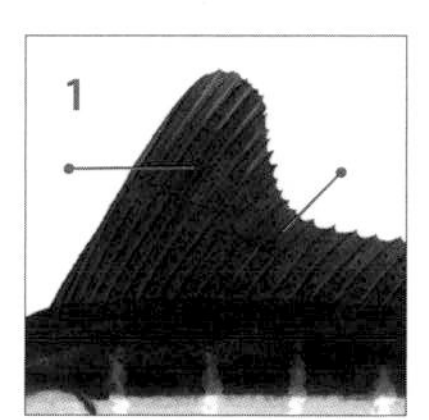

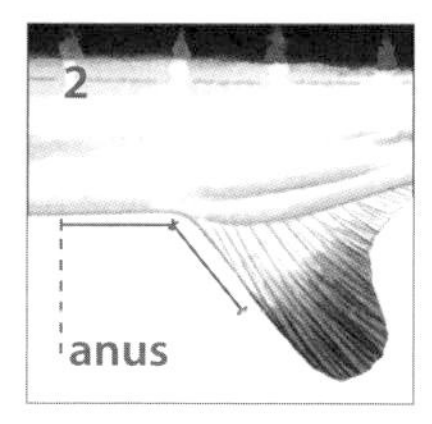

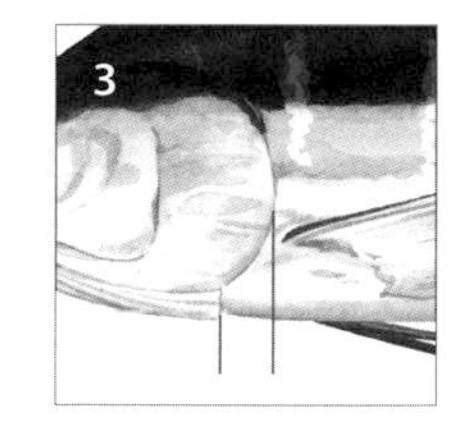

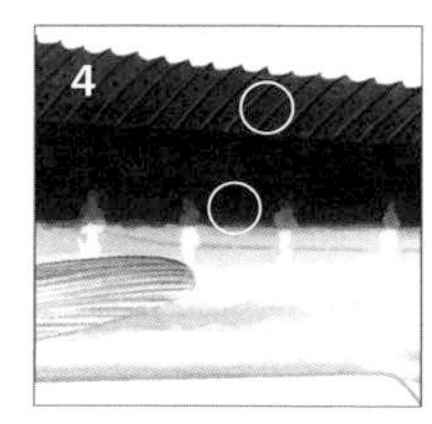

UNIQUE FEATURE: The only Atlantic billfish with generally rounded first dorsal and anal fins and an anal opening located close to the first anal fin.

DIAGNOSTIC FEATURES: The body is elongate and fairly compressed. The bill is stout and long, round in cross section. The nape is fairly elevated. The right and left branchiostegal membranes are completely united to each other but free from the isthmus. Gill rakers are absent. Both upper and lower jaws and the palatine bones (on the roof of the mouth) have small, file-like teeth. **Fins:** There are two dorsal fins, the first with 38–46 rays. The first dorsal fin usually has a rounded anterior lobe (although some may appear truncated); it is higher than body depth anteriorly, then abruptly decreases in height to about the 12th dorsal-fin ray and gently decreases further backward. The first dorsal-fin base is long, extending from above the posterior margin of the preopercle to near the origin of the second dorsal fin. The second dorsal fin, its position slightly backward with respect to the second anal fin, has 5 or 6 rays. There are two anal fins, the first with 12–17 rays, the second with 5 or 6 rays and very similar in size and shape to the second dorsal. The pectoral fins are long and wide, round-tipped, adpressible against sides of body, and with 18–21 rays. The pelvic fins are slender and almost equal to or slightly shorter than the pectoral fins. **Caudal peduncle:** The caudal peduncle is laterally compressed and slightly depressed dorsoventrally, with strong double keels on each side and a shallow notch on both the dorsal and ventral surfaces. **Anus:** The anus is situated just in front of the first anal-fin origin. **Lateral line:** The lateral line is single and obvious, curving above the base of the pectoral fin and then continuing in a straight line toward the caudal-fin base. **Scales:** The body is densely covered with elongate bony scales, each with 1 or 2 posterior points. Vertebrae number 24 (12 precaudal and 12 caudal). **Color:** The body is blue-black dorsally, silvery white laterally, and white ventrally. There are usually no blotches or marks on the body. When feeding, more than 15 whitish vertical bars are present, and brown blotches may appear in an exhausted fish. The first dorsal fin is dark blue with many black dots; the second dorsal fin is also dark blue; the pectoral fins are blackish brown, sometimes tinged with silvery white; the pelvic fins are blue-black with a black fin membrane; the caudal fin is blackish brown. The pectorals can turn neon blue when the fish is stimulated.

GEOGRAPHIC RANGE: White Marlin is found throughout warm waters of the Atlantic from 45° N to 45° S south to Capetown, South Africa, in the eastern Atlantic, including the Gulf of Mexico, Caribbean Sea, and Mediterranean Sea. However, records from the Mediterranean Sea and from Bretagne, France, likely represent a few straying individuals. The distribution shows some seasonal variation; it occurs in the highest latitudes only during warmer periods of the year. It is normally found in waters where surface temperature is above 22°C, in waters over 300 m in depth, and within a salinity range from 35–37 ppt but will move into shallower water for feeding.

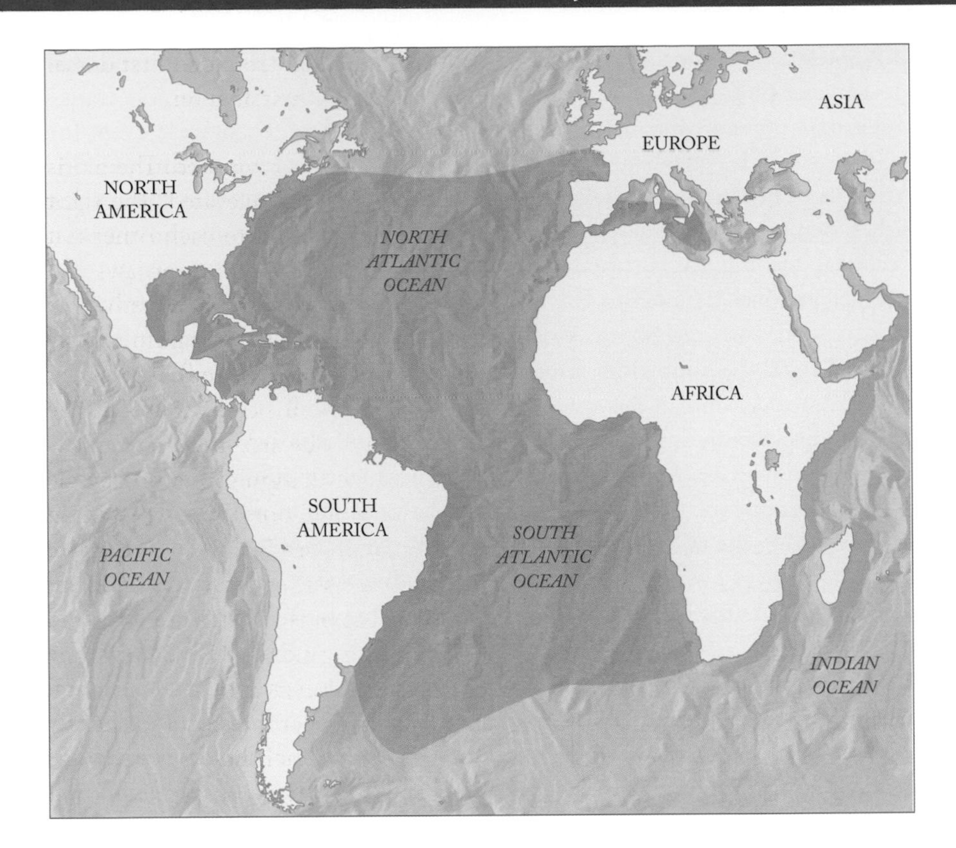

Left: White Marlin range map.

SIZE: White Marlin reach a maximum size of over 290 cm in total length and over 120 kg in weight. The IGFA all-tackle angling record is a fish weighing 181 lb, 14 oz (82.50 kg) caught off Vitoria, Brazil, in December 1979. The size of White Marlin caught by pelagic longline vessels ranges from 130 to 210 cm LJFL, with most fish between 150 and 180cm LJFL.

HABITAT AND ECOLOGY: White Marlin are pelagic and oceanodromous. Results from studies using pop-up satellite archival tags indicate that they spend the majority of their time in the upper 20 m of the water column and at temperatures within a few degrees of the surface temperature. More than 95% of the time is spent at depths <100 m and at temperatures within 8°C of the surface temperature. Vertical excursions, presumably for feeding, occur to depths below the thermocline and have been reported to almost 400 m. In general, White Marlin make deeper and longer vertical excursions during daylight hours. Throughout the western North Atlantic, habitat utilization varies among areas, with fish being more surface oriented in waters with a shallow thermocline depth.

MOVEMENTS: Conventional tagging data demonstrate that White Marlin are capable of traveling long distances, including trans-Atlantic and trans-equatorial movements; however, the majority of tag recaptures, even for individuals with considerable times at large, indicate limited net movements. Analysis of net movement versus time-at-large data suggests that the species either demonstrates seasonal site fidelity or is restricted in its movements. Seasonal site fidelity is supported by one individual tagged with a pop-up satellite archival tag off the US mid-Atlantic coast in the fall that travelled to waters off northern South America during the winter, returning to the US mid-Atlantic the following summer. Satellite tag studies also demonstrate that White Marlin can move long distances over a short time period. Mean 30-day displacements for 18 individuals tagged off the US mid-Atlantic in the late summer was 1,181 km, with one individual travelling 2,643 km in 30 days. There appears to be seasonal and spatial variation in movements as well. Fish tagged in the fall off the US mid-Atlantic coast demonstrated significant movements over a six-month period, while individuals tagged in the fall near the island of Aruba in the Caribbean had restricted movements.

FOOD: White Marlin have been seen to pick off individual fish by engulfing them, and whole specimens that appeared to be unscathed are sometimes found in the stomachs. But they also can kill or stun their food by spearing it or hitting it with their bill. White Marlin appear to be generalist feeders, with fish and squid comprising the majority of their diet. Squid (*Loligo pealei*) and Round Herring (*Etrumeus sadina*) were the most important food of White Marlin taken off New Jersey and Maryland in 1939 and 1959. In the Gulf of Mexico, the most consistently important food items observed from 1966 to 1971 were squids, Dolphinfish (*Coryphaena hippurus*), and Blue Runner (*Caranx crysos*). Mackerels were next in importance, followed by flyingfishes, and bonitos also play a big part. Along the central Atlantic coasts, Round Herring (*Etrumeus teres*) and squid (*Loligo pealei*) were the major prey items, but carangids are also well represented. Squids were found in 53.5% of the stomachs of 20 White Marlin caught in the mid-Atlantic bight. Off northeastern Brazil, the Oceanic Pomfret (*Brama brama*) and the Atlantic bird squid (*Ornithoteuthis antillarum*) comprised 67% of the diet of 105 White Marlin sampled. The most important food items found in a large sample of White Marlin from the southwestern equatorial Atlantic Ocean were juvenile Flying Gurnard (*Dactylopterus volitans*), with 27.9% of occurrence, and the Atlantic bird squid (*Ornithoteuthis antillarum*), with 21.2% of occurrence. Ten species of fishes were found in the stomachs of White Marlin from southern Brazil, with Flying Gurnard and Atlantic Cutlassfish (*Trichiurus lepturus*) being found the most frequently. Other food items included a broad array of epipelagic fishes and squids, including dolphinfish, Skipjack Tuna, Frigate Tuna, and even a Thorny Skate (*Amblyraja radiata*) from a White Marlin caught off Cape May, NJ.

REPRODUCTION: There is no apparent sexual dimorphism in White Marlin, but females attain larger sizes than males. Available estimates for size at 50% maturity for females vary from 149.0 cm LJFL (western equatorial Atlantic) to 160.4 cm LJFL (western central Atlantic) and 189.9 cm LJFL (Caribbean Sea). The information presently available indicates that White Marlin spawn once a year. Knowledge of spawning seasons, areas, and mating behavior is incomplete, because of difficulties in identifying the eggs and larvae and the lack of continuous and comprehensive gonad and ichthyoplankton surveys. White Marlin migrate into subtropical waters to spawn, with peak spawning occurring in early summer in both the western North Atlantic and western South Atlantic. Spawning occurs in deep and blue oceanic waters, generally at high temperatures ranging from 24–29°C and high surface salinities (over 35 ppt). Previous reports have mentioned spawning in the same area where Blue Marlin spawn in the Mona Passage (Caribbean Sea) in April and May and off southeast Brazil, but later in the year, from April to June. In the western North Atlantic, White Marlin have been reported to spawn in the Gulf of Mexico in June. Off southern Brazil (25–26° S and 40–45° W) White Marlin spawn from December to March.

AGE AND GROWTH: White Marlin ages of 1 to 13 years were estimated using counts of annular rings deposited on anal-fin spines. Longevity based on maximum time at liberty for fish tagged with conventional tags is estimated at 15 years. Given an observed longevity of 15 years and a length at maturity of approximately 145–60 cm LJFL (corresponding to an age at 50% maturity of 2.5–4 years for both sexes), the estimated generation length is 4.5–6 years.

EARLY LIFE HISTORY: A postlarval White Marlin, 124.9 mm BL, from the Florida Current was described and illustrated by de Sylva. Molecular techniques were used to identify larvae of three western Atlantic billfishes, including White Marlin, by Luthy et al. Their paper also included a key to identification of postlarvae and illustrations of the heads of the three species. A summary of characteristics of the early stages of most of the western Atlantic billfishes, with illustrations, is presented by Richards and Luthy.

STOCK STRUCTURE: Population structuring of White Marlin has been by conventional tagging studies that have demonstrated seasonal site fidelity for some individuals and the results of one long-term satellite tag that showed the individual made a large, circular movement throughout the western North Atlantic over the course of one year. However, recent genetic studies provide no evidence of significant population structure. ICCAT manages White Marlin as a single stock throughout the Atlantic Ocean.

FISHERIES INTEREST: White Marlin are taken in directed artisanal and recreational fisheries and as an incidental catch or bycatch in pelagic longline fisheries for tunas and swordfish and tuna purse seine fisheries. Over 90% of the reported landings are attributed to bycatch in longline fisheries, and there are also important directed recreational fisheries. The quality of the flesh is excellent. It is mostly marketed frozen in Japan.

White Marlin catch for the Atlantic has been recorded since 1956. Atlantic catches peaked in 1965 at nearly 5,000 mt, oscillated between 1,000 and 1,500 mt until 1993, increased to 1,900 mt in 1994, declined below 1,000 mt from 1995–2004, and have recently been below 500 mt. The decrease in catches over the past 10 years may be partially due to management measures requiring live release, reduced total allowable catches, and, more recently, country-specific quotas. It is recognized that there is significant underreporting of catches from small-scale artisanal fisheries. Also, catches of White Marlin have been inflated due to misidentification of Roundscale Spearfish.

THREATS: There is a lack of data reporting for the artisanal fisheries, and failure to record releases as well as catches has biased catch and effort data from the pelagic longline fleet. The observed distribution of several large pelagic predators, including the White Marlin, has significantly contracted from the 1960s to 2000. Petitions to declare White Marlin an endangered species in the United States were not accepted (White Marlin Review Team). There has been a continuous decline in the abundance of White Marlin since the beginning of exploitation of this species. Using a generation length of between 4.5 and 6.5 years, the estimated decline in overall stock abundance is 9–37% during the last 13 and 20 years, respectively. The most recent stock assessment shows a slight stabilization or increase in abundance in recent years; however, more data are needed to confirm whether this is accurate. Considering that this species is well below the biomass at maximum sustainable yield (BMSY) and is not considered to be well managed, it is listed as Vulnerable on the IUCN Red List.

CONSERVATION: White Marlin are currently considered to be severely overfished; however, the most recent assessment (ICCAT) indicated that overfishing is not likely occurring. This is a highly migratory species mentioned in Annex I of the 1982 Convention on the Law of the Sea. Mandatory live release of White Marlin caught on pelagic longlines or in purse seines was implemented by ICCAT in 2001. Subsequent management measures imposed country-specific catch limits, and in 2013 ICCAT adopted lower country-specific allocations. The US, Bermuda, and the Bahamas similarly do not allow commercial harvest of White Marlin, and in the US Atlantic, commercial harvest and import of marlins is prohibited. Brazil prohibits sale of White Marlin, and Mexico allows no commercial take within 50 miles of its coast. The US recreational fishery is limited to a total of 250 Blue Marlin and White Marlin (including Roundscale Spearfish) combined, and in 2013 ICCAT adopted a minimum size of 66 inches (167.6 cm) for all White Marlin recreational fisheries. In the United States, anglers are required to use circle hooks when fishing with natural baits in billfish tournaments. A meta-analysis of seven White Marlin studies showed high survival rates after release regardless of method of capture.

REFERENCES: Beerkircher and Serafy 2011; Beerkircher et al. 2009; Collette 2010; Collette et al. 2006, 2011a, 2011b; de Sylva 1963; de Sylva and Davis 1963; Die and Drew 2008; Drew et al. 2006; Ferreira and Hazin 2004; Graves 1988; Graves and McDowell 2006, 2012; Hanner et al. 2011; Horodysky et al. 2007; ICCAT 2003, 2007; IGFA 2018; Luthy et al. 2005; Mather et al. 1972, 1975; Musyl et al. 2014; Nakamura 1985; Nobrega et al. 2009; Oliveira et al. 2007; Orbesen et al. 2008; Ortiz et al. 2003; Pine et al. 2008; Pinheiro et al. 2010; Piva-Silva and Amorin 2014; Prince et al. 2005; Restrepo et al. 2003; Richards and Luthy 2005; Serafy et al. 2009; Staudinger et al. 2012; Talbot and Penrith 1963; Vaske Júnior et al. 2004; Wallace and Wallace 1942.

Striped Marlin

Kajikia audax (Philippi, 1877)

COMMON NAMES: English – Striped Marlin
French – Marlin Rayé
Spanish – Rayado
Japanese – Makajiki

ETYMOLOGY: The Striped Marlin was described by the Chilean naturalist Rudolph Philippi in 1887 using the species name *audax*, which comes from the Latin *audacia*, meaning "bold."

SYNONYMS: *Histiophorus audax* Philippi, 1887; *Tetrapturus mitsukurii* Jordan and Snyder, 1901; *Tetrapturus ectenes* Jordan and Evermann, 1926; *Makaira grammatica* Jordan and Evermann, 1926; *Makaira holei* Jordan and Evermann, 1926; *Makaira zelandica* Jordan and Evermann, 1926; *Kajikia formosana* Hirasaka and Nakamura, 1947; *Tetrapturus tenuirostratus* Deraniyagala, 1951; *Marlina jauffreti* Smith, 1956

TAXONOMIC NOTE: The Striped Marlin was previously included in the genus *Tetrapturus*. It is genetically similar to the White Marliln, *Kajikia albida,* but represents a distinct evolutionary unit.

FIELD MARKS:

1 The anterior lobe of the first dorsal fin is pointed and higher than the remainder of the fin, the height decreasing gradually backward.
2 The height of the dorsal fin is greater than the depth of the body.
3 The anus is situated near the origin of the first anal fin, the distance between them less than half of the first anal-fin height.
4 The tips of the pectoral and first anal fins are pointed.

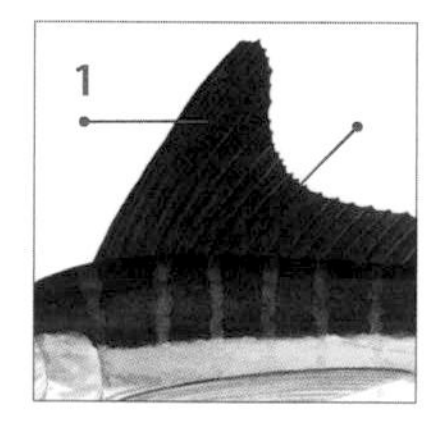

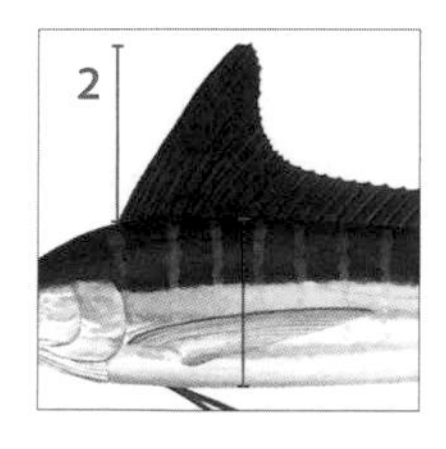

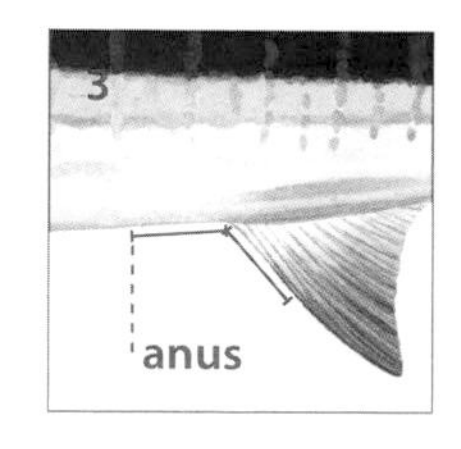

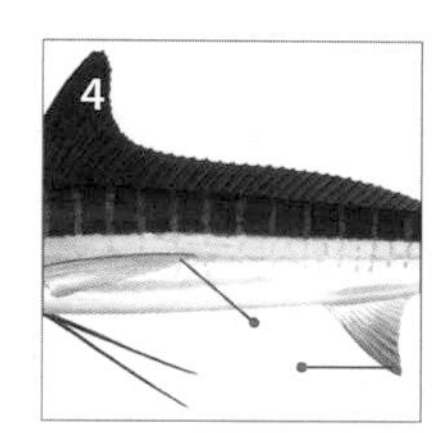

UNIQUE FEATURE: The only Pacific billfish with a pointed first dorsal fin with a height greater than the body depth and the anal opening located close to the first anal fin.

DIAGNOSTIC FEATURES: The body is elongate and moderately compressed. The bill is stout and long, round in cross section. The nape is fairly elevated. The right and left branchiostegal membranes are completely united to each other, but free from the isthmus. Gill lamellae are bound by an inter-lamellar fusion, which connects the leading edge of adjacent gill lamellae on the same filament. Gill rakers are absent. Both upper and lower jaws and the palatine bones (on the roof of the mouth) have small, file-like teeth. **Fins:** There are two dorsal fins, the first with 37–42 rays. The first usually has a pointed anterior lobe, higher than body depth anteriorly, which abruptly decreases in height to about the 10th dorsal-fin ray and gently decreases further backward. The first dorsal-fin base is long, extending from above the posterior margin of the preopercle to just in front of the second dorsal-fin origin. The second dorsal fin has 5 or 6 rays, its position slightly posterior with respect to the second anal fin. There are two anal fins, the first with 13–18 rays, the second with 5 or 6 rays and very similar in size and shape to the second dorsal fin. The pectoral fins are long and narrow, with pointed tips, adpressible against sides of body, and with 18–22 rays. The pelvic fins are slender and almost equal to or slightly shorter than the pectoral fins in large individuals, slightly longer in smaller individuals. **Caudal peduncle:** The caudal peduncle is laterally compressed and slightly depressed dorsoventrally, with a pair of keels on each side and a shallow notch on both the dorsal and ventral surfaces. **Anus:** The anus is situated just in front of the first anal-fin origin. **Lateral line:** The lateral line is single and obvious, curving above the base of the pectoral fin and then continuing in a straight line toward the caudal-fin base. **Scales:** The body is densely covered with elongate bony scales, each with 1 or 2 posterior points. **Color:** The body is blue-black dorsally and silvery white ventrally, with about 15 rows of cobalt-colored stripes, each consisting of round dots and/or narrow bands. The first dorsal fin is dark blue; the other fins are usually dark brown, sometimes tinged with dark blue. The bases of the first and second dorsal fins are tinged with silvery white.

GEOGRAPHIC RANGE: Striped Marlin are widely distributed in tropical and temperate waters of the Indo-Pacific, and strays are occasionally found in the eastern South Atlantic. They are strongly oceanic and rarely enter coastal waters. They are the most widely distributed of all Pacific istiophorids and are more abundant in the eastern Pacific and central North Pacific than the western Pacific.

Historic Japanese longline records indicate that adult Striped Marlin have a somewhat horseshoe-shaped distribution in the Pacific Ocean, which is unique among Pacific billfishes. The base of this distribution centers on the south Central American coast and runs on either side of the equator across to the Western Pacific Ocean. The central equatorial region in the western and central Pacific is characterized by very low and intermittent hook rate for Striped Marlin and thus is not considered part of the normal distribution. Early pelagic longline data indicated the areas of highest abundance were in the central and eastern North Pacific, while the waters of the southern and western Pacific were areas of lesser abundance. In the south and southwest Pacific, adults are most abundant during the austral spring and summer and least abundant in the winter. This region includes eastern Australia, where Striped Marlin are present from the Coral Sea to Tasmania (depending on currents and seasons).

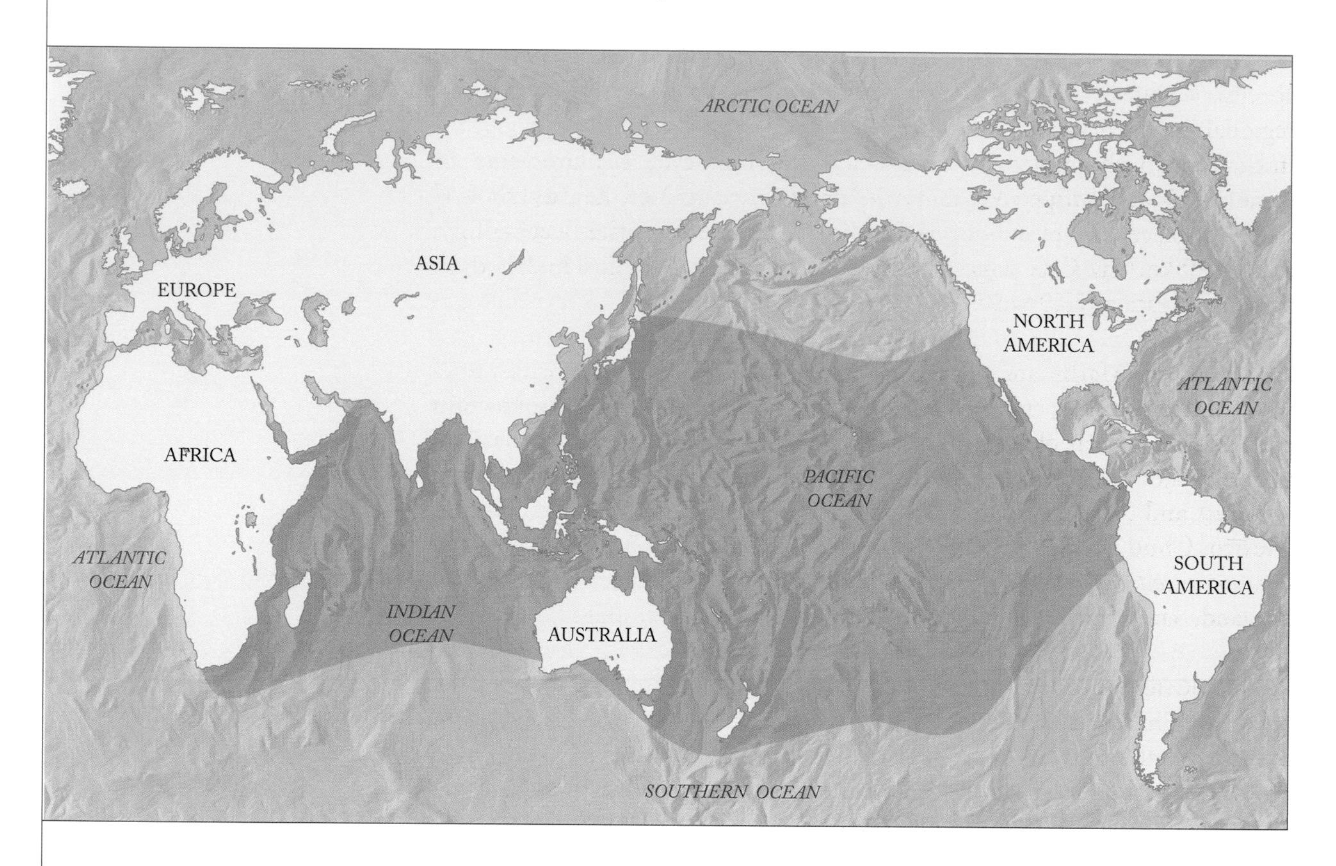

Above: Striped Marlin range map.

SIZE: Maximum size to 420 cm total length. The IGFA all-tackle game fish record is of a 494 lb (224.1 kg) fish caught off Tutukaka, New Zealand, in January 1986. The average size of Striped Marlin is much larger in the western South Pacific relative to the western, central, and eastern North Pacific.

HABITAT AND ECOLOGY: Striped Marlin are widely distributed, pelagic, and oceanodromous, usually found above the thermocline and shallower than 120 m, although they have been known to occur as deep as 532 m. They generally inhabit cooler water than either Black Marlin or Blue Marlin. Abundance increases with distance from the continental shelf; they are usually seen close to shore only where deep drop-offs occur. Striped Marlin are mostly solitary but form small schools by size during the spawning season and are also known to feed cooperatively on schooling prey. They are usually dispersed at considerably wide distances. Pop-up satellite archival tag deployments throughout the Pacific Ocean demonstrate that Striped Marlin spend more than 50% of their time in the upper 10 m of the water column and tend to be at shallower depths at night than during the day. Striped Marlin occur over a wide range of surface temperatures (15–31°C) but spend almost all of their time in waters within 8°C of the sea surface temperature.

MOVEMENTS: Conventional tag studies have shown Striped Marlin are capable of undertaking movements over 6,000 km. Trans-equatorial movements have been noted, but there have been no trans-Pacific recoveries. Unlike some species of billfish, Striped Marlin conventional tag recoveries do not suggest regional site fidelity or cyclical movements. In contrast, satellite tagging results indicate relatively limited movements and potential cyclical movements for some individuals. Striped Marlin tagged and tracked off New Zealand showed modal displacement rates of 20–30 km/day, but sometimes rates were as high as 50–120 km/day. One tagged Striped Marlin moved 1,461 km in 20.6 days, for a mean rate of 70.9 km/day.

FOOD: Striped Marlin are opportunistic feeders and are known to feed on a wide variety of fishes, crustaceans, and squids. They may use their bill to stun prey and have been observed to feed cooperatively on schooling fishes and squids. Major food items off the coast of California include Saury (*Cololabis adocetus*) and Northern Anchovy (*Engraulis mordax*), off the west coast of Mexico, Chub Mackerel, Pacific Sardine (*Sardinops sagax*), Round Herring (*Etrumeus sadina*), and Humbolt (jumbo) squid (*Dosidicus gigas*), and off New Zealand, Saury, Pacific Sardine, jacks, Chub Mackerel, and squid.

REPRODUCTION: Striped Marlin spawning is known to occur in several areas within the Indo-Pacific, evidenced by the seasonal occurrence of mature females and early life history stages. Larvae are most abundant in the respective local early summers. The lower temperature limit for the distribution of larvae is approximately 24°C, both in the Indian and Pacific oceans. Spawning sites are between 10° S and 30° S in the southwest Pacific and 10° S and 20° S in the northeastern Indian Ocean. A central Pacific spawning ground was documented by Hyde et al. based on identification of seven larvae from Hawaiian waters. Spawning activity in the southern Coral Sea, off the coasts of Fiji, and south of French Polynesia was confirmed by Kopf et al. Spawning occurs in concentrated aggregations across a broad longitudinal band. Ripe Striped Marlin were found from May through December in the southern Gulf of California. Batch fecundity ranges from about 2.2 to 4.1 million oocytes per spawning event for females ranging in size from 223–269 cm LJFL. An average 109 kg female may produce an annual reproductive output of 90–281 million hydrated oocytes spread across 27–90 spawning events per season. Age at 50% maturity is 2.5 years. Length at 50% maturity was 160 cm LJFL for males, 210 cm LJFL for females.

Sex ratios in three tropical areas of the eastern North Pacific were equal in the open ocean, biased to males in the near-continental area, and biased to females in the near-equatorial region.

AGE AND GROWTH: Striped Marlin have been aged using putative annular rings on fin spines. In the western South Pacific, age of females ranged up to 8.5 years at 287 cm lower jaw fork length (LJFL) and for males up to 7.0 years at 254 cm LJFL. In the eastern tropical Pacific, 10 age groups, 2 to 11, were found from the rings on the fourth dorsal spine, with the most abundant being age group 7. Striped Marlin reach 45% of their asymptotic growth (in length) in the first year of life; growth slows down to 10% in the second year and then averages 4% per year over the next eight years.

EARLY LIFE HISTORY: Ueyanagi reported on 42 larval Striped Marlin ranging in size from 2.9 mm to 21.2 mm standard length. He illustrated a developmental series of seven specimens, 2.9 to 21.2 mm. They resemble Sailfish but can be distinguished by the eye position being relatively low and the tip of the snout and center of the eye being at about the same level.

STOCK STRUCTURE: Early genetic studies showed significant stock structuring of Striped Marlin in the Pacific Ocean, an observation that was consistent with conventional and satellite tagging studies that show relatively restricted movement and the existence of geographically separated spawning grounds. Expanded genetic studies support the existence of at least four different genetic stocks: western Indian Ocean, Oceania, North Pacific Ocean, and eastern Pacific Ocean, with the possibility of additional structure within the North Pacific Ocean. The Indian Ocean stock structure and the relationships of those fishes to the stocks in the Pacific have not been investigated.

FISHERIES INTEREST: Striped Marlin is an important commercial and recreational resource throughout its range, with bycatch from pelagic longline fisheries targeting tunas and swordfish accounting for the greatest reported landings. Commercial catch is taken mostly by pelagic longlining, with less than 1% of the total catch taken by harpooning. The quality of the flesh is among the best of the billfishes for sashimi and sushi. It is marketed mostly frozen, sometimes fresh. FAO-reported landings of Striped Marlin increased to 16,000 mt in the 1950s and peaked at over 26,000 mt in the 1960s. Landings varied from 12,000 to 20,000 mt through the late 1990s and have ranged between 8,000 and 12,000 mt since that time.

In the western and central Pacific Ocean (WCPO), catches of Striped Marlin were dominated by the Japanese longline fleet until the early 1990s. Taiwanese and Korean fleets reported relatively small catches of striped marlin since the mid-1960s and mid-1970s, respectively. However, Taiwanese catches have increased in recent years, mainly due to the high effort of this fleet in the eastern temperate WCPO, targeting albacore tuna. Longline fleets of Pacific Island

Countries and Territories, and of Australia and New Zealand, have reported increasing catches of Striped Marlin since the early 1990s, mainly due to the development of these domestic fleets. Catches by Australian longline fleets have rapidly increased in recent years due, at least in part, to specific targeting of Striped Marlin by some vessels during some periods. For the period 1991–2002, Striped Marlin represented about half of the istiophorid billfish catch by the US Hawaii-based longline fleet.

Extensive recreational fisheries for Striped Marlin exist throughout the eastern North Pacific and western South Pacific Ocean, although total landings by recreational fisheries are very small relative to commercial landings. In addition, a high proportion (up to 60%) of Striped Marlin are (tagged and) released by recreational fisheries in the WCPO.

THREATS: This species may be threatened by the expansion of longline fisheries and also increased artisanal fisheries in the tropical eastern Pacific region. Globally, it is estimated that Striped Marlin biomass has declined 20–25% over three generation lengths (16 years). This species is not considered to be well managed, and stock assessments are needed, especially in the Indian Ocean and the western North Pacific. Overfishing is currently occurring relative to FMSY, and the western and central Pacific stock is in an overfished state. Striped Marlin was listed as Near Threatened on the IUCN Red List.

CONSERVATION: Striped Marlin is a highly migratory species listed in Annex I of the 1982 Convention on the Law of the Sea and managed by three different international regional fisheries management organizations: the IATTC, the WCPFC, and the IOTC.

Western South Pacific: The most recent assessment considered a single population of Striped Marlin within the region from the equator to latitude 40° S and from 140° E to 130° W. The results of the model were considered preliminary, as there remains a great deal of uncertainty regarding some of the key parameters included in the assessment model, in particular natural mortality and growth rates. The assessment indicates that current levels of fishing mortality may approximate or exceed FMSY and current biomass levels may approximate or be below BMSY. The estimated decline in spawning biomass over a three-generation-length period (from 1991 to 2006) was 8%. The fishery is not considered to be well managed. Since 1987, longline fleets operating in the New Zealand exclusive economic zone (EEZ) have been prohibited from landing Striped Marlin, in an attempt to support recreational fisheries in the north of the country.

Eastern North Pacific: The most recent assessment considered a single stock of Striped Marlin in the eastern North Pacific Ocean. The results of the Stock Synthesis model indicated the stock is not overfished and that overfishing is not occurring. Over a three-generation-length period (1992–2008), the population decline was calculated to be 0%, as the spawning biomass has been increasing since 2003. However, mean size modes of longline-caught Striped Marlin have declined dramatically in the eastern North Pacific over the past 30 years. It is highly likely that the stock has been experiencing excessive fishing mortality since the 1980s. There are no conservation measures for this species in the eastern Pacific, but it is recommended that there should be no increase in fishing for this species' northern population in the Eastern Pacific.

Western North Pacific: Previous stock assessments have been conducted, combining data from both the eastern and western North Pacific. Using the available data from Brodziak and Piner, a decline of 55–60% was estimated for the North Pacific stock as a whole. It is likely that declines are in the order of 40–50% for the Northwest Pacific. The International Scientific Committee for Tuna and Tuna-Like Species in the North Pacific Ocean (ISC) recommended that fishing mortality for Striped Marlin in the North Pacific not be permitted to exceed current levels. There is a need for a new stock assessment for the northwest Pacific. This stock is not considered to be well managed.

Indian Ocean: Catch rates of Striped Marlin in the Indian Ocean have decreased dramatically over time. Nominal CPUE of Japanese longliners operating off northwest Australia dropped 40.1% over a three-generation time period, while the CPUE decline for Japanese longliners in the Sychelles was 95% over the same period. The most recent assessment for Striped Marlin in the Indian Ocean indicates that the stock may be overfished, but overfishing is not occurring. There was considerable uncertainty with the assessment due to a lack of knowledge of biological parameters and the length-frequency composition for several fisheries. The IOTC recently adopted management measures encouraging parties to reduce catches to levels below the average of levels from 2009–14.

REFERENCES: Abitia-Cardenas et al. 1997, 2002; Baker 1966; Brill et al. 1993; Brodziak and Piner 2010; Bromhead et al. 2004; Collette et al. 2006, 2011; Domeier et al. 2003; González-Armas et al. 2006; Graves 1998; Graves and McDowell 1994; Hinton and Bayliff 2002; Hinton et al. 2010; Holdsworth et al. 2008; Hubbs and Wisner 1953; Hyde et al. 2006; IATTC 2008; IGFA 2017; IOTC 2009; ISC 2007; Kailola et al. 1993; Kopf 2005; Kopf et al. 2011, 2012; Lam et al. 2015; Langley et al. 2006; Lee et al. 2012; Matsumoto et al. 2001; McDowell and Graves 2008; Melo-Barrera and Uraga 2004; Melo-Barrera et al. 2003; Morrow 1952; Musyl et al. 2014; Nakamura 1985; Ortiz et al. 2003; Pepperell and Davis 1999; Pillai and Ueyanagi 1977; Pine et al. 2008; Piner et al. 2013; Purcell and Edmands 2011; Shimose et al. 2013; Sippel et al. 2007, 2011; Squire and Suzuki 1990; Ueyanagi 1959; Ueyanagi and Wares 1975; Uozumi 1999; Walsh et al. 2005; Wang et al. 2006; Wegner et al. 2006; Whitelaw 2001; Worm and Tittensor 2011.

Blue Marlin

Makaira nigricans Lacepède, 1802

COMMON NAMES: English – Blue Marlin
French – Empereur, Empereur Bleu
Spanish – Marlin Azul

ETYMOLOGY: The Blue Marlin was described by the French ichthyologist Bernard Lacepéde from the Bay of Biscay, France, with no etymology for the name he chose, but *nigricans* means "blackish," from the Latin *niger*, for "black."

SYNONYMS: *Makaira nigricans* Lacepède, 1802; *Tetrapturus herschelii* Gray, 1838; *Tetrapturus amplus* Poey, 1860; *Tetrapturus mazara* Jordan and Snyder, 1901; *Makaira bermudae* Mowbray, 1931; *Makaira perezi* de Buen, 1950; *Eumakaira nigra* Hirasaka and Nakamura, 1947; *Istiompax howardi* Whitley, 1954

TAXONOMIC NOTE: Blue Marlin in the Indo-Pacific were previously known as *Makaira mazara*, but based on genetic data, they are conspecific with Atlantic Blue Marlin, *Makaira nigricans*, which name has priority.

FIELD MARKS:
1 The pectoral fins are long and narrow and adpressible to sides of body.
2 The body is not very compressed laterally.
3 The nape is highly elevated.
4 The height of the anterior lobe of the first dorsal fin is less than the body depth.
5 The position of the second dorsal fin is slightly posterior to that of the second anal fin.

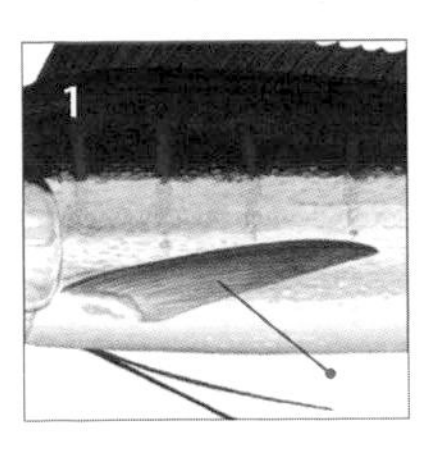

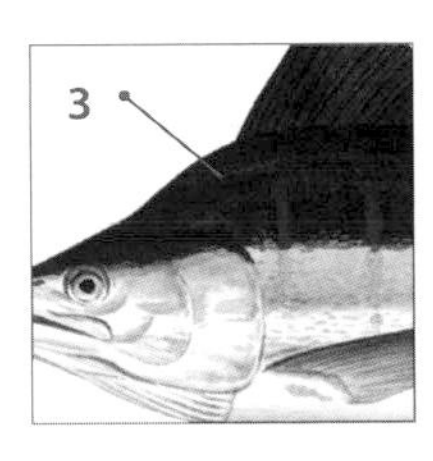

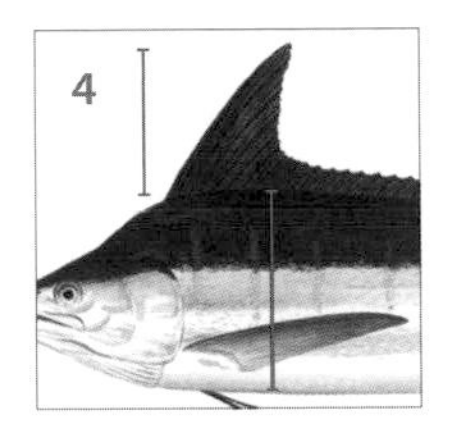

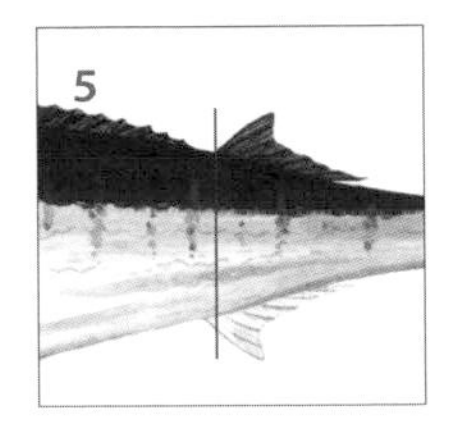

UNIQUE FEATURE: The Blue Marlin is the largest billfish, reaching 1,800–2,000 lb (816–907 kg).

DIAGNOSTIC FEATURES: The body is not strongly compressed laterally. The bill is long, very stout, and round in cross section. The nape is conspicuously elevated. The left and right branchiostegal membranes are completely united to each other but free from the isthmus. Gill rakers are absent. Both upper and lower jaws and the palatine bones (on the roof of the mouth) have small file-like teeth. **Fins:** There are two dorsal fins, the first with 39–43 rays and a pointed anterior lobe, the height of which is less than the body depth in adults. The dorsal fin originates above the posterior margin of the preopercle and ends near the second dorsal-fin origin. The second dorsal fin has 6 or 7 rays, and its position is slightly posterior to that of the second anal fin. There are two anal fins, the first with 13–16 rays, the second with 6 or 7 rays, very similar in size and shape to the second dorsal fin. The pectoral fins are long and narrow, adpressible to sides of body, with 19–22 rays. The pelvic fins are shorter than the pectoral fins, with a poorly developed membrane, and depressible into deep ventral grooves. **Caudal peduncle:** The caudal peduncle is compressed laterally and slightly depressed dorsoventrally, with strong double keels on each side and a poorly developed notch on both the dorsal and ventral surfaces. **Anus:** The anus is situated just in front of the origin of the first anal fin. **Lateral line:** The lateral line forms a complicated network pattern, obvious in immature fish but obscure in adults, as it becomes progressively imbedded in the skin. However, the lateral line always becomes clearly visible when the epidermis is removed. **Scales:** The body is densely covered with thick, bony scales, each with usually 1 or 2 (mostly 1) and sometimes with 3 posterior points. **Vertebrae:** Vertebrae number 24 (11 precaudal and 13 caudal). **Color:** The body is blue-black dorsally and silvery white ventrally, with about 15 rows of pale cobalt-colored stripes, each consisting of round dots and/or narrow bars. The first dorsal-fin membrane is blackish or dark blue, without any dots or marks. The other fins are usually brown-black, sometimes tinged with dark blue. The bases of the first and second anal fins are tinged with silvery white.

GEOGRAPHIC RANGE: Blue Marlin primarily inhabits tropical and subtropical pelagic waters of the Atlantic and Indo-Pacific oceans with surface temperatures between 22°C and 31°C. It is the most tropical of the billfishes. Its latitudinal range changes seasonally, expanding into higher latitudes in the warmer months and contracting toward the equator in colder months. Its latitudinal range in the Atlantic, based on data from the pelagic longline fishery, extends from about 40–45° N in the North Atlantic to 40° S in the western South Atlantic, 30° S in the central South Atlantic, and 35° S in the eastern South Atlantic. In the Indo-Pacific, its latitudinal range extends to about 45° N in the western North Pacific Ocean, 35° N in the eastern North Pacific, 35° S in the western South Pacific, 25° S in the eastern South Pacific, 40–45° S in the southwestern Indian Ocean, and 35° S in the southeastern Indian Ocean. In both the eastern tropical Pacific and the eastern tropical Atlantic, Blue Marlin concentrates in shallower waters than in the western part of both oceans, due to hypoxia-based habitat compression over oxygen-minimum zones in the eastern tropical seas.

Japanese pelagic longline catch data show two main seasonal concentrations of Blue Marlin in the western Atlantic: one from January through April in the western South Atlantic, between 5° and 30° S, and the other from June through October, in the western North Atlantic, between 10° and 35° N. May, November, and December appear to be transitional months. In the eastern Atlantic, where Blue Marlin are less abundant, they occur mostly off the east coast of Africa between 25° N and 25° S. In the Pacific, the monthly distribution of catches shows two main seasonal concentrations: one from December through March in the western and central South Pacific between 8° and 26° S, and the other from May through October in the western and central North Pacific between 2° and 24° N. In the remaining two months (April and November) the fish tend to concentrate in the equatorial Pacific between 10° N and 10° S. Blue Marlin become less abundant toward the eastern Pacific. In the Indian Ocean, they are known to be relatively abundant around Sri Lanka and Mauritius. Off the east coast of Africa, they are apparently abundant between the equator and 13° S during the southeast monsoon period (from April to October).

Below: Blue Marlin range map.

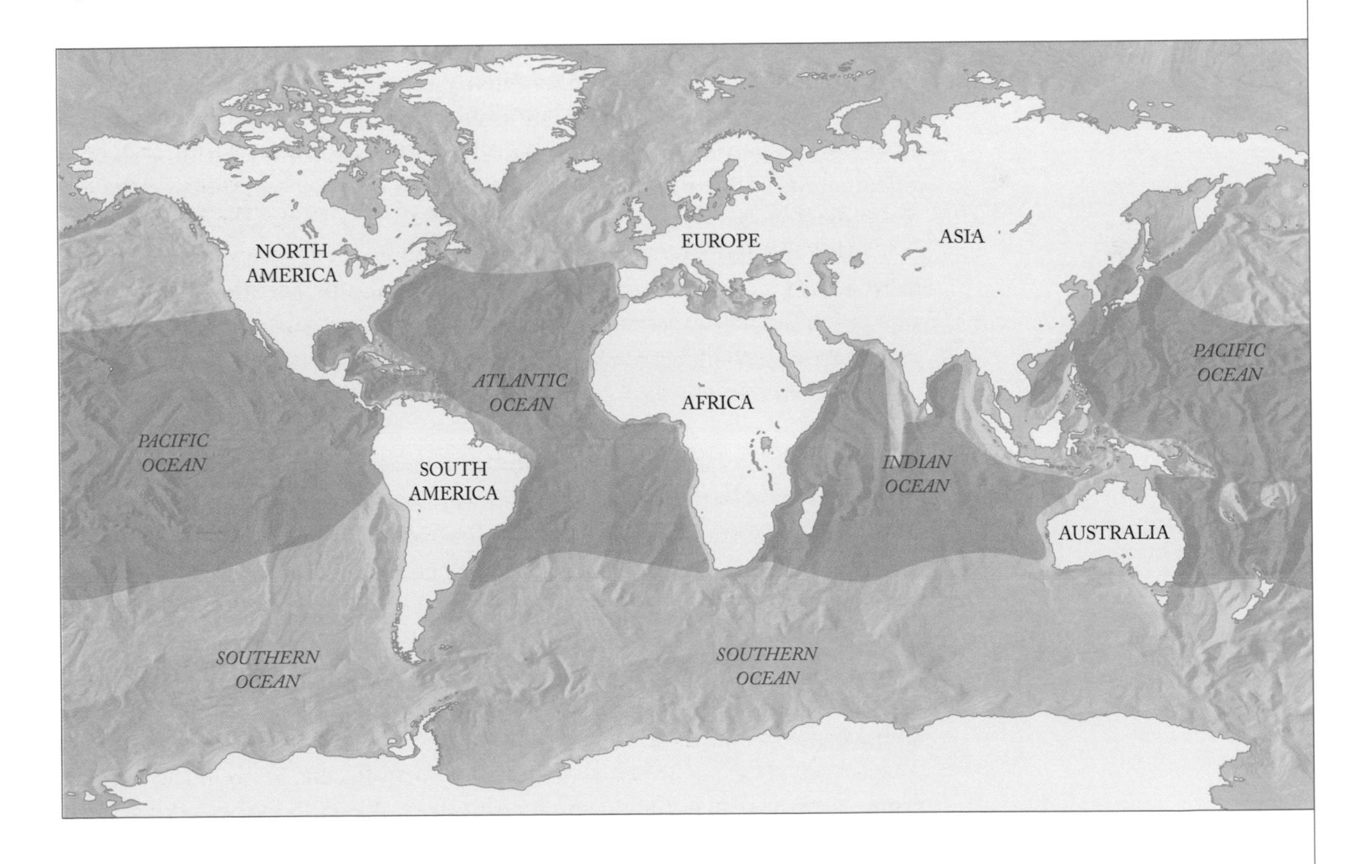

SIZE: Blue Marlin are reported to attain sizes over 2,000 lb (907 kg). The heaviest Blue Marlin taken on sport fishing gear weighed 1,805 lb (818 kg) and was caught off Waikiki, Hawaii. However, this was an unofficial game fish record, as more than one person fought the fish. The official IGFA all-tackle angling record in the Pacific is a 1,376 lb (624 kg) fish caught at Kona, Hawaii, in May 1982, and the IGFA all-tackle game fish angling record in the Atlantic is a 1,402 lb (636 kg) fish caught off Vitoria, Brazil, in February 1992. Blue Marlin caught by pelagic longliners average approximately 200–275 cm LJFL in the Atlantic, 200–85 cm LJFL in Pacific equatorial waters, and 215–300 cm LJFL in the Indian Ocean.

HABITAT AND ECOLOGY: Blue Marlin are epipelagic and oceanic, typically occurring in clear (blue) surface waters above 25°C. They are primarily daylight sight feeders, and while the majority of their time is spent in the upper 10 m, they regularly dive into and below the thermocline, as deep as 1,000 m, presumably in search of prey. They are not usually seen close to land masses or islands, unless there is deep water (>1,000 m) close to shore, such as in the waters around oceanic islands like Hawaii. Blue Marlin are typically solitary, but may form small aggregations for feeding or possibly mating.

Acoustic and electronic tagging of Blue Marlin have provided considerable information on habitat utilization. The fish spend the majority of their time in the upper 10 m but are capable of making dives below 500 m. In many areas mean depths occupied at night are shallower than those during the day, as excursions to depth are more frequent during the day. Blue Marlin spend the vast majority of their time in water temperatures within 1 or 2 degrees of the sea surface temperature.

MOVEMENTS: Conventional tagging studies indicate that Blue Marlin are capable of long-distance dispersal, including trans-oceanic, trans-equatorial, and inter-ocean movements. Blue Marlin have the longest movements, inferred from conventional tagging, of any of the istiophorid billfishes. Unlike some other species of billfishes, plots of time at large versus minimum travel distances do not suggest cyclical annual movements or seasonal site fidelity. In the western North Pacific, north-south migration is related to temperature and feeding activities and differs between males and females. Like all billfishes, reporting rates for conventional tags are low, generally less than 1.5%, but many recaptures in the western Atlantic demonstrate seasonal connectivity between the US mid-Atlantic and Mexican Caribbean coasts with northern Venezuela. Acoustic and electronic tagging studies suggest mean minimum straight-line daily displacements of approximately 20 km, with some individuals exceeding 90 km/day.

FOOD: Blue Marlin are opportunistic feeders consuming a wide variety of prey, particularly fishes and squids. They feed mostly in near-surface waters but sometimes make trips to relatively deep water for feeding, as evidenced by the presence of deep-sea fishes such as *Pseudoscopelus* and a deep-dwelling squirrelfish in stomachs. Blue Marlin feed primarily during the day, but nocturnal feeding activity is reported to increase around the full moon. The size range of the prey taken by adult Blue Marlin is relatively wide, ranging from postlarval surgeonfishes to large (>50 kg) tunas. Blue Marlin have been observed to use their bill to stun larger prey, and Skipjack, Yellowfin Tuna, and Bigeye Tuna found in stomachs often show deep slashes on their bodies, presumably caused by the bill of the marlin. In the Gulf of Guinea, a 290 kg Blue Marlin swallowed a Bigeye Tuna weighing about 50 kg. Stomach content analyses demonstrate that Blue Marlin feeding varies by region and season, probably reflecting the availability of prey. Small tunas, especially frigate tunas (*Auxis*) and dolphinfish, are common food items, but in some studies squids and various pelagic fishes have comprised a large fraction of the diet. Twelve species of fishes were found in the stomachs of Blue Marlin from Brazil, with scombrids again being the most important component, especially Frigate Tuna (24%), Little Tunny (20%), and Blackfin Tuna (13.6%); other prey included Skipjack Tuna, Bigeye Tuna, dolphinfishes, flyingfishes, and squids. In the eastern Atlantic, off the coast of Portugal, Blue Marlin fed on pelagic fishes, especially Atlantic Chub Mackerel, Scad (*Trachurus*), and Bullet Tuna. Off Mazatlán, Mexico, in the eastern tropical Pacific, the main food items, based on the index of relative importance, were frigate tunas (*Auxis*, 52%) and jumbo squid (*Dosidicus gigas*). Off Baja California, after a Blue Marlin about 3 m body length found a school of jumbo squids measuring about 40 cm mantle length that were gathering under the night-light of a squid-fishing boat, it approached the school at almost full speed with its fins completely held back in the grooves, then suddenly hit the squids with its bill, subsequently nudging the stunned prey and eating it head first. Prey species in Okinawa included Slender Mola (*Ranzania laevis*), Frigate Tuna, and Skipjack Tuna.

REPRODUCTION: In the Pacific, size at first maturity of males is thought to range from 130 to 140 cm eye–fork length (86.8 to 87.8% of body length), with age at maturity estimated as two years. Blue Marlin spawning probably takes place year-round in equatorial waters to 10° latitude and during summer periods in both hemispheres to 30° N/S, in all oceans. Based on the capture of gravid females or early life history stages, Blue Marlin are known to spawn in several locations within the western North Atlantic Ocean, including the central and northern Caribbean Sea, the Gulf of Mexico, and even as far north as Bermuda. Off Brazil, spawning occurs February to March from 20–23° S, primarily in the Abrolhos Archipelago. In the Pacific Ocean, concentrations of spawning fish occur around French Polynesia, and gravid females, eggs, and larvae have been collected off Hawaii. In the Indian Ocean larvae have been collected around the Mascarene Islands and off the south coasts of Java and Sumatra.

AGE AND GROWTH: Blue Marlin females attain larger sizes than males; around the Bonin Islands, fish over 200 cm eye–fork length are all females; around

Taiwan Island males attain less than 120 kg weight, while females grow to over 300 kg. Maximum age is estimated to be at least 20 years, with estimates up to 27 years for females and 18 years for males. Age estimation in marlins is problematic, and longevity information from the Pacific has also been applied to the Atlantic. The generation length was estimated to be between 4.5 and 6 years. The maximum time at large for a conventionally tagged Blue Marlin is 11 years.

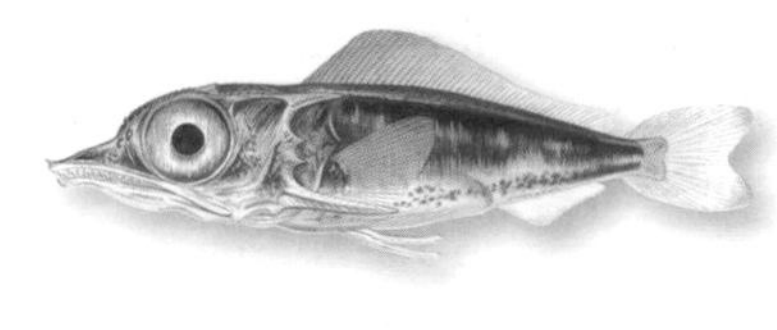

Above left: Juvenile Blue Marlin approx. 63 mm LJFL

Above right: Larval Blue Marlin, approx. 5 mm LJFL

EARLY LIFE HISTORY: Unfertilized eggs from running ripe female Blue Marlin are relatively colorless and spherical and range from 0.8 to 0.9 mm in diameter, while fertilized eggs, identified on shipboard with molecular genetic techniques, ranged from 1.24–1.30 mm in diameter. Pigmentation of early-stage eggs seems restricted to a single dense patch of melanophores on the oil globule. A summary of characteristics of the early stages of Blue Marlin, White Marlin, and Sailfish is presented by Richards and Luthy. Blue Marlin larvae grow fast. Growth rates were comparable to those for other pelagic species such as Sailfish and Dolphinfish and much higher than for several species of mackerels and tunas. Maximum (ca. 16 mm/day at 50 days) and sustained length growth rates (ca. 10 mm/day) during the first 100 days indicate that Blue Marlin are one of the fastest growing of all teleosts, and growth rates can vary regionally. Blue Marlin larvae begin to feed on larval pelagic fishes and zooplankton at 6 mm but switch almost entirely to fish larvae by 12 mm. Larval fish prey include Blue Marlin, Albacore, snake mackerels, and flyingfishes. Feeding is most intensive in the morning (0700–1000 hours) and evening (1600–1900 hours) and decreases sharply or ceases at night.

FISHERIES INTEREST: Blue Marlin are taken in directed artisanal and sport fisheries and also as incidental catch in pelagic longline fisheries targeting tunas and swordfish. The primary source of fishing mortality results from the pelagic longline fishery, although catches from artisanal fisheries are not well known and may represent a significant source of mortality in several coastal areas. Catches of Blue Marlin in the Atlantic Ocean increased dramatically in the late 1950s and early 1960s with the global expansion of the Japanese longline fleet, reaching peak values of 9,000 mt in the early 1960s and averaging 4,000–5,000 mt from the late 1980s until the early 2000s, when catch limits were established. Catches in the Pacific Ocean reached 20,000 mt in the early 1960s and varied without much of a trend since that time, reaching a peak of about 25,000 mt. In the Indian Ocean, catches of Blue Marlin increased from 3,000–5,000 mt in the 1980s until the present level of 9,000–13,000 mt.

THREATS: Fishing effort in the Pacific and Indian oceans is increasing as a result of longline fisheries and in the Atlantic due to expansion of artisanal fleets. In many instances, landed istiophorid billfishes are not identified to species, contributing to uncertainty with catch data. In addition, there is generally poor reporting from artisanal fisheries, some of which may have very large catches of Blue Marlin. In the Atlantic there has been poor compliance by several fleets to record regulatory live releases of Blue Marlin, which have affected long-term catch-per-unit-effort time series used in the assessments, and the increasing use of anchored FADs by various artisanal and sport fisheries is causing greater vulnerability of these stocks. For the species on a global basis, using generation lengths of 4.5 and 6 years, a decline in overall stock abundance was estimated at 31% and 38% over 14 years and 18 years, respectively. Global declines were calculated as a weighted average of the declines for each stock using maximum historical catch as a proxy for the stock's contribution to the global population. Based on the available data, this species is listed as Vulnerable on the IUCN Red List.

CONSERVATION: This is a highly migratory species listed under Annex I of the 1982 Convention on the Law of the Sea. The species is managed by four different RFMOs: ICCAT, IATTC, WCPFC, and IOTC. As of the 2011 assessment in the Atlantic, Blue Marlin were overfished, and overfishing was occurring. Catch limits and mandatory live release from pelagic longline and purse seine gear were implemented in 2001, and country-specific quotas were implemented in 2012. In addition, ICCAT has implemented a minimum size for recreationally caught Blue Marlin of 251 cm LJFL, and the US Atlantic recreational fishery is limited to a total of 250 Blue Marlin, White Marlin, and Roundscale Spearfish combined.

In the Pacific Ocean, the 2013 stock assessment concluded that Blue Marlin were not overfished and that overfishing was not occurring. However, assessments that allow for environmental factors, movement dynamics, and sexual dimorphism indicate that the Pacific population is in an overexploited state, with current spawning stock biomass below the level corresponding to maximum sustainable yield. Several nations have restricted pelagic longlining in coastal regions to prevent gear conflicts with recreational anglers. Catches of Blue Marlin in the Indian Ocean are very poorly reported, and there is need of better data to evaluate the condition of the Indian Ocean stock. At the time of the 2013 assessment, Blue Marlin in the Indian Ocean were considered to be slightly overfished, but overfishing was not occurring.

SELECTED REFERENCES: Abitia-Cárdenas et al. 2010; Buonaccorsi et al. 1999, 2001; Collette et al. 2006; Kraus et al. 2011; Kraus and Rooker 2007; Nakamura 1985; Ortiz et al. 2003; Prince et al. 1991; Shimose et al. 2006; Su et al. 2012.

Shortbill Spearfish

Tetrapturus angustirostris Tanaka, 1915

COMMON NAMES: English – Shortbill Spearfish
French – Makaire à Rostre Court
Spanish – Marlin Trompa Corta
Japanese – Fûraikajikii

ETYMOLOGY: Described by the Japanese ichthyologist Shigeho Tanaka (1915). The species name *angustirostris* comes from a combination of the Latin *angustus*, meaning "narrow," and the Latin *rostrum*, meaning "beak," in allusion to the short bill of this species.

SYNONYMS: *Tetrapturus angustirostris* Tanaka, 1915; *Tetrapturus illingworthi* Jordan and Evermann, 1926; *Tetrapturus kraussi* Jordan and Evermann, 1926; *Pseudohistiophorus angustirostris* (Tanaka, 1915); *Pseudohistiophorus illingworthi* (Jordan and Evermann, 1926)

TAXONOMIC NOTE: It would have been more descriptive if Tanaka had named his new species *brevirostris*, "short nose," rather than *angustirostris*, "narrow nose."

FIELD MARKS:

1 The bill is very short, usually less than 15% of the body length.
2 The pectoral fins are narrow and short, less than 15% of the body length.
3 The distance between the anus and the origin of the anal fin is nearly equal to the anal-fin height.

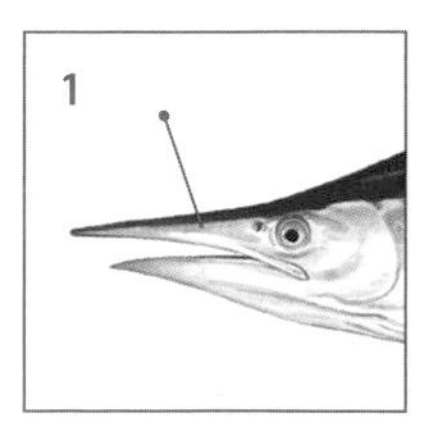

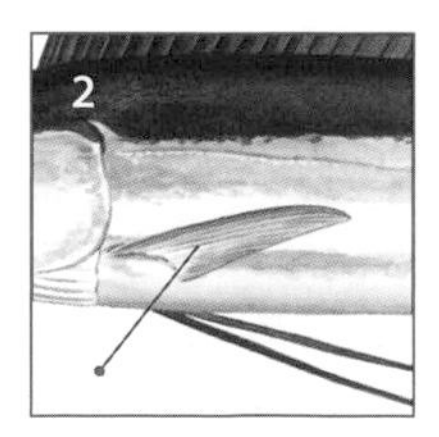

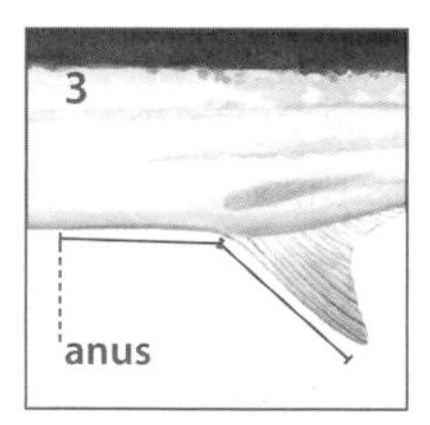

UNIQUE FEATURE: The billfish with the shortest bill.

DIAGNOSTIC FEATURES: The body is elongate and compressed. The bill is short and slender and round in cross section. The right and left branchiostegal membranes are completely united to each other, but free from the isthmus. No gill rakers are present. Both upper and lower jaws and the palatine bones (on the roof of the mouth) have small, file-like teeth. **Fins:** There are two dorsal fins, the first with 45–50 spines and a pointed anterior lobe that is higher than the body depth anteriorly. The first dorsal fin abruptly decreases in height to about the 19th dorsal spine and then keeps the same height posteriorly. The first dorsal-fin base is long and extends from above the posterior margin of the preopercle to near the second dorsal-fin origin. The second dorsal fin has 6 or 7 rays and is located slightly posterior to the second anal fin. There are two anal fins; the first has 12–15 rays and the second has 6 to 8 rays and is very similar in size and shape to the second dorsal fin. The pectoral fins have 17–19 rays. The pelvic fins are slender and depressible into deep ventral grooves. **Caudal peduncle:** The caudal peduncle is laterally compressed and slightly depressed dorsoventrally, with strong double keels on each side. **Anus:** The anus is situated far anterior to the origin of the first anal fin, a distance usually greater than the height of the first anal fin. **Lateral line:** The lateral line is single and obvious, with its arch ending near the midpoint of the pectoral fin. **Scales:** The body is densely covered with elongate bony scales, each with 3–5 posterior points. **Vertebrae:** Vertebrae number 24 (12 precaudal and 12 caudal). **Gonads:** The gonads are Y shaped. **Color:** The body is dark blue dorsally, blue splattered with brown laterally, and silvery white ventrally, without dots or stripes. The dorsal fins are dark blue and lack dots or spots.

GEOGRAPHIC RANGE: Shortbill Spearfish are widely distributed offshore throughout the tropical and temperate Indo-Pacific and enter the eastern Atlantic (around the Cape of Good Hope) but do not spawn there. Putative vagrants have been recorded in the Mediterranean Sea, but these likely represent misidentified Mediterranean Spearfish (*T. belone*). In the eastern Pacific they are found from California and the mouth of the Gulf of California to Peru, including all of the oceanic islands.

SIZE: Maximum size is about 2 m in total length and 52 kg in weight. The IGFA all-tackle game fish record is of a 110 lb, 3 oz (50 kg) fish taken off Botany Bay, Sydney, Australia, in May 2008. Average length of fish caught in the eastern Pacific pelagic longline fishery is 135 cm eye–fork length. Females are on average slightly larger than males.

HABITAT AND ECOLOGY: This species is oceanic and epipelagic, and is generally found above the thermocline. It is found well offshore and rarely enters coastal waters.

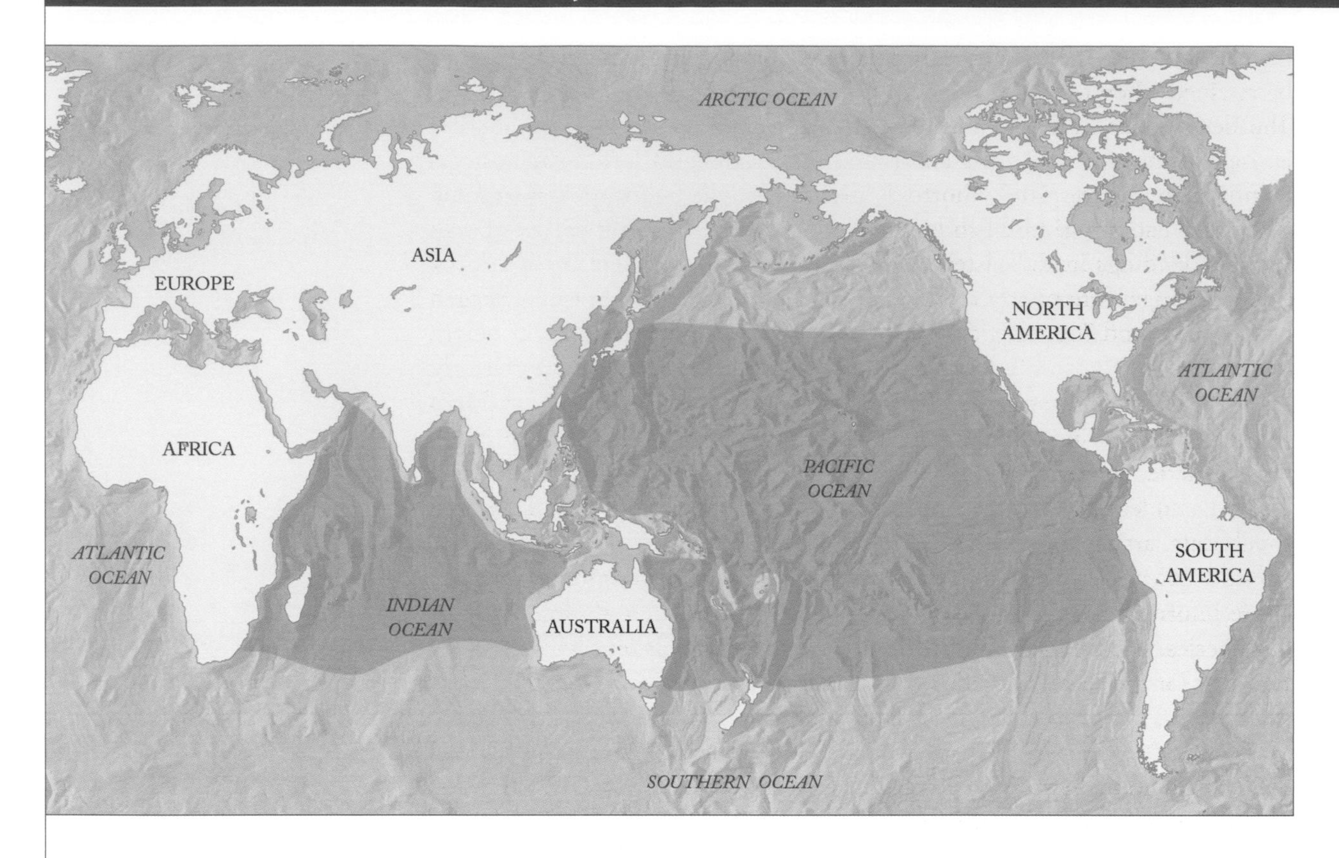

Above: Shortbill Spearfish range map.

FOOD: Shortbill Spearfish feed on fishes such as small tunas, flyingfishes, and bramids, and also on cephalopods and crustaceans.

REPRODUCTION: Spawning is believed to occur mainly during winter months in both hemispheres, especially in warm offshore currents with surface temperatures of about 25°C.

EARLY LIFE HISTORY: Diameters of eggs shed range from 1.3 to 1.6 mm (mean 1.44 mm) in the equatorial western Indian Ocean. Off Hawaii, Hyde et al. found higher concentrations of Shortbill Spearfish eggs and larvae in May relative to July. Ueyanagi reported on 84 larval specimens of Shortbill Spearfish from tropical and subtropical areas of the Pacific ranging in size from 2.5 to 82 mm standard length. A developmental series of 11 specimens 2.4 to 76 mm standard length was described and illustrated. Larvae below 6 mm standard length resemble larvae of Black Marlin, while those over 10 mm resemble the larvae of Sailfish. Larvae of Shortbill Spearfish can be distinguished from larvae of other species of billfishes by the presence of pigment on the branchiostegal membranes. Early-stage embryos have limited pigmentation, with two series of melanophores on either side of the developing embryo.

FISHERIES INTEREST: This species is not typically targeted in fisheries, but may be retained when caught incidentally. It is marketed, mostly frozen, in Japan. The flesh is not of high value compared with that of other billfishes and is used mainly for fish cakes and sausages. Billfish landings are not always identified to species, and Shortbill Spearfish landings may be aggregated with other Istiophorid billfish landings as "billfish unidentified." From 1994 to 2004, landings in the eastern Pacific varied between 100 and 300 mt per year. Landings were poorly reported prior to the late 1990s, but have ranged between 400 and 800 mt without trend over the past 10 years.

THREATS: There are no directed fisheries for the Shortbill Spearfish, and it is primarily taken as incidental catch in pelagic longline and secondarily in purse seine fisheries for tunas and swordfish. It is also caught by sports fishermen. There is underreporting and no reporting from some fisheries, especially small-scale artisanal fisheries. The lack of reliable catch data introduces considerable uncertainty into stock assessments and management advice. More information is needed on catch landings, discards, and effort for this rare species. The Shortbill Spearfish is listed as Data Deficient on the IUCN Red List for the eastern Pacific and globally.

CONSERVATION: This is a highly migratory species listed under Annex I of the 1982 Convention on the Law of the Sea. There have been no stock assessments for this species, and only very limited catch data is available. Given that it is caught with the same gears as the Blue Marlin and Striped Marlin, it is likely that the biomass of this species is declining as well, but there are insufficient data to detect such a trend. There is very little information on Shortbill Spearfish life history and stock structure.

REFERENCES: Collette et al. 2006, 2011; Hanner et al. 2011; Hyde et al. 2005; IATTC 2008; IGFA 2018; Kikawa 1975; Nakamura 1985; Nakamura and Nakano 1978; Pepperell 2010; Serafy et al. 2009; Ueyanagi 1962; Watanabe and Ueyanagi 1963.

Mediterranean Spearfish

Tetrapturus belone Rafinesque, 1810

COMMON NAMES: English – Mediterranean Spearfish
French – Auggia Imbriale, Aguglia Impériale, Marlin de la Méditerranée, Poisson-pique
Spanish – Marlín del Mediterráneo

ETYMOLOGY: The species name *belone* comes from the Greek *belone*, which means "needle," and is also the generic and specific name of an eastern Atlantic needlefish, *Belone belone*.

SYNONYMS: *Tetrapturus belone* Rafinesque, 1810; *Skeponopodus typus* Nardo, 1833; *Tetrapterurus belone* (Rafinesque, 1841); *Histiophorus belone* (Rafinesque, 1810); *Scheponopodus prototypus* Canestrini, 1872; *Tetrapterurus belone* (Rafinesque, 1810); *Makaira belone* (Rafinesque, 1810)

TAXONOMIC NOTE: Western Atlantic records of *Tetrapturus belone* through the early 1960s are of *T. pfluegeri* (Robins and de Sylva 1963). Genetically, the Mediterranean Spearfish is closely related to the Longbill Spearfish (Collette et al. 2006, Hanner et al. 2011).

FIELD MARKS:
1 The bill is very short, about 18% of the body length.
2 The pectoral fins are narrow and short, less than 15% of the body length.
3 The distance between the anus and the origin of the anal fin is nearly equal to the anal-fin height.

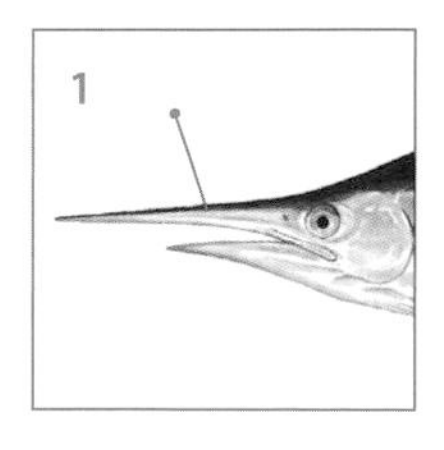

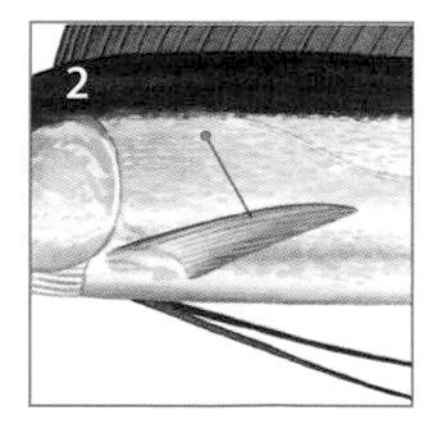

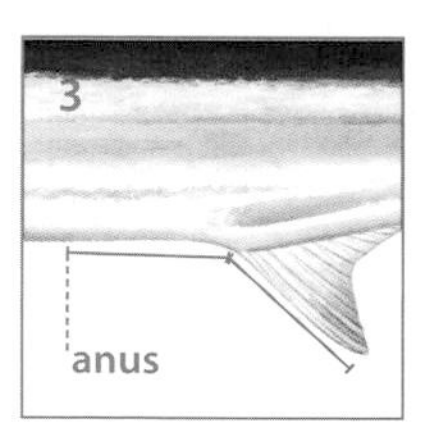

UNIQUE FEATURE: The only billfish confined to the Mediterranean Ocean.

DIAGNOSTIC FEATURES: The body is elongate and compressed. The bill is short and slender and round in cross section. The nape is almost straight. The right and left branchiostegal membranes are completely united to each other, but free from the isthmus. No gill rakers are present. Both upper and lower jaws and the palatine bones (on the roof of the mouth) have small, file-like teeth. **Fins:** There are two dorsal fins, the first with 39–46 rays and with a rounded anterior lobe that is higher than the body depth anteriorly. The first dorsal fin abruptly decreases in height to about the 10th dorsal spine and then maintains the same height posteriorly. The first dorsal-fin base is long, extending from above the posterior margin of preopercle to just in front of second dorsal-fin origin. The second dorsal fin has 5–7 rays and is located posterior to the second anal fin by half the length of the anal-fin base. There are two anal fins: the first has 11–15 rays, the second has 6 or 7 rays and is very similar in size and shape to the second dorsal fin. The pectoral fins are short, 10–13% of body length, adpressible against the sides of the body, their upper margins curved, their lower margins nearly straight and the tips pointed, and have 16–20 rays. The pelvic fins are long and slender, slightly shorter than twice the pectoral fins in length, and depressible into deep ventral grooves. **Caudal peduncle:** The caudal peduncle is laterally compressed and slightly depressed dorsoventrally, with strong double keels on each side. **Anus:** The anus is situated far anterior to the origin of the first anal fin. **Lateral line:** The lateral line is single and obvious, with its arch ending between the midpoint and the tip of the pectoral fin. **Scales:** The body is densely covered with elongate bony scales, each with 3–5 posterior points. **Vertebrae:** Vertebrae number 24 (12 precaudal and 12 caudal). **Color:** The body is dark bluish gray to nearly black dorsally and silvery white ventrally. There are usually no blotches or marks on the body or fins.

GEOGRAPHIC RANGE: Generally restricted to the Mediterranean Sea, although there is one, apparently valid, record from Madeira. It is abundant around Italy and has been reported from Tunisia. There are no confirmed reports from the Black Sea. Catches have been reported from the southern Aegean Sea, but no adults have been reported in the northern region.

SIZE: Exceeds 240 cm in total body length and 70 kg in weight. The IGFA all-tackle game fish record is 90 lb, 13 oz (41.2 kg) for a fish caught off Madeira in June 1980, outside the usual range of the species.

HABITAT AND ECOLOGY: The Mediterranean Spearfish is epipelagic. It is the most common istiophorid in the central basin of the Mediterranean Sea. Little is known about its biology. Mediterranean Spearfish undertake limited vertical movements, ranging between 0 and 200 m depths, generally above or within the thermocline. They often travel in pairs.

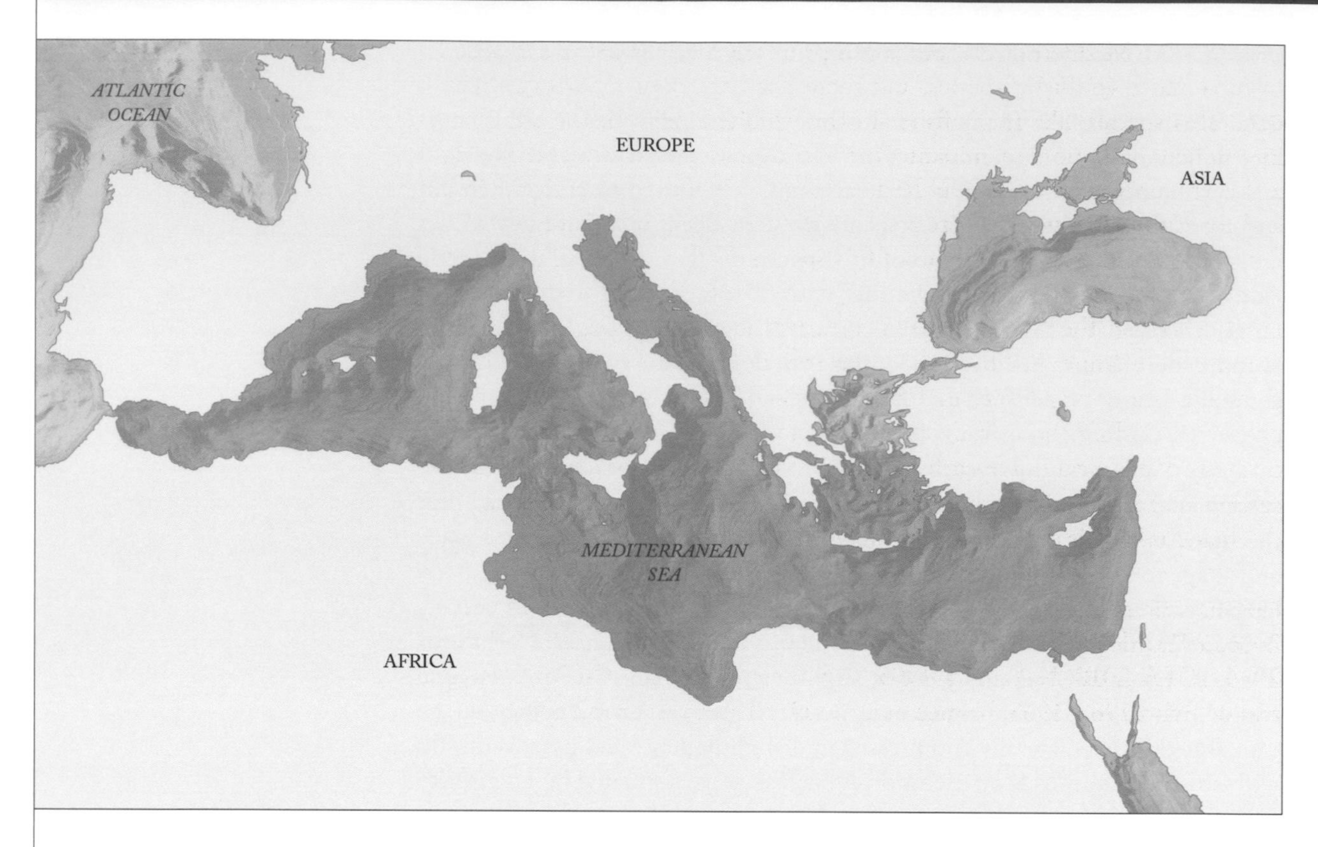

Above: Mediterranean Spearfish range map.

FOOD: The Mediterranean Spearfish feeds mainly on pelagic fishes and cephalopods. The most important fishes in the diet of the Mediterranean Spearfish in the Strait of Messina, Italy, are needlefishes, herrings, and sauries (Belonidae, Clupeidae, and Scomberesocidae), and there is little dietary overlap with Swordfish caught in the same area.

REPRODUCTION: Spawning occurs from spring to winter.

EARLY LIFE HISTORY: Two larvae, 29 and 54 mm, were described and illustrated by Sparta, and a 37 cm juvenile was described and illustrated by Cavaliere.

FISHERIES INTEREST: Mediterranean Spearfish is a species with minor commercial importance. It is targeted in a small, traditional harpoon fishery from boats known as "passerelle," but most are taken as incidental catch in pelagic longline, drift net, and set net fisheries for Bluefin Tuna, Albacore, and Swordfish. Over the past 10 years reported catches of Mediterranean Spearfish have approached 400 mt, but few countries report catches. Mediterranean Spearfish are also caught in directed sport fisheries in several Mediterranean countries. Up to 40 fish were caught in the sport fishery over three years in Majorca.

THREATS: The Mediterranean Spearfish is primarily taken as an incidental catch in tuna and Swordfish fisheries, and there has been poor reporting of catch data. This species, like many fishes in the Mediterranean Sea, is exposed to lipophilic xenobiotic contaminants. The Mediterranean Spearfish was assessed as Least Concern on the IUCN Red List for the Mediterranean Sea, globally, and for Europe. However, more research is necessary as little is known about the biology, ecology, and fisheries of this species.

CONSERVATION: The Mediterranean Spearfish is a highly migratory species mentioned in Annex I of the 1982 Convention on the Law of the Sea. While generally not a target species for commercial fleets, spearfish and billfish catches, including those from the recreational fishery, should be monitored carefully. Mediterranean Spearfish is common and locally abundant. Although catches seem to be increasing, there is no directed commercial fishery, and catch and release is commonly encouraged in the recreational fishery.

REFERENCES: Abdul Malak et al. 2011; Castriota et al. 2008; Collette et al. 2006, 2011; de Sylva 1975; Fossi et al. 2002; Hanner et al. 2011; Hattour 2996; IGFA 2018; Nakamura 1985; Pepperell 2010; Potoschi 2000; Robins and de Sylva 1961, 1963; Romeo et al. 2009, 2014; Sparta 1953, 1961.

Roundscale Spearfish

Tetrapturus georgii Lowe, 1840

COMMON NAMES: English – Roundscale Spearfish
French – Makaira Epée
Spanish – Marlin Peto

ETYMOLOGY: The species was described from Madeira by the British ichthyologist Richard Lowe to commemorate "by its specific name the valuable assistance rendered to the cause of ichthyology by Mr. George Butler Leacock" (Lowe 1840).

SYNONYMS: There are no synonyms for *T. georgii*, although it has sometimes been misspelled as *T. georgei*.

TAXONOMIC NOTE: Although the most distinct genetically and basal to the other three species of spearfishes (Collette et al. 2006, Hanner et al. 2011), the Roundscale Spearfish was only recently confirmed as a valid species (Shivji et al. 2006).

FIELD MARKS:
1 First dorsal fin is unspotted.
2 Tips of first dorsal and anal fins are truncated.
3 The anal opening in the Roundscale Spearfish is located far from the anal fin, separated by a distance of more than half the anal-fin height.
4 Branchiostegals extend almost to edge of operculum.
5 Scales on mid-body are soft and round anteriorly.

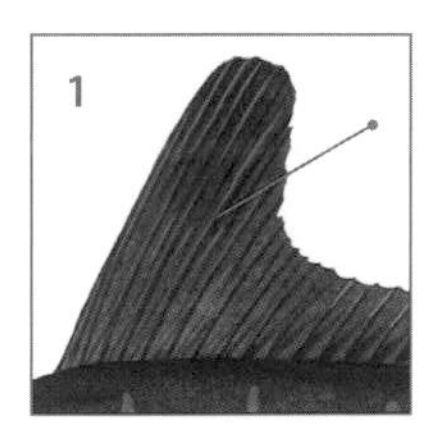

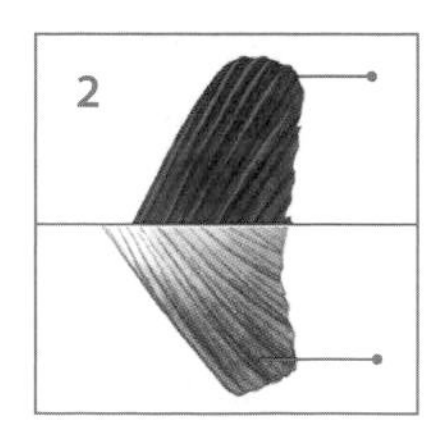

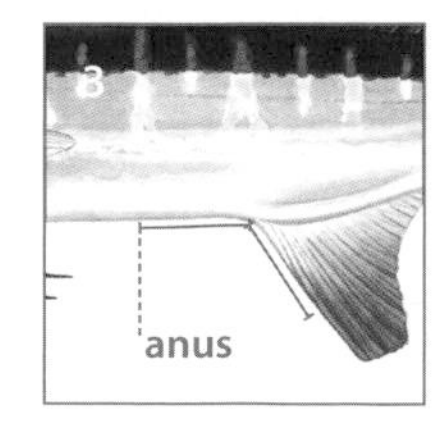

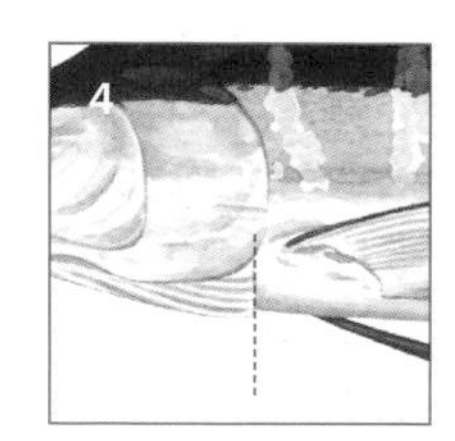

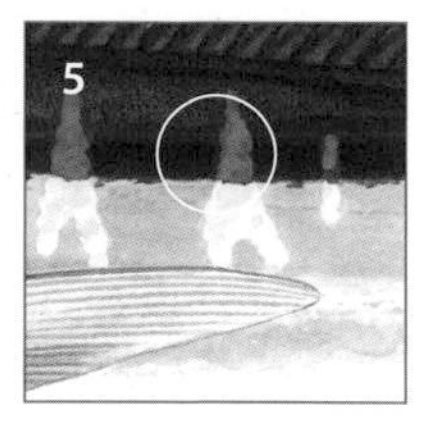

UNIQUE FEATURE: The most recently recognized species of billfish.

DIAGNOSTIC FEATURES: The body is fairly robust and compressed. The bill is long, slender, and round in cross section. The nape is moderately humped. The right and left branchiostegal membranes are completely united to each other, but free from the isthmus. No gill rakers are present. Both jaws and palatines (on the roof of the mouth) have small, file-like teeth. **Fins:** There are two dorsal fins, the first with 43–48 rays. The first is higher than the maximum body depth anteriorly and lower posteriorly, with a truncate anterior lobe and a long base extending from above the posterior margin of the preopercle to just in front of the second dorsal-fin origin. The second dorsal fin has 6 or 7 rays and is located slightly posterior to the second anal fin. There are two anal fins, the first high and truncate with 14–16 rays, the second with 5–7 rays and very similar in size to the second dorsal. The pectoral fins are long, subequal to pelvic fins, reaching beyond the curve of the lateral line, adpressible against sides of body, and with 19 or 20 rays. The pelvic fins are long and slender. **Caudal peduncle:** The caudal peduncle has a pair of keels on each side. **Anus:** The anus is moderately far from the origin of the first anal fin, a distance equal to about half the height of the first anal fin. **Lateral line:** The lateral line is single and simple. **Scales:** Scales on sides of the body are rounded anteriorly and only slightly imbricated and stiff. **Vertebrae:** Vertebrae number 24 (12 precaudal and 12 caudal). **Color:** When dead, there are no bars on the body. However, of all the billfish observed by Graves (pers. com.), Roundscale Spearfish have the brightest coloration when excited, presenting very distinct, upside-down Y-shaped bars along the dark-blue upper body. The first dorsal fin is dark blue and completely unspotted.

GEOGRAPHIC RANGE: This species is known from the entire tropical and subtropical Atlantic Ocean and the Mediterranean Sea. Originally described from Madeira and reported from several other eastern Atlantic localities but only recently known with certainty from the western Atlantic. Analysis of 85 White Marlin/spearfish samples from 18 landings by the Santos-based longliners off southern Brazil in 2009 showed Roundscale Spearfish represented 10.6% of the samples, White Marlin represented 84.7%, and Longbill Spearfish 4.7%.

SIZE: Maximum LJFL is at least 73.5 in. (187 cm) and weight is to 76 lb (34.5 kg) in samples from mid-Atlantic billfish tournaments observed by Graves. In Brazil, maximum size recorded is 180 cm LJFL. The IGFA all-tackle game fish record is of a 70 lb (31.75 kg) fish caught in Baltimore Canyon off Maryland in August 2010.

HABITAT AND ECOLOGY: This species is pelagic, oceanodromous, and found in subtropical waters. Because the existence of this species was only recently validated, little is known of its movements or habitat utilization. One Roundscale Spearfish carried a satellite tag for 34 days off the US mid-

Atlantic coast, moving more than 1,300 km due east during the tagging period, spending 74.7% of its time in the top 10 m and 87% of its time in waters of 24–28°C.

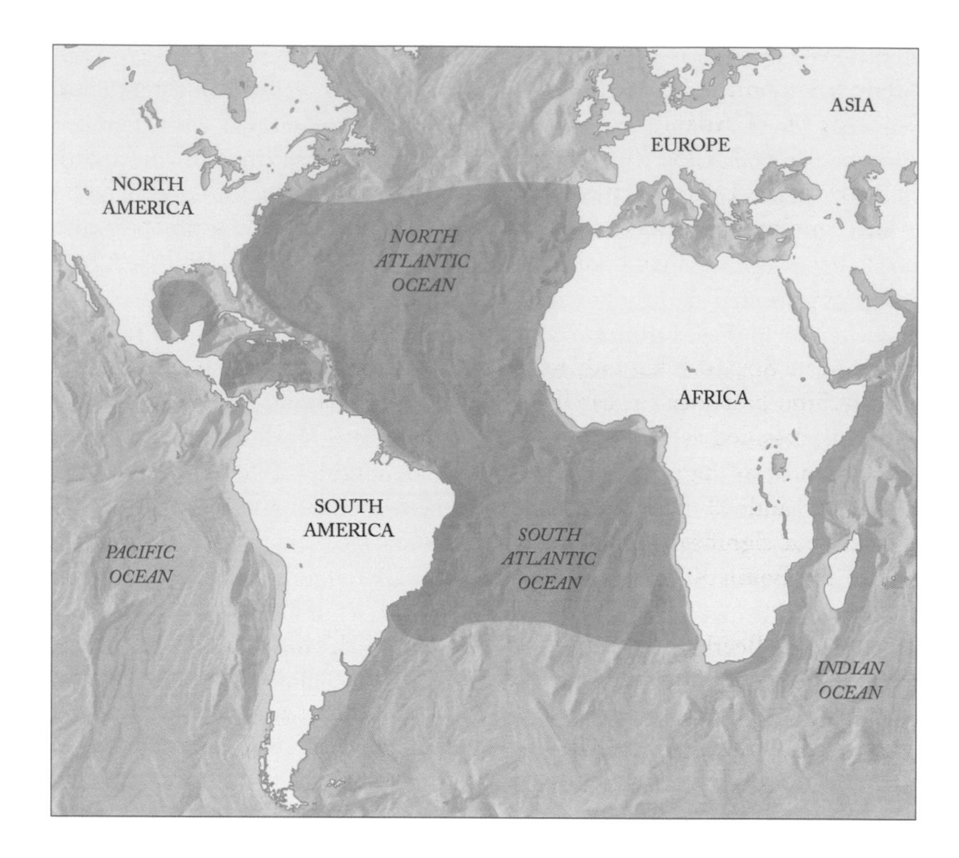

Right: Roundscale Spearfish range map.

FOOD: Squids form an important part of the diet of Roundscale Spearfish. They were found in 55% of nine stomachs sampled from the Mid-Atlantic Bight. Fishes, including Frigate Tuna, are also eaten off the coast of the United States.

REPRODUCTION: Very little is known of its reproductive biology. Graves received and identified a photo of a ripe female Roundscale Spearfish caught by a recreational angler off the Azores in May a few years ago.

FISHERIES INTEREST: Roundscale Spearfish are primarily taken as bycatch by longline fisheries but also by purse seines, by some artisanal gears, and by various sport fisheries located on both sides of the Atlantic. This species is easily misidentified in the field, and it is likely that most captures of this species were classified as White Marlin or Longbill Spearfish until only recently. The proportion of this species in landings classified as Longbill Spearfish or White Marlin is temporally unstable, and the relative abundance of Roundscale Spearfish and White Marlin in specific locations changes seasonally and interannually.

THREATS: The lack of landings data (due to the misidentification problem) and information on population structure and biology led to its being listed as Data Deficient on the IUCN Red List.

CONSERVATION: This is a highly migratory species listed in Annex I of the 1982 Convention on the Law of the Sea. The United States prohibits commercial landings of all Atlantic istiophorid billfish, and the recreational fishery is limited to a total of 250 Blue Marlin, White Marlin, and Roundscale Spearfish combined. The United States has a minimum size (66 inches LJFL), and a similar minimum size was adopted by ICCAT in 2012. The United States also requires the use of circle hooks in natural baits for anglers fishing in billfish tournaments.

Roundscale Spearfish has only recently been recognized as distinct, and there is very little biological or catch information available for it. Landings have been misclassified as Longbill Spearfish and White Marlin, confounding an evaluation of its status. Analyses of the mitochondrial DNA control-region sequences and 13 nuclear microsatellite loci provide mixed evidence for shallow but significant structuring between Roundscale Spearfish from the North and South Atlantic.

REFERENCES: Beerkircher et al. 2008, 2009; Beerkircher and Serafy 2011; Beerkircher et al. 2011; Bernard et al. 2013, 2014; Collette 2010; Collette et al. 2006, 2011; Graves and McDowell 2012; Hanner et al. 2011; IGFA 2018; Loose 2014; Lowe 1840; Nakamura 1985; Piva-Silva et al. 2009; Robins 1974; Serafy et al. 2009; Shivji et. 2006; Staudinger et al. 2012; White Marlin Biological Review Team 2007.

Longbill Spearfish

Tetrapturus pfluegeri Robins & de Sylva, 1963

COMMON NAMES: English – Longbill Spearfish
French – Makaire bécune
Spanish – Aguja picuda

ETYMOLOGY: The Longbill Spearfish was "dedicated to the memory of Albert Pflueger, Sr., Miami taxidermist, who was a steady source of valued information and who first called to our attention the presence of a spearfish in western Atlantic waters" (Robins and de Sylva 1963).

SYNONYMS: *Tetrapturus belone* Fowler, 1936 (by many authors who did not distinguish this species from *Tetrapturus belone* Rafinesque, 1810, until *Tetrapturus pfluegeri* was described by Robins and de Sylva, 1963); *Tetrapterus belone* La Monte, 1940

TAXONOMIC NOTE: Confused with the eastern Atlantic *T. belone* until described as a distinct species by Robins and de Sylva (1963).

FIELD MARKS:
1 The bill is long, its length usually equal to or slightly longer than head length.
2 The pectoral fins are wide, long, and rounded, longer than 18% of body length.
3 The anus is situated far anterior to first anal-fin origin; the distance between the anus and the anal-fin origin is nearly equal to anal-fin height.

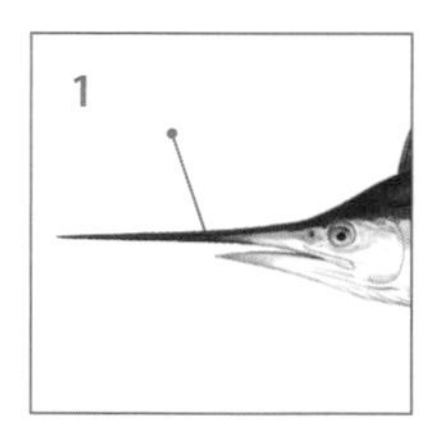

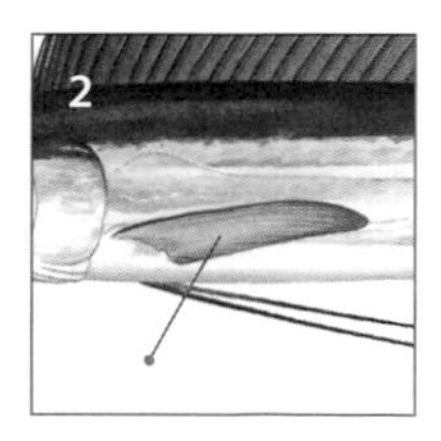

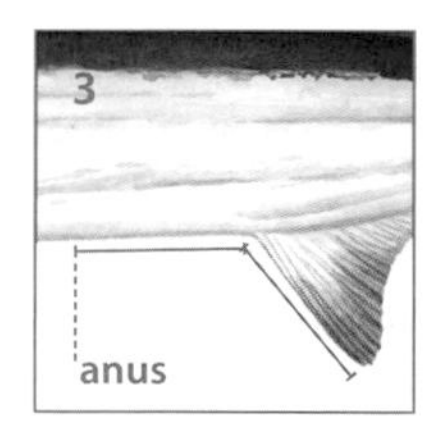

UNIQUE FEATURES: The only Atlantic istiophorid with a long bill and an anal opening situated well in front of the origin of a triangular first anal fin. The most recently described istiophorid.

DIAGNOSTIC FEATURES: The body is elongate and remarkably compressed, its depth very low. The bill is slender, rather long, and round in cross section. The nape is nearly straight. The right and left branchiostegal membranes are completely united to each other, but free from the isthmus. Gill rakers are absent. Both upper and lower jaws and the palatine bones (on the roof of the mouth) with small, file-like teeth. **Fins:** There are two dorsal fins, the first with 44–50 rays and a rounded anterior lobe higher than the body depth anteriorly. The first dorsal fin then abruptly decreases in height to about the ninth dorsal-fin ray and maintains almost the same height further backward, except at the posterior end. The first dorsal-fin base is long, extending from above the posterior margin of the preopercle to just in front of the second anal fin by a distance of one-third the length of the anal-fin base. The second dorsal fin is small, with 6 or 7 rays. It is located posterior to the second anal fin by a distance of one-third of the length of the anal-fin base. There are two anal fins, the first with 12–17 rays, the second with 6 or 7 rays and very similar in size and shape to the second dorsal fin. The pectoral fins are long and wide, round-tipped, adpressible against sides of body, and with 18–21 rays. The pelvic fins are slender, almost equal to or slightly longer than the pectoral fins, and depressible into deep ventral grooves. **Caudal peduncle:** The caudal peduncle is laterally compressed and slightly depressed dorsoventrally. It has strong double keels on each side and a shallow notch on both the dorsal and ventral surfaces. **Anus:** The anus is situated far anterior to the origin of the first anal fin. **Lateral line:** The lateral line is single and obvious, curving above the base of the pectoral fin and then continuing in a straight line toward the caudal-fin base. **Scales:** The body is densely covered with elongate bony scales, each with 2 to 5 posterior points. **Vertebrae:** Vertebrae number 24 (12 precaudal and 12 caudal). The gonad is Y shaped. **Color:** The body is blue-black dorsally, silvery white splattered with brown laterally, and silvery white ventrally. The first dorsal fin is dark blue without dots or blotches. The second dorsal fin is dark blue. The pectoral fins are blackish brown, sometimes tinged with grayish white. The pelvic fins are blue-black with a black fin membrane. The first anal fin is dark blue, its base tinged with silvery white. The second anal fin is blackish brown.

GEOGRAPHIC RANGE: The Longbill Spearfish is an epipelagic and oceanic species chiefly found in offshore waters, usually above the thermocline. It was described rather recently as a new species by Robins and de Sylva (1963). At that time, it was known with certainty only from the western North Atlantic, where it occurs from Georges Bank to Puerto Rico and from the Gulf of Mexico and the Caribbean Sea. Recent pelagic fishery research surveys have clearly shown that it is also widely distributed in Atlantic offshore waters, much more densely so in the western than in the eastern Atlantic. Its latitudinal range extends from approximately 40° N to 35° S. In Brazil, it has been frequently recorded between 20–33° S and 40–50° W. It has also been recorded in northeast Brazil from 57–240 m depth. Analysis of 85 samples of istiophorid billfishes from 18 landings by the Sao Paulo–based longliners off southern Brazil in 2009 showed that Longbill Spearfish represented only 4.7% of the samples, White Marlin represented 84.7%, and Roundscale Spearfish 10.6%.

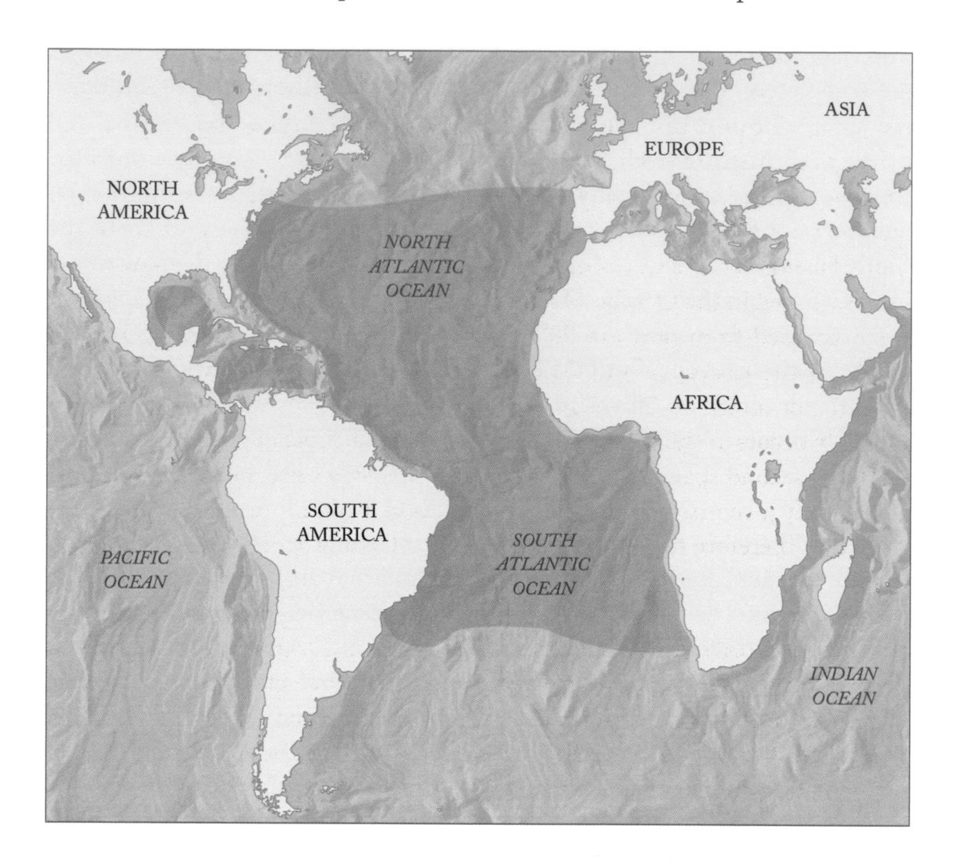

Right: Longbill Spearfish range map.

SIZE: The maximum recorded size of the Longbill Spearfish is 161 cm LJFL, but it probably exceeds 200 cm in body length and 58 kg in weight. The IGFA all-tackle angling record is of a 127 lb, 13 oz. (58 kg) fish caught off Puerto Rico, Gran Canaria, Canary Islands, in May 1999. Weights of fish typically taken by recreational anglers range from 9 to 36.5 kg. The most common size caught by surface longlines is about 165 cm body length throughout the Atlantic fishing grounds. Longevity is probably four years.

HABITAT AND ECOLOGY: Little information is available on the movements and habitat utilization of Longbill Spearfish. Data from satellite tags attached to two individuals released near Ascension Island in the South Atlantic for 11 and 45 days indicated the fish spent the vast majority of time in waters 22–26°C. The fish exhibited a bimodal depth distribution, spending the majority of their time at depths less than 25 m, with a smaller mode at depths of 50–100 m.

FOOD: Like other billfishes, Longbill Spearfish feed chiefly on pelagic fishes and squids. Off northeastern Brazil, Atlantic Pomfret (*Brama brama*) and the squid *Ornithoteuthis antillarum* comprised 63% of the diet of 47 Longbill Spearfish sampled. Another study from Brazil also found Atlantic Pomfret and *Ornithoteuthis* as well as Snake Mackerel (*Gempylus serpens*) to be important components of the diet.

REPRODUCTION: Based on occurrence of larvae and mature fish, spawning of the Longbill Spearfish takes place throughout wide areas of the tropical and subtropical Atlantic from late November to early May. Mature individuals are caught only in the January–March quarter (mostly) and in the April–June quarter (less frequently), with the exception of the areas around the Cape Verde Islands and the Caribbean Sea where some mature individuals have also been recorded in the October–December quarter. Mature individuals have not been recorded from north of 20° N in the western North Atlantic, north of 30° N in the eastern North Atlantic, south of 10° S in the western South Atlantic, or south of 30° N in the eastern South Atlantic. First spawning probably occurs at the end of the first year, and few females apparently survive beyond a second spawning. Females probably spawn once a year. In northeast Brazil, only juveniles and adults with gonads in a resting condition have been collected. Therefore the region is considered a feeding ground, not a spawning area.

EARLY LIFE HISTORY: Eggs are undescribed by Robins, Nakamura, de Sylva and Breder, and Richards and Luthy. A summary of characteristics of the early stages of three of the western Atlantic billfishes is presented by Richards and Luthy, but only one larval Longbill Spearfish has been recorded. It had pigmentation similar to that of the Indo-Pacific Striped Marlin.

FISHERIES INTEREST: Longbill Spearfish are taken incidentally by pelagic longline vessels targeting tuna and swordfish. Historical Japanese statistical data combine this species with the Atlantic Sailfish and other spearfish species as "spearfish and sailfish." It was assumed that most nearshore records applied to Sailfish and offshore records to Longbill Spearfish. However, it has been demonstrated that Sailfish occur continuously across the Atlantic, and it is likely that Roundscale Spearfish were also included in the "spearfish and sailfish" category. In this category, the recorded catch declined from 52,000–118,000 fishes in the early years from 1962 to 1980 to only 1,000–10,000 fishes in the later years, with an effort ranging from 22 million to 97.5 million hooks per year. In the ICCAT data base, annual reported catches peaked in the 1960s at more than 1,000 mt and quickly dropped to lower levels. For the past 20 years reported catches have averaged over 200 mt per year.

The Longbill Spearfish is taken by recreational anglers with the same methods used for other billfishes. The species is not generally targeted, as it is relatively rare. Gear and tackle preferences vary with fishermen, but all are well described in the sport fishing literature, such as Migdalski, Tinsley, Rybovich, Goadby, etc. Prior to the mid-1990s, as many as ten Longbill Spearfish were sent to each of the two principal taxidermy shops in southeastern Florida. Catches everywhere are low; probably fewer than 100 fish are caught per year by sport fishermen in the western Atlantic.

THREATS: The Longbill Spearfish is primarily taken as an incidental catch by pelagic longline fisheries but also by purse seines, trolling, artisanal gears, and various sport fisheries located in both sides of the Atlantic. The EU Scientific, Technical and Economic Committee for Fisheries (STECF) is concerned about the lack of attention given to this species, because it might be affected by the same problems as other billfish species. However, it is widely distributed and common, and there are no directed fisheries for it. Catches have been at the same moderate level for a long period of time, and there is no evidence of substantial decline over the past several generations, so it is listed as Least Concern on the IUCN Red List.

CONSERVATION: Longbill Spearfish are listed as a highly migratory species in Annex I of the 1982 Convention on the Law of the Sea. The Unites States prohibits commercial landings of all Atlantic isotiophorid billfishes, and recreational landings of Longbill Spearfish are also prohibited.

REFERENCES: Amorim and Arfelli 1979; Amorin et al. 2011; Arocha et al. 2007; Collette 2010; Collette et al. 2011; de Sylva and Breder 1997; Hanner et al. 2011; IGFA 2018; Kerstetter et al. 2009; Nakamura 1985; Nóbrega et al. 2009; Piva-Silva and Amorin 2014; Richards and Luthy 2005; Robins 1975; Robins and de Sylva 1961, 1963; Souza et al. 1994; Vaske et al. 2004; Wor et al. 2010.

GLOSSARY of TERMS

Adipose eyelid: Translucent fold covering the anterior and posterior margins of the eye in mackerels (*Scomber* and *Rastreliger*).

Anal fin: The fin on the ventral surface of fishes located between the anus and the caudal fin.

Axillary branches: Fine branches that extend dorsally and ventrally from the anterior portion of the lateral line in *Scomberomorus guttatus* and *S. koreanus*.

Base: The point or area where a fin emerges from the body.

Bifid: Divided by a deep notch into two parts.

Body length (BL): The length measured from the tip of the snout to the hypural plate at the base of the caudal fin.

Branchiostegal membrane: The membrane supported by and connecting the branchiostegals and enclosing the ventral gill chamber in some fishes.

Branchiostegals: Series of long, curved bony rays below the operculum that support the branchiostegal membrane.

Caudal fin: The fin at the rear end of the body.

Caudal keel: Raised ridge on either side of the caudal peduncle. All scombrids have a small, obliquely oriented pair at the caudal-peduncle base. More advanced scombrids also have a large median keel anterior to the small keels.

Caudal peduncle: The narrow portion of the body just anterior to the caudal fin.

Caudal vertebrae: Vertebrae that lack ribs and instead bear a ventral spine. The first caudal vertebra is located near the anal-fin origin.

Corselet: Region behind the head and surrounding the pectoral fins that is covered in large, thick scales in advanced scombrids.

Dorsal: Refers to the back or upper portion of a fish.

Dorsal fin: A fin located on the back of a fish that may be single or separated into two or more fins.

Epipelagic: The open ocean habitat from the surface to a depth of about 200 m.

Fin groove: The groove on the dorsal surface of the body in all scombrids that the first dorsal fin folds down into when they are swimming rapidly.

Fork length (FL): The length measured from the tip of the snout or lower jaw to the ends of the middle caudal-fin rays.

Gill arch: The J-shaped structure under the gill cover that bears the gill filaments. There are four gill arches on each side in scombrids.

Gill teeth: Short and flattened structures on the gill arch located medially from the gill rakers. Counted like gill rakers. Also called inner gill rakers.

Gill rakers: The stiff and pointed structures that extend dorsally or anteriorly from the first gill arches and toward the mouth. Counts are usually given as the number on the upper plus the number on the lower limbs of the first arch, e.g., 4 + 12 = 16. In scombrids there is usually one at the arch angle that is not clearly on the upper or lower arch and is counted 4 + 11 + 1 = 16. If not counted separately, this gill raker is included with those on the lower arch.

Interpelvic process: Fleshy process between the bases of the pelvic fins—it may be single or bifid, small or large.

Isthmus: The triangular, front-most part of the underside of the body, largely separated from the head, in most bony fishes, by the gill opening.

Laminae: Fleshy folds containing odor-detecting cells arranged in a circular or rosette pattern under the area between the anterior and posterior nostril opening.

Lateral line: A series of sense organs enclosed in tubular scales along the sides of the body that perceives vibrations. Most scombrids have a single lateral line; *Grammatorcynus* species have two lateral lines.

Lower jaw fork length (LJFL): The length measured in a straight line from the tip of the lower jaw to the end of the middle caudal-fin rays; used for billfishes.

Maxilla: The supporting bone for the premaxilla, the bone in the upper jaw that bears teeth.

Membrane: The thin skin between spines or rays of fins.

Mesopelagic: Inhabiting the intermediate ocean depths, about 200 to 1,000 m.

GLOSSARY of TERMS / ACRONYMS and ABBREVIATIONS

Palatine: The plow-shaped bone, the ventral margin of which lies in the roof of the mouth and may bear conical or fine villiform teeth.
Parapophysis: Transverse projection from the vertebral centrum, or oval-shaped central part.
Pelagic: Living in open waters or open ocean, away from the bottom.
Preopercle: The portion of the side of the head composed of a bone lying anterior to the opercle.
Preorbital bone: The largest of the infraorbital series of bones.
Oceanic: Living in open ocean waters beyond continental shelves.
Oceanodromous: Migrating within ocean waters, typically between spawning and feeding areas.
Opercle: The portion of the side of the head that covers the gills and is typically composed of large, flattened bones.
ppt: Parts per thousand; the units of measurement for salinity.
Sagittal otolith: The calcium carbonate structure in the inner ear of a fish.
Standard length (SL): The length measured from the tip of the snout or lower jaw to the rear end of the vertebral column.
Thermocline: The depth or layer of water at which the rate of temperature decline is the greatest and which separates warm surface waters from cooler midwaters.
Total length (TL): The length measured from the tip of the snout or lower jaw to the tip of the longest caudal-fin rays, which may be compressed along the midline.
Ventral: Refers to the abdominal or lower portion of a fish.
Vomer: A median skull bone, the ventral surface of which lies in the roof of the mouth and may bear teeth.

ACRONYMS and ABBREVIATIONS

BMSY: The biomass that enables a fish stock to deliver the maximum sustainable yield. In theory, the population size at the point of maximum growth.
CCSBT: Commission for the Conservation of Southern Bluefin Tuna
CPUE: Catch per unit of fishing effort; the total catch divided by total amount of effort used to harvest the catch.
FADs: Fish Aggregating Devices. Floating or anchored objects that are strategically placed to attract pelagic fishes.
FAO: Food and Agriculture Organization of the United Nations.
FMSY: The maximum rate of fishing mortality (the proportion of a fish stock caught and removed) resulting eventually in a population size of BMSY.
IATTC: Inter-American Tropical Tuna Commission.
ICCAT: International Commission for the Conservation of Atlantic Tunas.
IGFA: International Game Fish Association.
IOTC: Indian Ocean Tuna Commission
IUCN: International Union for the Conservation of Nature.
IUCN Red List: Provides taxonomic, conservation status, and distribution information on plants, fungi, and animals that have been globally evaluated for the purpose of guiding conservation activities of governments, NGOs, and scientific institutions.
MSY: Maximum sustainable yield; the longest long-term average catch or yield that can be taken from a stock or stock complex under prevailing ecological and environmental conditions.
PCR: Polymerase Chain Reaction. A technique used in molecular biology to amplify, or make many copies of, a specific target region of DNA.
RFMO: Regional fishery management organization
SSB: Spawning stock biomass
SSBMSY: Spawning stock biomass that would produce the maximum sustainable yield.
TAC: Total Allowable Catch.
WCPFC: Western and Central Pacific Fisheries Commission

SELECT BIBLIOGRAPHY

NOTE: The entire bibliography can be found online at: www.press.jhu.edu.
In the search box, type: Tunas and Billfishes of the World.

Two early references that were part of the basis for this book were issued as reports from FAO, the Food and Agriculture Organization of the United Nations.

Collette, B.B. and C.E. Nauen. 1983. Scombrids of the world: An annotated and illustrated catalogue of tunas, mackerels, bonitos and related species known to date. Food and Agriculture Organization of the United Nations (FAO) Fisheries Synopsis number 125, volume 2.

Nakamura, I. 1985. Billfishes of the world: An annotated and illustrated catalogue of marlins, sailfishes, spear fishes and swordfishes known to date. Food and Agriculture Organization of the United Nations (FAO) Fisheries Synopsis number 125, volume 5.

The Tuna and Billfish Specialist Group, Species Survival Commission, International Union for the Conservation of Nature, conducted a series of workshops around the world to evaluate the threat status of all 61 species of mackerels, tunas, and billfishes. Results of the assessments, including information on the distribution and biology of each species, were published in the IUCN Red List in 2011 (http://www.iucnredlist.org/) and summarized in:

Collette, B.B., K.E. Carpenter, B.A. Polidoro, M.J. Juan-Jorda, A. Boustany, et al. 2011. High-value and Long-lived: Double jeopardy for threatened tunas, mackerels and billfishes. Science 333: 291-292.

Here we provide some of the most useful additional references, first for the tunas and mackerels (family Scombridae) and then for the billfishes (families Istiophoridae and Xiphiidae):

SCOMBRIDAE

Abdussamad, E.M., N.G.K. Pillai, H.M. Kasim, O.M.M.J. Habeeb Mohamed, and K. Jeyabalan. 2010. Fishery, biology and population characteristics of the Indian Mackerel, *Rastrelliger kanagurta* (Cuvier) exploited along the Tuticorin coast. Indian Journal of Fisheries 57(1):17–21.

Abdussamad, E.M., K.P. Said Koya, S. Ghosh, P. Rohit, K.K. Joshi, B. Manojkumar, D. Prakasan, S. Kemparaju, M.N.K. Elayath, H.K. Dhokia, M. Sebastine, and K.K. Bineesh. 2012. Fishery, biology and population characteristics of Longtail Tuna, *Thunnus tonggol* (Bleeker, 1851) caught along the Indian coast. Indian Journal of Fisheries 59(2):7–16.

Aguirre-Villaseñor, H., E. Morales-Bojórquez, R.E. Morán-Angulo, J. Madrid-Vera, M.C. Valdez-Pineda. 2006. Biological indicators for the Pacific sierra (*Scomberomorus sierra*) fishery in the southern Gulf of California, Mexico. Ciencias Marinas 32(3):471–484.

Ancieta Calderón, F. 1964. Sinopsis sobre la biología y pesquería del "bonito" *Sarda chilensis* (Cuvier y Valenciennes) frente a la cost del Perú. Revista de la Facultad de Ciencias Biologicas 1(1):17–49.

Banford, H.M., E. Bermingham, B.B. Collette, and S.S. McCafferty. 1999. Phylogenetic systematics of the *Scomberomorus regalis* (Teleostei: Scombridae) species group: molecules, morphology and biogeography of Spanish mackerels. Copeia 1999(3):596–613.

SELECT BIBLIOGRAPHY

Begg, G.A., and G.A. Hopper. 1997. Feeding patterns of School Mackerel (*Scomberomorus queenslandicus*) and Spotted Mackerel (*S. munroi*) in Queensland east-coast waters. Marine and Freshwater Research 48:565–571.

Begg, G.A., and M.J. Sellin. 1998. Age and growth of School Mackerel (*Scomberomorus queenslandicus*) and Spotted Mackerel (*S. munroi*) in Queensland east-coast waters. Marine and Freshwater Research 49:109–120.

Bentley, B.P., E.S. Harvey, S.J. Newman, D.J. Welch, A.K. Smith, and W.J. Kennington. 2014. Local genetic patchiness but no regional differences between Indo-West Pacific populations of the Dogtooth Tuna *Gymnosarda unicolor*. Marine Ecology Progress Series 506:267–277.

Cheng, J., T. Gao, Z. Miao, and T. Yanagimoto. 2011. Molecular phylogeny and evolution of *Scomber* (Teleostei: Scombridae) based on mitochondrial and nuclear DNA sequences. Chinese Journal of Oceanology and Limnology 29(2):297–310.

Collette, B.B., and C.R. Aadland. 1996. Revision of the frigate tunas (Scombridae, *Auxis*), with descriptions of two new subspecies from the eastern Pacific. Fishery Bulletin 94(3):423–441.

Collette, B.B., and L.N. Chao. 1975. Systematics and morphology of the bonitos (Sarda) and their relatives (Scombridae, Sardini). Fishery Bulletin 73(3):516–625.

Collette, B.B., and G.B. Gillis. 1992. Morphology, systematics, and biology of the double-lined mackerels (*Grammatorcynus*, Scombridae). Fishery Bulletin 90(1):13–53.

Collette, B.B., and J.L. Russo. 1985. Morphology, systematics, and biology of the Spanish mackerels (*Scomberomorus*, Scombridae). Fishery Bulletin 82(4):545–692.

Collette, B.B., C. Reeb, and B.A. Block. 2001. Systematics of the Tunas and Mackerels (Scombridae). In: Block, B.A., and Stevens, E.D. (eds), Tuna: Physiology, ecology and evolution, pp. 1–33. Academic Press, San Diego, CA.

D'Aubenton, F., and Blanc, M. 1965. Etude systématique et biologique de *Scomberomorus sinensis* (Lacepède, 1802), poisson des eaux douces du Cambodge. Bulletin du Muséum National d'Histoire Naturelle 2e ser., 37(2):233–243.

Devaraj, M. 1976. Discovery of the scombrid *Scomberomorus koreanus* (Kishinouye) in India, with taxonomic discussion on the species. Japanese Journal of Ichthyology 23(2):79–87.

Devaraj, M. 1977. Osteology and relationships of the Spanish mackerels and seerfishes of the tribe Scomberomorini. Indian Journal of Fisheries 22(1–2):1–67.

Devaraj, M. 1986. Maturity, spawning and fecundity of the Streaked Seer, *Scomberomorus lineolatus* (Cuvier & Valenciennes) in the Gulf of Mannar and Palk Bay. Indian Journal of Fisheries 33(3):293–319.

Devaraj, M. 1998. Food and feeding habits of the Spotted Seer, *Scomberomorus guttatus* (Bloch and Schneider), in the Gulf of Mannar and Palk Bay. Journal of the Marine Biological Association of India 40(1–2):105–124.

SELECT BIBLIOGRAPHY

Devaraj, M. 1998. Food and feeding habits of the Streaked Seer, *Scomberomorus lineolatus* (Cuvier and Valenciennes) in the Gulf of Mannar and Palk Bay. Journal of the Marine Biological Association of India 40(1–2):91–104.

Devaraj, M., and H. Mohamad Kasim. 1998. Seerfish fishery of the world. Journal of the Marine Biological Association of India 40(1–2):51–68.

Dominguez-López, M., P. Díaz-Jaimes, M. Uribe-Alcocer, and C. Quiñónez-Velázquez. 2015. Post-glacial expansion of the Monterey Spanish Mackerel, *Scomberomorus concolor*, in the Gulf of California. Journal of Fish Biology 86(3):1153–1162.

Fable, W.A., Jr., A.G. Johnson, and L.E. Barger. 1987. Age and growth of Spanish Mackerel, *Scomberomorus maculatus*, from Florida and the Gulf of Mexico. Fishery Bulletin 85(4):777–783.

Farley, J.H., J.P. Eveson, T.L.O. Davis, R. Andamari, C.H. Proctor, B. Nugraha, and C.R. Davies. 2014. Demographic structure, sex ratio and growth rates of Southern Bluefin Tuna (*Thunnus maccoyii*) on the spawning ground. PloS ONE 9(5):e96392.

Finucane, J.R., and L.A. Collins. 1984. Reproductive biology of Cero, *Scomberomorus regalis*, from the coastal waters of south Florida. Northeast Gulf Science 7(1):101–107.

Finucane, J.H., L.A. Collins, H.A. Brusher, and C.S. Saloman. 1986. Reproductive biology of King Mackerel, *Scomberomorus cavalla*, from the southeastern United States. Fishery Bulletin 84(4):841–850.

Garber A.F., M.D. Tringali, and J.S. Franks. 2005. Population genetic and phylogeographic structure of Wahoo, *Acanthocybium solandri*, from the western central Atlantic and central Pacific oceans. Marine Biology 147(1): 205–214.

Gibbs, R.H., Jr., and Collette, B.B. 1967. Comparative anatomy and systematics of the tunas, genus *Thunnus*. Fishery Bulletin 66(1):65–130.

Godsil, H.C. 1955. A description of two species of bonito, *Sarda orientalis* and *S. chiliensis*, and a consideration of relationships with the genus. California Department of Fish and Game, Fish Bulletin 99, 43 pp.

Godsil, H.C., and R.B. Byers. 1944. A systematic study of the Pacific tunas. California Fish and Game Fish Bulletin 60, 131 pp.

Godsil, H.C., and E.K. Holmberg. 1950. A comparison of the bluefin tunas, genus *Thunnus* from New England, Australia and California. California Division of Fish and Game, Fish Bulletin No. 77, 55 pp.

Graham, J.B., and K.A. Dickson. 2000. The evolution of thunniform locomotion and heat conservation in scombrid fishes: New insights based on the morphology of *Allothunnus fallai*. Zoological Journal of the Linnean Society 129:419–466.

Grandcourt, E.M. 2013. A review of the fisheries, biology, status and management of the Narrow-barred Spanish Mackerel (Scomberomorus commerson) in the Gulf Cooperation Council countries (Bahrain, Kuwait, Oman, Qatar, Saudi Arabia and the United Arab Emirates. Third Working Party on Neritic Tunas, Bali, Indonesia. IOTC-2013-WPNT03-27, 17 pp.

Hattour, A. 2000. Contribution à l'étude des poissons pélagiques des eaux tunisiennes. Thèse de Doctorat, Faculté des Sciences de Tunis, Université d'El Manar II.

Joshi, K.K., E.M. Abdussamad, K.P. Said Koya, M. Sivadas, S. Kuriakose, D. Prakasan, M. Sebastine, M. Beni, and K.K. Bineesh. 2012. Fishery, biology and dynamics of Dogtooth Tuna, *Gymnosarda unicolor* (Rüppell, 1838) exploited from Indian seas. Indian Journal of Fisheries 59(2):75–79.

Kikawa, S., and staff of the Nankai Regional Fisheries Laboratory. 1963. Synopsis of biological data on bonito *Sarda orientalis* Temminck and Schlegel 1842. FAO Fisheries Report No. 6, 2:147–156.

Kishinouye, K. 1923. Contributions to the comparative study of the so-called scombroid fishes. Journal of the College of Agriculture, Imperial University of Tokyo 8(3):293–475.

Manacop, P.R. 1958. A preliminary systematic study of the Philippine chub mackerels, family Scombridae, genera *Pneumatophorus* and *Rastrelliger*. Philippine Journal of Fisheries 4(2):79–101.

Manooch, C.S., III, E.L. Nakamura, and A.B. Hall. 1978. Annotated bibliography of four Atlantic scombrids: *Scomberomorus brasiliensis*, *S. cavalla*, *S. maculatus*, and *S. regalis*. NOAA Technical Report, NMFS Circular 418, 166 pp.

Matsui T. 1967. Review of the mackerel genera *Scomber* and *Rastrelliger* with description of a new species of *Rastrelliger*. Copeia 1967(1):71–89.

Matsumoto, W.M., R.A. Skillman, and A.E. Dizon. 1984. Synopsis of biological data on Skipjack Tuna, *Katsuwonus pelamis*. NOAA Technical Report, NMFS Circular 451.

McBride, R.S., A.K. Richardson, and K.L. Maki. 2008. Age, growth, and mortality of Wahoo, *Acanthocybium solandri*, from the Atlantic coast of Florida and the Bahamas. Marine and Freshwater Research 59:799–807.

Muhling, B.A., J.T. Lamkin, F. Alemany, A. Garcia, J. Farley, G.W. Ingram, Jr., D. Alvarez Berastegui, P. Reglero, and R. Laiz Carrion. 2017. Reproduction and larval biology in tunas, and the importance of restricted area spawning grounds. Review of Fish Biology and Fisheries 27:697–732.

Muto, N., U.B. Alama, H. Hata, A.M.T. Guzman, R. Cruz, A. Gaje, R.F.M. Traifalgar, R. Kakioka, H. Takeshima, H. Motomura, F. Muto, R.P. Babaran, and S. Ishikawa. 2016. Genetic and morphological differences among the three species of the genus *Rastrelliger* (Perciformes: Scombridae). Ichthyological Research 63(2):275–287.

Nakajima, K., S. Kitada, Y. Habara, S. Sano, E. Yokoyama, T. Sugaya, A. Iwamoto, H. Kishino, and K. Hamasaki. 2014. Genetic effects of marine stock enhancement: a case study based on the highly piscivorous Japanese Spanish Mackerel. Canadian Journal of Fisheries and Aquatic Sciences 71(2):301–314.

SELECT BIBLIOGRAPHY

Nikolic, N., G. Morandeau, L. Hoarau, W. West, H. Arrizabalaga, S. Hoyle, S.J. Nicol, J. Bourjea, A. Puech, J.H. Farley, A.J. Williams, and A. Fonteneau. 2016. Review of Albacore Tuna, *Thunnus alalunga*, biology, fisheries and management. Reviews of Fish Biology and Fisheries DOI 10.1007/s11160-016-9453-y.

Nishikawa, Y. 1979. Early development of the Double-lined Mackerel, *Grammatorcynus bicarinatus* (Quoy and Gaimard) from the western tropical Pacific. Bulletin, Far Seas Fisheries Research Laboratory, Shimizu 17:125–140.

Oxenford, H.A., P.A. Murray, and B.E. Luckhurst. 2003. The biology of Wahoo (*Acanthocybium solandri*) in the western Central Atlantic. Gulf and Caribbean Research 15:33–49.

Pecoraro, C., I. Zudaire, N. Bodin, H. Murua. P. Taconet, P. Díaz-Jaimes, A. Cariani, F. Tinti, and E. Chassot. 2016. Putting all the pieces together: integrating current knowledge of the biology, ecology, fisheries status, stock structure and management of Yellowfin Tuna (*Thunnus albacares*). Review of Fish Biology and Fisheries DOI 10.1007/s11160-016-9460-z.

Postel, E. 1955. Contribution à l'étude de la biologie de quelques Scombridae de l'Atlantique tropico-oriental. Annales Station Océanographique de Salammbô 10:1–168.

Rey, J.C., E. Alot, and A. Ramos. 1984. Sinopsis biologica del bonito, *Sarda sarda* (Bloch), del Mediterraneo y Atlantico este. ICCAT Collective Volume of Scientific Papers 20(2):469–502.

Schaefer, K.M. 2001. Reproductive biology of tunas. In: Block, B.A., and Stevens, E.D. (eds), Tuna: Physiology, ecology, and evolution. pp. 225–270. Academic Press, San Diego, CA.

Scoles, D.R., B.B. Collette, and J.E. Graves. 1998. Global phylogeography of mackerels of the genus *Scomber*. Fishery Bulletin 96:823–842.

Uchida, R.N. 1981. Synopsis of biological data on Frigate Tuna, *Auxis thazard*, and Bullet Tuna, *A. rochei*. NOAA Technical Report NMFS Circular 436, 63 pp.

Van der Elst, R., and B.B. Collette. 1984. Game fishes of the east coast of southern Africa. 2. Biology and systematics of the Queen Mackerel, *Scomberomorus plurilineatus*. South African Association Marine Biological Research Oceanographic Research Institution Investigation Report No. 55:1–12.

Warashina, I., and K. Hisada. 1972. Geographical distribution and body length composition of two tuna-like fishes, *Gasterochisma melampus* Richardson, and *Allothunnus fallai* Serventy, taken by Japanese tuna longline fishery. Far Seas Fisheries Research Laboratory Bulletin 6:51–74.

Welch, D.J., and A.C. Ballagh. 2009. Northern Australia Grey Mackerel biology. Fishing & Fisheries Research Centre Technical Report No. 4:27–55. James Cook University, Townsville, Australia.

Yoshida, H.O. 1979. Synopsis of biological data on tunas of the genus *Euthynnus*. NOAA Technical Report NMFS Circular 429, 57 pp.

Yoshida, H.O. 1980. Synopsis of biological data on bonitos of the genus *Sarda*. NOAA Technical Report NMFS Circular 432, 50 pp.

XIPHIIDAE and ISTIOPHORIDAE

Beardsley, G.L., Jr., N.R. Merrett, and W.J. Richards. 1975. Synopsis of the biology of the Sailfish, *Istiophorus platypterus* (Shaw and Nodder, 1791). NOAA Technical Report NMFS SSRF 675(3):95–120.

Beerkircher, L.R., D.W. Lee, and G.F. Hinteregger. 2008. Roundscale Spearfish *Tetrapturus georgii*: morphology, distribution, and relative abundance in the western North Atlantic. Bulletin of Marine Science 82(1):155–170.

Bromhead, D., J. Pepperell, B. Wise, and J. Finlay. 2004. Striped Marlin: Biology and ecology. Bureau of Rural Sciences, Canberra.

Buonaccorsi, V.P., J.R. McDowell, and J.E. Graves. 2001. Reconciling patterns of inter-ocean molecular variance from four classes of molecular markers in Blue Marlin (Makaira nigricans). Molecular Ecology 10:1179–1196.

Collette, B.B., J.R. McDowell, and J.E. Graves. 2006. Phylogeny of Recent billfishes (Xiphioidei). Bulletin of Marine Science 79(3):455–468.

de Sylva, D.P. 1975. Synopsis of biological data on the Mediterranean Spearfish, *Tetrapturus belone* Rafinesque. NOAA Technical Report, NMFS Special Scientific Report Fisheries 675(3):121–131.

Hanner, R., R. Floyd, A. Bernard, B.B. Collette, and M. Shivji. 2011. DNA barcoding of billfishes. Mitochondrial DNA 22(Sl):1–10.

Kikawa, S. 1975. Synopsis of biological data on the Shortbill Spearfish, *Tetrapturus angustirostris* Tanaka, 1914 in the Indo-Pacific area. NOAA Technical Report, NMFS Special Scientific Report Fisheries 675(3):39–54.

Mather, F.J., III, Clark, H.L., and Mason, J.M., Jr. 1975. Synopsis of the biology of the White Marlin, *Tetrapturus albidus* Poey (1861). NOAA Technical Report, NMFS Special Scientific Report Fisheries 675(3):55–94.

Palko, B.J., G.L. Beardsley, and W.J. Richards. 1981. Synopsis of the biology of the Swordfish, *Xiphias gladius* Linnaeus. NOAA Technical Report, NMFS Circular 441.

Robins, C.R. 1975. Synopsis of biological data on the Longbill Spearfish, *Tetrapturus pfluegeri*. In: R.S. Shomura, F. Williams (eds.). NOAA Technical Report, NMFS Special Scientific Report Fisheries 675(3):28–38.

Shimose, T., K. Yokawa, and H. Saito. 2010. Habitat and food partitioning of billfishes (Xiphioidei). Journal of Fish Biology 76(10):2418–2433.

Ueyanagi, S., and P.G. Wares. 1975. Synopsis of biological data on Striped Marlin, *Tetrapturus audax* (Philippi), 1887. NOAA Technical Report, NMFS Special Scientific Report Fisheries 675(3):132–139.

INDEX of SCIENTIFIC and COMMON NAMES

ABOUT the AUTHORS and ILLUSTRATOR

In 1960, Bruce accepted a position as an ichthyologist at the National Systematics Laboratory in what is now NOAA's National Marine Fisheries Service, housed in the Smithsonian's National Museum of Natural History. Bruce's research focuses on the anatomy, systematics, evolution, and biogeography of tunas and their relatives plus other fishes such as halfbeaks, needlefishes, and toadfishes. Results of his research have been published in nearly 400 papers in many scientific journals plus two regional fish books, *The Fishes of Bermuda* and *Bigelow and Schroeder's Fishes of the Gulf Of Maine*. His ichthyology textbook *The Diversity of Fishes* is now in its second (2009) edition and is the most widely used college-level ichthyology text. He is Chair of the IUCN Species Survival Tuna and Billfish Specialist Group. He is a past President of the American Society of Ichthyologists and Herpetologists and has earned many honors, such as the first Robert H. Gibbs Memorial Award for an outstanding body of published work in systematic ichthyology, the Joseph S. Nelson Lifetime Achievement Award in Ichthyology, and a Gold Medal from the US Department of Commerce for his leadership in using IUCN Red List standards to assess extinction risk of tunas, mackerels, and billfishes.

John Graves is Chancellor Professor of Marine Science at the Virginia Institute of Marine Science (VIMS), College of William & Mary, and served as Chair of the Department of Fisheries Science for 20 years. John's research focuses on fisheries genetics and the fate of fishes released from recreational and commercial fisheries, with an emphasis on highly migratory species. He has been the advisor or co-advisor for 39 graduate students at VIMS and authored more than 120 papers. John is currently serving his twelfth consecutive two-year term as Chair of the US ICCAT Advisory Committee, and he has been a member of the NMFS Highly Migratory Species Advisory Panel since its inception in 1997. John's awards include the Virginia Outstanding Faculty Award, William & Mary's Graves Award for Sustained Excellence in Teaching, an NOAA Fisheries Special Recognition Award, and *Sportfishing* magazine's Making a Difference Award, and he was inducted into the International Game Fish Association's Hall of Fame in 2016. He is an avid angler and wishes for more time to be on the water.

Val Kells is a widely recognized freelance Marine Science Illustrator. She completed her B.S. in Environmental Studies and her formal training in Scientific Communication and Illustration at the University of California, Santa Cruz, in 1985. She has worked with designers, educators, and curators to develop a wide variety of illustrations for interpretive and educational displays in over 30 public aquariums, museums, and nature centers. Val also collaborates with editors, ichthyologists, and fisheries biologists, and her work has been published in many books and periodicals. She recently illustrated and co-authored *A Field Guide to Coastal Fishes: From Maine to Texas* and its companion, *A Field Guide to Coastal Fishes: From Alaska to California*. She also illustrated *Field Guide to Fishes of the Chesapeake Bay* and illustrated and co-authored *Field Guide to Freshwater Fishes of Virginia*. All five books were published by Johns Hopkins University Press. Val is an avid fisherman and naturalist and spends her off-time exploring, fishing, and documenting aquatic habitats along the coasts.